W9-DEU-906

FOUNDATIONS OF
PHYSICAL GEOGRAPHY

Air & wind   Chap 3   Page 55-57, 63-67

FOUNDATIONS OF
PHYSICAL GEOGRAPHY

MICHAEL
BRADSHAW
&
RUTH
WEAVER

WCB

Wm. C. Brown Publishers

Dubuque, IA   Bogota   Boston   Buenos Aires   Caracas   Chicago
Guilford, CT   London   Madrid   Mexico City   Sydney   Toronto

Book Team

Editor *Lynne Meyers*
Developmental Editor *Tom Riley*
Production Editor *Daniel Rapp*
Designer *Lu Ann Schrandt*
Art Editor *Joyce Watters*
Photo Editor *Janice Hancock*
Permissions Coordinator *Gail I. Wheatley*
Visuals/Design Developmental Coordinator *Donna Slade*

Wm. C. Brown Publishers
A Division of Wm. C. Brown Communications, Inc.

Vice President and General Manager *Beverly Kolz*
Vice President, Publisher *Jeffrey L. Hahn*
Vice President, Director of Sales and Marketing *Virginia S. Moffat*
Vice President, Director of Production *Colleen A. Yonda*
National Sales Manager *Douglas J. DiNardo*
Marketing Manager *Amy Halloran*
Advertising Manager *Janelle Keeffer*
Production Editorial Manager *Renée Menne*
Publishing Services Manager *Karen J. Slaght*
Royalty/Permissions Manager *Connie Allendorf*

Wm. C. Brown Communications, Inc.

President and Chief Executive Officer *G. Franklin Lewis*
Senior Vice President, Operations *James H. Higby*
Corporate Senior Vice President, President of WCB Manufacturing *Roger Meyer*
Corporate Senior Vice President and Chief Financial Officer *Robert Chesterman*

Cover photos: Background: Yellowstone National Park, © Stephen
Simpson/FPG International Corp.; Inset: New Britain Island Papua,
New Guinea, © Telegraph Colour Library/FPG International Corp.

The credits section for this book begins on page 433 and
is considered an extension of the copyright page.

Copyright © 1995 by Wm. C. Brown Communications, Inc. All
rights reserved

A Times Mirror Company

Library of Congress Catalog Card Number: 94–72286

ISBN 0–697–25075–X

No part of this publication may be reproduced, stored in a retrieval
system, or transmitted, in any form or by any means, electronic,
mechanical, photocopying, recording, or otherwise, without the
prior written permission of the publisher.

Printed in the United States of America by Wm. C. Brown Communications, Inc.,
2460 Kerper Boulevard, Dubuque, IA 52001

10 9 8 7 6 5 4 3 2 1

C Brief ONTENTS

CONTENTS

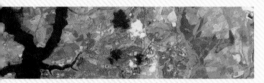

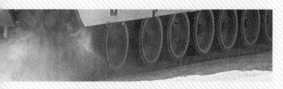

PREFACE

Foundations in Physical Geography is for introductory courses in physical geography. It attempts to cover the needs of students who study the natural environment from a geographic point of view and of those who study it as a science elective.

PHYSICAL GEOGRAPHY

The geographic thrust is met in a number of ways. The spatial focus of the subject is emphasized by the use of maps and three-dimensional diagrams. Examples and cases from across the United States and from other parts of the world illustrate distributions of phenomena and differences among places.

A second aspect of geographic studies is the focus on interactions between the natural environment and human activities. This book purposefully includes the human factor in physical geography because people continue to be affected by natural events and because people are an increasingly potent factor of change in many natural situations.

SCIENTIFIC CONCEPTS AND METHODS

Physical geography and other sciences share common characteristics. Those who take a physical geography course as a science element will find a range of scientific concepts, knowledge, and methods illustrated in the text. Terms such as atom, molecule, heat, and temperature are defined and examples are provided. Scientific concepts are explained succinctly and clearly. Physical geography uses similar working methods to other sciences, based on observation, classification, devising hypotheses for testing, and deriving theories that form the basis of prediction.

A range of sciences figures throughout the book, including basic physics and chemistry in meteorology together with geology, and ecology. In addition, the book demonstrates how new technology, such as remote sensing and geographic information systems, is used in scientific enquiry.

All scientific results are prone to blind acceptance when simplified. It is emphasized, therefore, that the study of physical geography, like that of any science, is a continuing search for a fuller understanding of how the natural environment works. Current theories may be replaced by new ones when new evidence becomes available. Some of the points made and conclusions drawn in this text gloss over a fuller debate which might confuse introductory students. The temptation to oversimplify has been resisted and checked by our reviewers. The reader should, however, regard this text as one statement of current ideas on the topics concerned.

THEMES

Five main themes are developed in the text. Each brings out aspects of physical geography and scientific concepts and methods.

EARTH ENVIRONMENTS

The study of the physical nature of our planet is approached through the concept of environment. An environment is the sum of conditions in which an organism lives. For example, the varied conditions of weather, soils, and other organisms in the environment affect the way in which an organism acts. In physical geography, human beings are the central organisms within Earth's physical environments.

The concept of environment is extended to cover the contexts in which processes act in both the living and nonliving worlds. Thus, there are climatic environments and environments in which mountains form and continents move. Four major Earth environments are used as the structure of this text, each based on the sources of energy and materials that determine how they function.

The *atmosphere-ocean environment* is powered by energy from the sun. Movements within the environment composed largely of gases and water produce short-term weather and longer-term climatic conditions and changes.

The *solid-earth environment* derives its energy from within the planet, and its materials are rocks and minerals. Earth's internal processes cause continents to move and mountains to rise.

The *surface-relief environment* is a narrow zone on the continental surfaces where the first two major environments interact. Running water, moving ice, and other agents dependent on the workings of the atmosphere-ocean environment carry out the detailed molding of the land produced by processes in the solid-earth environment.

The *living-organism environment* also gains its energy from the sun but interacts with all the other major environments in cycling chemical elements.

Energy and material pathways associated with living organisms provide the basic conceptual framework for this all-embracing capstone of physical geography—the ecosystem.

As well as providing an organizing structure linked closely to a systems approach, the focus on Earth environments also makes it possible to highlight and comment on a set of environmental issues that may provide a motivation for students' interest in physical geography.

ONENESS OF EARTH'S ENVIRONMENTS

Although the major Earth environments are studied separately in order to appreciate the workings of individual aspects, a consciousness of the wholeness of environmental conditions at a place is vital in considering environmental issues. This oneness is emphasized throughout this book by highlighting in the summaries linkages and interactions among the four major environments.

CHANGING EARTH ENVIRONMENTS

Change over time is a feature of all Earth environments. Weather changes each day; climate changes over longer periods; earthquakes and volcanoes are short-lived events that signal longer-term changes within the solid-earth environment; continents move around on Earth's surface over millions of years; mountains rise and are torn down as rivers, glaciers, and wind etch their distinctive features on the surface; plants and animals migrate and ecosystems develop. Human beings must know that environments change. Change is emphasized in this text through a discussion of climate change and studies of the evolution of Earth surface features and living communities.

HUMANS CHANGE ENVIRONMENTS

Human activities affect the operation of environmental processes, change the composition of air and ocean, alter the shape of Earth's surface, and have major impacts on living organisms. Human intervention in natural environments is an increasing concern in physical geography. In this book, the human factor is regarded as important in the workings of the four major environments. Particular attention is paid to the role of humans in climate change, landform changes, and ecological changes.

ORGANIZATION

After Chapter 1, which provides an introduction to physical geography, the rest of the book is ordered around the four major environments. Chapters 2 to 6 present the *atmosphere-ocean environment*, beginning with the materials that constitute the atmospheric and oceanic realms and proceeding to studies of the processes—heating, movement, and cycling of water—that cause weather and climate. Chapter 7 is about the *solid-earth environment* and how the internal Earth processes affect major surface features. Chapters 8 to 11 show how atmospheric and marine processes carve land into distinctive features in the *surface-relief environment*. Chapters 12 and 13 focus on ecosystem processes and the major groupings of vegetation and soil types of the *living-organism environment*.

Each chapter provides a variety of teaching and learning resources. The main text is organized around a clear and logical development of the material and is interspersed with internal summaries. Within each chapter the *Environmental Issue* sections emphasize the significance of physical geography in society's environmental concerns and in so doing provide illustrations of applied geography. At the end of each chapter there is a *chapter summary*, a list of *key terms, chapter review questions,* and *suggested activities.* At the end of the book, there is a reference section that includes a *glossary* of the key terms, information on *map projections, symbol keys* for weather and topographic maps, a *soil classification* summary, and an *index.*

PRESENTATION

The art program for this book has been a major priority. The *diagrams* are clearly linked to the text. Color and three-dimensional representations are used to help students understand the concepts and geographic relationships involved. The *photographs* and *satellite images* have been selected for similar reasons. A special feature of the book is the utilization of results from the latest work at NASA, including photos from space shuttle missions.

Additional instructor materials are available that will enrich the teaching and learning that arises from using this book. They include:

Instructor's Manual and Test Item File. The Instructor's Manual includes helpful supplementary resources, chapter synopses, and suggested activities. The Test Item File provides hundreds of questions in several formats. Use these questions or modify them for fresh test materials.

MicroTest test-generating software. Available in DOS, Windows and Macintosh formats, this easy-to-use testing software comes complete with all the questions from the Test Item File.

Student Study Guide. This useful guide offers the student chapter overviews, learning objectives, key terms and concepts, and sample questions with answers.

Student Study Art/Map Notebook. Included free with every new copy of the text, this is a bound version of the transparency set, which students use to take notes directly on the illustration while concentrating on what is being discussed in the lecture. This unique ancillary also includes a variety of base maps for use in quizzes or outside work and is offered exclusively with *Foundations of Physical Geography.*

100 Transparencies. Full-color transparencies of the most pedagogically important line art from the text.

100 Slides. Full-color slides of the outstanding photos from the text.

AGI Videodisc. This videodisc contains a comprehensive visual collection of over 3,500 photographs, animations, and videos, designed for classroom presentation and reference work. Ask your Wm. C. Brown representative for details.

Computer Tutorials for Physical Geology. This interactive software is designed to help students understand and explore earth processes through graphics, animations, and moving footage. From volcanoes to cloud formation, students can learn about processes and respond to questions to reinforce their understanding. Available on diskette or CD-ROM, Macintosh or Windows. Ask your Wm. C. Brown representative for details.

Exercises in Physical Geography, second edition, by Don W. Duckson. This fully customizable lab manual allows instructors to combine the outstanding general coverage from Duckson's lab manual with their own or regional materials. Ask your Wm. C. Brown representative for details.

ACKNOWLEDGMENTS

Mary Catherine Prante
Western Kentucky University

James R. Powers
Pasadena City College

Rudi H. Kiefer
The University of North Carolina at Wilmington

Percy H. Dougherty
Kutztown University

Stanford E. Demars
Rhode Island College

Thomas S. Krabacher
California State University

Perry J. Hardin
Brigham Young University

George A. Schnell
SUNY at New Paltz

Mark W. Williams
University of Colorado

James F. Petersen
Southwest Texas State University

Roberto Garza
San Antonio College

Joan C. Stover
South Seattle Community College

Richard A. Crooker
Kutztown University

Elliot G. McIntire
California State University

Steven Jennings
Texas A & M University

Stephen H. Sandlin
San Bernardino Valley College

Mark R. Anderson
University of Nebraska

Roger K. Sandness
South Dakota State University

Jerry E. Green
Miami University of Ohio

Donald E. Petzold
University of Wisconsin—River Falls

THE LEARNING SYSTEM

CHAPTER

11

COASTAL LANDFORMS

THIS CHAPTER IS ABOUT:

- Coastal landform environments
- Waves and tides
- Cliffs
- Beaches
- Deltas and estuaries
- Salt marshes and mangroves
- Reefs
- Climate and coastal landforms
- Human impacts on coastal landforms
- Environmental Issue: The Aral Sea—Is It Dead?

Land-based and ocean processes meet at coasts. Sand brought to the coast by rivers builds up to form beaches that protect wetland areas of marsh grasses, like those on the shores of southern Mozambique in Africa (inset). Ocean currents smooth the outer margins and give the beaches a straight line.

These **icons,** representing the four major environments, further alert students to the interaction among the physcial environments.

"Environmental Issue" boxes emphasize the environmental impact of human activity and natural occurrences on physical geography. All boxes end with critical thinking questions.

ENVIRONMENTAL ISSUE:

Acid Rain—Who Is to Blame?

Some of the issues involved include:
  What causes acid rain?
  What are its effects?
  Can it be stopped?

Acid precipitation is rain or snow that has an enhanced acidity caused by atmospheric pollution. The increasing acidity of precipitation in the northeastern United States is shown on the series of maps. Similar increases over the same period of time occurred in northwestern Europe. Acid fog occurs around industrial areas.

The acidity of water governs its ability to react chemically with substances in soils and rocks. Acidity is usually measured on the pH scale, which registers the proportion of hydrogen ions (components of acids). The scale goes from 1—highly acid, with many hydrogen ions—to 14—strongly alkaline, the opposite of acid. Normal rainwater usually has a pH of 5.5, which is slightly acid because it contains dissolved carbon dioxide (weak carbonic acid). Acid rain has a pH between 5.5 and 2.5.

Acid rain is caused by sulfur and nitrogen compounds being emitted into the atmosphere following the burning of coal in power plants and of gasoline in automobiles. The sulfur and nitrogen unite with oxygen in the presence of sunlight, forming oxides that may become dry deposition near the source, as the diagram shows. If they travel farther from the source, further chemical reactions convert the oxides of sulfur and nitrogen into sulfuric and nitric acids. These acids fall to the ground in rain or snow several hundred kilometers downwind of their source. The building of high chimneys at coal-fired power stations since the early 1970s reduced dry deposition but has increased wet deposition at a distance.

Acid rain is blamed for fish kills in upland lakes, loss of tree leaves, destruction of building materials, and hazards to human health. The effects on upland lakes in the Adirondack Mountains of upper New York State, parts of eastern Canada, and Scandinavia are most obvious. The thin soils of these areas, which have an acidic chemical environment themselves, rapidly pass the acidity in rain and snow to the lakes. The greatest effect of acid rain may be on crop yields, but little data are available.

While the urban-industrial atmospheric sources of acidity have received most blame for the declining quality of soils and lakes, other human activities also cause or enhance the condition. Adding lime to poor soils makes them less acid, but farm management practices that do not replace the lime after crops have extracted lead to the impoverishment of soils by increasing acidity. In 1993 it was shown that another source of acidity in the atmosphere comes from nitrate fertilizer used on farm fields. This washes into rivers and the sea. The tiny organisms living near the ocean surface around northern Europe grow and reproduce at increased rates because of the added nitrate, and they release more sulfur compounds into the atmosphere. It is estimated that between one-third and three-quarters of acid rain falling on less industrialized countries, such as Ireland and Norway, could be due to this source of sulfur.

Acid rain has become a public issue at an international level between the United States (source) and Canada (recipient), and between other European countries, such as Great Britain and Germany, (source) and Scandinavia (recipient). Plans are now under way to reduce emissions of sulfur and nitrogen in the source countries by using low-sulfur fuels, installing filters at power plants, or by installing catalytic devices to cars. The more developed countries bear the additional expense, but the expansion of industrialization in the developing world may spread acid rain sources to countries that cannot afford the costs of control.

Linkages

Acid rain demonstrates the linkages among Earth environments. Human activities alter the composition of the atmosphere and precipitation brings particles and acids down to the soil and surface waters. The effect of increased acidity is greatest in those soils and waters overlying rocks with a high natural acidity.

Questions for Debate

1. Should all sources of sulfur and nitrogen oxides in the atmosphere be banned?
2. Do we need to find more conclusive evidence before taking further (costly) action to reduce sulfur and nitrogen emissions?
3. What should be done about such emissions in developing world countries?

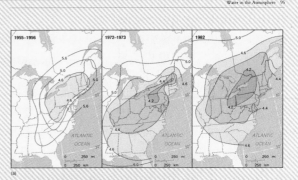

(a)

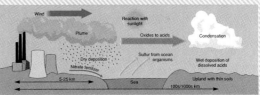

Gases (sulphur and nitrogen oxides)
Solid particles

(b) The distribution of acid precipitation (pH values) in eastern North America, 1955–82, showing the increase in intensity and geographic coverage. (b) The processes that give rise to acid rain.
Source: (a) Top left and center: Park after Likens; top right, U.S. National Atmosphere Deposition Project.

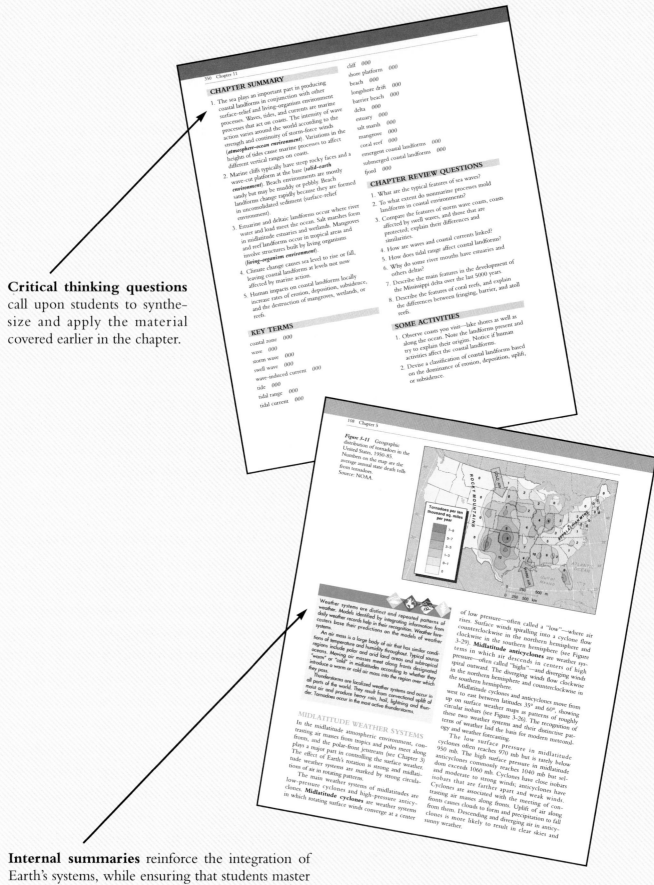

Critical thinking questions call upon students to synthesize and apply the material covered earlier in the chapter.

Internal summaries reinforce the integration of Earth's systems, while ensuring that students master concepts as they proceed through the chapter.

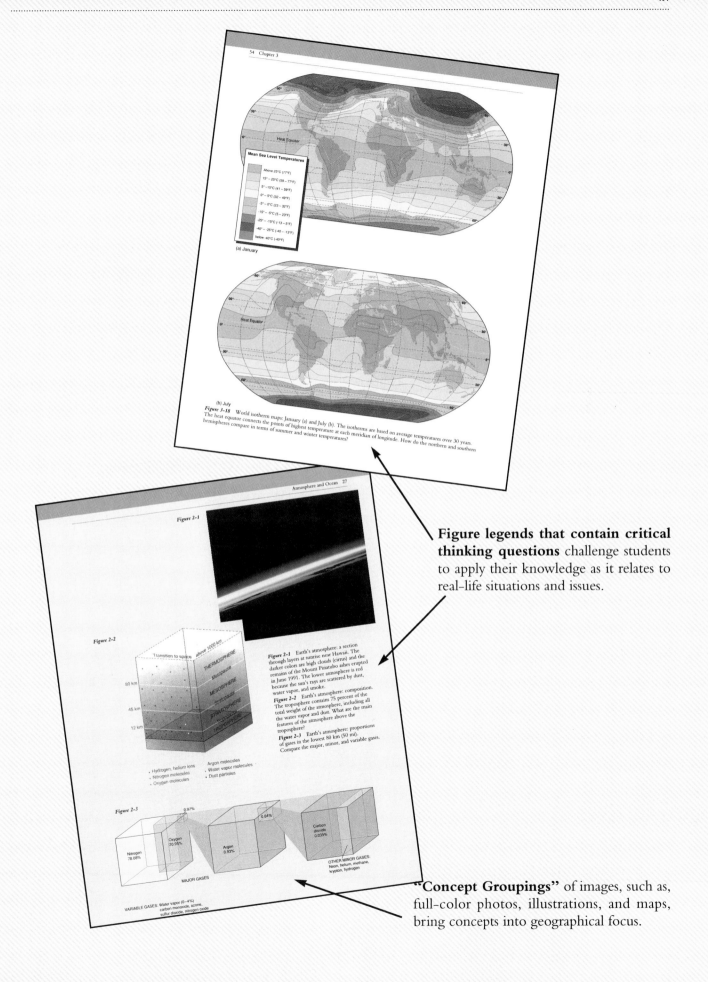

54   Chapter 3

Mean Sea Level Temperatures

Above 25°C (77°F)
15° – 25°C (58 – 77°F)
5° – 15°C (41 – 59°F)
0° – 5°C (32 – 49°F)
-5° – 0°C (23 – 32°F)
-15° – -5°C (5 – 23°F)
-25° – -15°C (-13 – 5°F)
-40° – -25°C (-40 – -13°F)
below -40°C (-40°F)

Heat Equator

(a) January

Heat Equator

(b) July

Figure 3–18   World isotherm maps: January (a) and July (b). The heat equator connects the points of highest temperature at each meridian of longitude. The isotherms are based on average temperatures over 30 years. How do the northern and southern hemispheres compare in terms of summer and winter temperatures?

Figure legends that contain critical thinking questions challenge students to apply their knowledge as it relates to real–life situations and issues.

Atmosphere and Ocean   27

Figure 2–1

Figure 2–2

Transition to space

above 1000 km

THERMOSPHERE

Mesopause

80 km

MESOSPHERE

Stratopause

45 km

STRATOSPHERE

Tropopause

12 km

TROPOSPHERE

· Hydrogen, helium ions
· Nitrogen molecules
· Oxygen molecules

Argon molecules
· Water vapor molecules
· Dust particles

Figure 2–1   Earth's atmosphere: a section through layers at sunrise near Hawaii. The darker colors are high clouds (cirrus) and the remains of the Mount Pinatubo ashes erupted in June 1991. The lower atmosphere is red because the sun's rays are scattered by dust, water vapor, and smoke.

Figure 2–2   Earth's atmosphere: composition. The troposphere contains 75 percent of the total weight of the atmosphere, including all the water vapor and dust. What are the main features of the atmosphere above the troposphere?

Figure 2–3   Earth's atmosphere: proportions of gases in the lowest 80 km (50 mi). Compare the major, minor, and variable gases.

Figure 2–3

0.97%

0.04%

Nitrogen
78.08%

Oxygen
20.95%

Argon
0.93%

Carbon
dioxide
0.035%

OTHER MINOR GASES:
Neon, helium, methane,
krypton, hydrogen

MAJOR GASES

VARIABLE GASES: Water vapor (0–4%)
carbon monoxide, ozone,
sulfur dioxide, nitrogen oxide

"Concept Groupings" of images, such as, full-color photos, illustrations, and maps, bring concepts into geographical focus.

FOUNDATIONS OF PHYSICAL GEOGRAPHY

CHAPTER

1

PHYSICAL GEOGRAPHY AND EARTH ENVIRONMENTS

THIS CHAPTER IS ABOUT:

- Physical geography
- Natural environment
- Physical geography and environmental issues
- Physical geographers at work
- Maps and mapmaking

The study of physical geography is based on relationships between four major environments—the linked atmosphere and oceans, the rocky solid Earth, the surface land features, and the plants and animals. The Hawaiian lava flow comes from inside Earth (solid-earth environment), the water flowing down the river in California is carving Earth's surface (surface-relief environment), the rock pool is part of the atmosphere-ocean environment, and the meadow illustrates the living-organism environment.

Surface-relief environment

Living-organism environment

Geomorphology

Biogeography

Pedology

Geology

Atmosphere-ocean environment

Meteorology

Climatology

Solid-earth environment

Oceanography

PHYSICAL GEOGRAPHY

The natural environment affects every part of our lives. Daily weather changes, local soil conditions, or extreme events such as hurricanes, floods, earthquakes, and volcanic eruptions influence us all from time to time. Human activities also affect the natural environment. In some cases the result of human actions is clear; rain forest destruction, acid rain, and soil erosion are three good examples. In cases like global warming, the vanishing ozone layer, and expanding deserts, we are still not sure what effects human activities have and how long they will last. These issues and their possible results all concern physical geographers, who study patterns and processes in the natural environment and their interactions with human activities.

Physical geographers ask several sets of questions about the natural environment. First, physical geographers try to explain why certain features—such as glaciers, volcanoes, or tropical forests—exist in some places and not others. They want to know why it is cold in Alaska and warm in California, why kangaroos live in Australia but nowhere else, and why Hawaii has many volcanoes and Iowa has none.

Second, physical geographers want to know how and why these features change over time. For example, they aim to discover how long it takes for rainfall to drain into a river, how quickly a hurricane can develop, and how long it takes to build up a mountain range and wear it down.

Third, physical geographers study human interactions with the natural environment. How, for example, does extracting irrigation water from a river affect flooding downstream? Why do certain farming practices lead to soil erosion? Can an early warning system be developed to save lives when a tornado strikes? In short, physical geographers aim to understand how the natural environment works in a certain place, how it changes over time, and how it affects and is affected by human activity.

NATURAL ENVIRONMENT

An **environment** is the set of surrounding conditions that acts on a place or person and gives it a certain character. A body of air, for example, receives energy from the sun and Earth's surface and water from the oceans, lakes, soils, and plants over which it passes. Bodies of air from contrasting environments have different characteristics of temperature and moisture. In another example, farmers are partly influenced in choosing their crops by the farm's environment: the temperature and humidity of the surrounding air, the amount of rain or snow that falls, the character of the underlying soil, and the steepness of the terrain.

Earth's **natural environment** is the set of physical and biologic conditions that surround human beings at Earth's surface (Figure 1-1). It can be studied at three levels of geographic, or spatial, scale: global, regional, and local. The global scale deals

(a)   (b)

Figure 1-1   Contrasting natural environments. (a) A tropical island with wooded hills and vigorous tree growth. (b) The arctic with ice-covered sea and snow-covered hills. (c) A desert with sand dunes, hills of bare rock—little soil and few plants only where water is available. (d) A coastal area where the ocean forms beaches and cliffs at its margin with the land. (e) High mountains, with bare rock and snow.

(c)

(d)

(e)

with the whole Earth. The regional scale involves large sections of Earth's surface, many thousands of square kilometers in size. Major regions are areas the size of continents; smaller regions may be the size of several U.S. states. The local scale refers to areas from a few kilometers to a few hundred square kilometers. Physical geographers study processes that create spatial patterns at any or all of these scales. They discover where a feature exists, how much of Earth's surface it covers, and the causes and significance of such a distribution.

Issues that arise from human activities in the natural environment occur at a range of spatial scales and change rapidly over time. Because physical geographers understand the complex processes that link together Earth's major environments, they can contribute to current political and scientific debates about environmental issues.

FOUR MAJOR EARTH ENVIRONMENTS

Earth's natural environment can be divided into four major parts (Figure 1-2). Each major Earth environment has unique characteristics, but they all overlap and interact to create dynamic and living systems at all geographic scales.

In the **atmosphere-ocean environment,** the atmosphere and oceans are inextricably linked by exchanges of heat and materials (Figure 1-3a). The atmosphere is a thin blanket that both traps energy from the sun's rays and protects Earth, particularly plants and animals, from the harmful elements of those rays. The oceans act as a giant store of water and heat. Exchanges between the atmosphere and oceans help to maintain the composition of the atmosphere. The atmosphere-ocean environment transports energy from the sun around the globe and creates weather and climate. Chapters 2

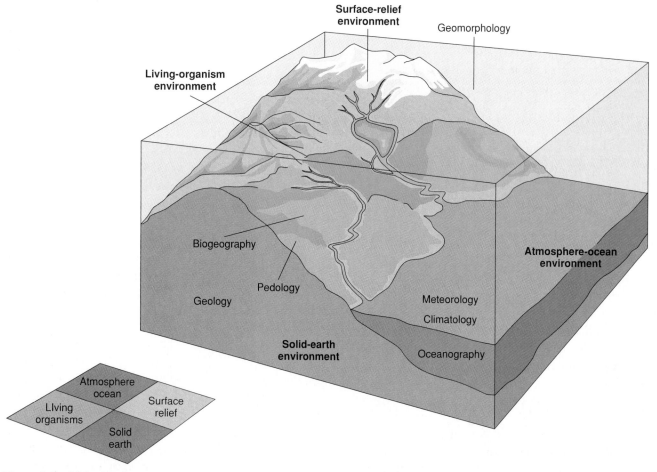

Figure 1-2 Major Earth environments and the sciences that study them. Geomorphology is the study of surface landforms; biogeography is the study of the distributions of plants and animals; pedology is the study of soils; geology is the study of rocks; meteorology is the study of the atmosphere and weather; climatology is the study of long-term conditions in the atmosphere; oceanography is the study of oceans. The small diagram is used to highlight linkages among the four major environments throughout this book.

through 6 of this book describe the features and workings of the atmosphere-ocean environment.

The **solid-earth environment** is very different. It consists of rocks that are hot and fluid near Earth's center and relatively cool and rigid near the surface (Figure 1-3b). Gravity and the heat that keeps rocks melted inside our planet are the sources of energy for moving surface rocks. Activity in the solid-earth environment is most apparent in volcanic eruptions and earthquakes. Such sporadic surface events are a powerful reminder of continuous activity deep within Earth. Over the longer term, volcanoes and earthquakes are part of global-scale movements that raise mountain ranges and cause continents to drift across Earth's surface. Chapter 7 of this book describes the features and workings of the solid-earth environment.

The atmosphere-ocean environment and the solid-earth environment interact on the surfaces of continents to form the hills, plains, and valleys—or landforms—of the **surface-relief environment** (Figure 1-3c). The solid-earth environment forms the broad features of continents and mountain chains, while the reactions of different rocks to atmospheric agents such as wind, ice, and running water control the detailed appearance of land surfaces. Chapters 8 through 11 describe the features and workings of the surface-relief environment.

The **living-organism environment** (Figure 1-3d) is shaped by the other three Earth environments. Plants and animals need water and the sun's energy from the atmosphere-ocean environment; plant nutrients come from the rock minerals of the solid-earth environment. The distribution of plants and animals around the world depends on how water and nutrients are delivered to the organisms by processes in the surface-relief environment. Chapters 12 and 13 of this book describe the features and workings of the living-organism environment.

These four major Earth environments create a dynamic and living planet as they constantly interact and adjust to each other. The atmosphere-ocean environment is normally in a state of organized turmoil, as shown by the patterns of wind and weather experienced around the globe. Weather patterns change daily and hourly; climate changes over thousands of years. The solid-earth environment slowly but steadily adjusts the positions of the continents, ocean basins, mountain ranges, and volcanoes over millions of years. Activity in the surface-relief environment gradually wears down land above sea level. Changes over time in the atmosphere-ocean environment mean that the wearing down of land started

by glaciers in cold conditions, many thousands of years ago, is continued in today's warmer conditions by rivers. In the living-organism environment, animals migrate between different areas. Plant distribution changes as the atmospheric temperature and moisture conditions change and as plants change their surroundings. Evolution over millions of years leads to new forms of plants and animals.

The four major Earth environments operate simultaneously at global, regional, and local scales. Processes acting at a global scale affect those at the local or regional scale and vice versa. For example, activity in the solid-earth environment creates volcanoes that strongly influence living organisms and landforms locally, while dust and gases from the eruption enter the atmosphere-ocean system and affect weather around the globe.

Amongst the intense activity caused by the interaction and changes in Earth environments, living things, including humans, find temporary lodgings. The special nature of Earth environments allows organisms to live and makes our planet a unique place. For all practical purposes, Earth is our only home in the solar system. This makes it important for us to understand how Earth's environments work and how human activities can exist in harmony with them.

STRUCTURE AND PROCESS IN EARTH ENVIRONMENTS

Physical geographers seek to understand how a specific natural environment works and how the environment changes over time. This understanding makes it possible to assess how the natural environment affects and is affected by human activity.

Earth environments are understood by studying their **structure**—the observed patterns within the environment, the types, amounts, and distributions of things—and the **processes** that link these components together into a working system. For example, the structure of a forest, part of the living-organism environment, includes the plants and animals that live there, the nature of the soil, the local climate, and the steepness of the hill slope on which the forest lives. The processes at work in a forest environment include the trees and other plants in the forest taking energy from the sun and nutrients from the soil. The energy and nutrients pass to grazing animals when they eat the plants. Animals return the nutrients in their droppings and when their bodies decompose after death.

All Earth's major environments have distinctive structures. The atmosphere-ocean environment is composed mainly of gases in the atmosphere and water in the oceans. Both atmosphere and oceans

(a)

(b)

(c)

(d)

Figure 1-3   Major Earth environments: some of their features. (a) The atmosphere-ocean environment. The water that forms the clouds comes from the ocean water as energy from the sun causes evaporation, heating, and movement of the water and gases. Olympic National Park, Washington. (b) The solid-earth environment. The high temperature of melted rock in a volcanic eruption on Hawaii gives some idea of the heat within Earth's interior. (c) The surface-relief environment. Glaciers carved the details of the high mountains—Morain Lake, Banff National Park, Canadian Rockies. (d) The living-organism environment at close-up scale. The plant absorbs solar energy and takes food from the soil; the butterfly depends on the plant for its food.

are layered by temperature variations. The solid-earth environment comprises rocks and minerals organized into layers by the heaviness of the materials. The surface-relief environment consists of the surface features of land areas—such as hills, valleys, and stream channels. The structure of the living-organism environment is based around groupings of trees, shrubs, and grasses, the soils in which they root, and the animals that feed on the plants.

Processes are also distinctive in each major environment. The sun daily replenishes the energy supply of the atmosphere-ocean environment. This energy drives movements of the air and water that transport heat around the globe to produce our weather. In the solid-earth environment, heat energy from Earth's interior and gravity scatter the continents and raise up mountain systems. In the surface-relief environment, wind, rain, snow, and ocean waves caused by movements in the atmosphere wear down the land surfaces raised by internal Earth movements. The living-organism environment is marked by processes that take solar energy and convert it into chemical energy in living matter, and by processes that circulate mineral nutrients from soil and air through plants and animals. Such transfers of energy and materials link the four major Earth environments.

PHYSICAL GEOGRAPHY AND ENVIRONMENTAL ISSUES

The effect of humans on the natural environment is of increasing concern to scientists and politicians alike, as shown by the following excerpt.

> In a very short time, human activity has become so varied and complex that it is having effects not only at local and national levels, but on the whole world itself. Having discovered only the other day that our world is round, we are suddenly finding it uncomfortably small and fragile.
>
> Mankind has always been capable of great good and great evil. This is certainly true of our role as custodians of our planet. We have a moral duty to look after our planet and to hand it on in good order to future generations.

("This Common Inheritance," UK Government document 1991)

Some environmental issues are straightforward, especially at the local scale. For example, it is possible to identify the polluters when a farmer uses excessive amounts of fertilizers or pesticides, or a factory discharges toxic substances into a river. The leakage of harmful chemicals from a landfill site into groundwater affects the community dependent on the water source. A solution to this problem involves city authorities, planners, and water engineers.

Many current environmental issues, however, are complex and poorly understood. Their processes operate at all spatial and temporal scales and involve all Earth environments. Their study often needs collaboration across countries and traditional academic boundaries. The thinning of Earth's protective ozone layer, for example, involves the whole of the global atmosphere and is studied by meteorologists, oceanographers, physicists, and chemists in many different countries. Desertification affects several million square kilometers of the Earth's surface and brings misery and starvation to millions of people. To understand desertification, physical geographers examine the climate, water resources, and soils of large regions in partnership with botanists, hydrologists, and soil scientists. Geographers, anthropologists, and sociologists must also understand the structure of local societies and the social and economic reasons for certain farming practices.

Because physical geographers often integrate results from several sciences and undertake interdisciplinary research that might otherwise be overlooked, they are well placed to contribute to discussions about global, regional, and local environmental problems. The study of physical geography also provides a starting point for personal involvement in the issues. Each person has a responsibility to ensure that the Earth's natural environment is not degraded or destroyed, and that we can "hand it on in good order to future generations." Citizens educated in physical geography should have the knowledge to take part in, for example, effective land development or lobbying their political representatives to vote funds for environmental research and monitoring.

The consideration of environmental issues by physical geographers leads to a realization that some of the issues involved are complex, that it is important to have a full understanding of the workings of the natural environment, and that there are different views about the decisions to be made. Some environmental issues are investigated in this book through a series of *Environmental Issues* boxes.

Earth can be divided into four dynamic and interacting environments. The atmosphere-ocean environment consists of the atmosphere and all the oceans. The solid-earth environment comprises the rocks, solid and molten, that make up the framework of the planet. The atmosphere-ocean environment and the solid-earth environment interact to form the surface-relief environment. Plants and animals form the living-organism environment, involving interactions among the other three environments.

Physical geography is the study of Earth's environments, interactions among them, and changes in conditions over time. Physical geographers also study how Earth environments affect and are affected by human activities.

PHYSICAL GEOGRAPHERS AT WORK

Physical geography is, therefore, the study of Earth environments, the interactions among them, and changes in their conditions over time. In total this study is an immense and complex undertaking. Most physical geographers focus what they do more narrowly and make a detailed study of only one Earth environment. For instance, a geomorphologist studies the surface-relief environment, while a biogeographer studies the living-organism environment. There are further specialties within these areas, such as fluvial geomorphology (the study of running water within the surface-relief environment) and glacial geomorphology (the study of the effects on surface rocks of ice and glacier movement). It takes many geographers working together to gain a full view of the Earth environments.

The early physical geographers were explorers. Their job was to visit uncharted territory and bring home diaries, specimens, and drawings of their new findings. Now, however, little of the world remains to be explored. Humans have visited most places on the planet, including the highest mountains and the deepest oceans. Modern geographers still photograph and describe their findings. Data collection remains important, but they spend more time and energy trying to understand and explain their results. Explanations in physical geography and other sciences can lead to predictions. For example, a study of rainfall and streamflow might result in a prediction of how quickly a stream will flood after a storm. Explanations and predictions help the physical geographer understand the processes that drive and link the Earth environments.

DATA AND MODELS IN PHYSICAL GEOGRAPHY

Physical geographers test their ideas about the natural environment by gathering data in several forms and at different scales. For many researchers, data collection means fieldwork, such as digging out glacial deposits from remote mountain areas; measuring streamflow before, during, and after a storm; monitoring the buildup of sediment on a beach; or surveying the vegetation in an ancient woodland (Figure 1-4).

In many cases, more detailed work on the field samples is carried out in the laboratory using specialized equipment that cannot be carried into the field. A common example is the particle size analysis of sediments from glacial deposits, streambeds, or soils (Figure 1-5). The proportions of large,

Figure 1-4   Fieldwork in physical geography. Making observations in the natural environment. Students assessing the amount of footpath erosion in a popular tourist area in England.

Figure 1-5   Particle size analysis of glacial deposits. The equipment is a series of sieves with progressively finer meshes from the top downward. Each size class of particle separated in the sieves is expressed as a percent of the total amount of sediment.

medium, and small particles in a sample tell the researcher about the origin of the sediment and how it was deposited. Other examples of laboratory work include measuring the amount of moisture and nutrients in soils or the acidity of lake or rainwater.

A different type of laboratory work shrinks the real world to a controlled model that simplifies and imitates some of the features of the natural environment. Models work on a scale that can be grasped by the human observer. An example is the environmental cabinet shown in Figure 1-6. The cabinet can simulate a year's fluctuations in climate in a matter of hours and, at the flick of a switch, can move from creating the climatic conditions in a desert environment to those on a snowy mountain peak. The cabinet accelerates the natural processes of weathering (see Chapter 8) so that rocks disintegrate before the researchers' eyes.

Other models are based on computers. Data gathered in the field or laboratory "set up" the model by describing an initial state of some geographic process. For example, when a certain amount of beach sand is moved by offshore currents, it creates a sandbank of a particular size, shape, and position. The researcher can alter the amount and type of sand being moved or the strength and direction of the currents and rerun the model to discover the outcome. Specialized mathematical models, called general circulation models, summarize patterns in the atmosphere and help in weather forecasting and climate prediction (see Chapters 5 and 6).

Not all geographic data can be measured directly in the field or the laboratory. A researcher often builds on data gathered by earlier experiments or uses historical records of rainfall, temperature, or land use. It is also difficult for a research team to cover more than a small ground area in detail. To help them find the right spot to take samples and to compare one site with other sites, geographers use a range of data sources that show spatial information, including maps, aerial photographs, and satellite images.

MAPS AND MAPMAKING

Maps are vital to the physical geographer. Everybody uses a mental map of home, stores, school, and work to organize their days. Physical geographers use more specialized maps to organize their investigations of Earth environments and to guide, extend, or replace fieldwork and give researchers a first guess at the structure and processes of Earth environments in the study area.

Figure 1-6 Environmental cabinet being prepared for use. Conditions of temperature and humidity inside are adjusted to simulate different climates.

Not everything in a landscape appears on its map. The *cartographer,* or mapmaker, chooses which features to show and which to leave out. Maps made for different purposes, therefore, show different features. For instance, a standard topographic map includes the shape of the land surface, the courses of rivers, and coastal features, plus other data such as roads and towns. A geologic map may include some of this information, but it emphasizes the surface rock types. Weather maps show the distribution of atmospheric pressure, temperature, or rainfall. There are also soil maps, vegetation maps, and land-use maps.

To read a map, a physical geographer must understand three important sets of information. First, the coordinate system locates every spot on the map. Second, the scale relates distance on the map to distance at the ground. Third, the legend explains the symbols used to represent landscape features on the map.

COORDINATE SYSTEMS

Latitude and longitude provide the international reference system that locates any place on Earth's surface (Figure 1-7). Other coordinate systems are used locally. **Latitude** describes how far north or south of the equator a place is and is measured in degrees. The **equator** is at 0° of latitude. A place at 30° of latitude lies where a line from the surface to Earth's center intersects the

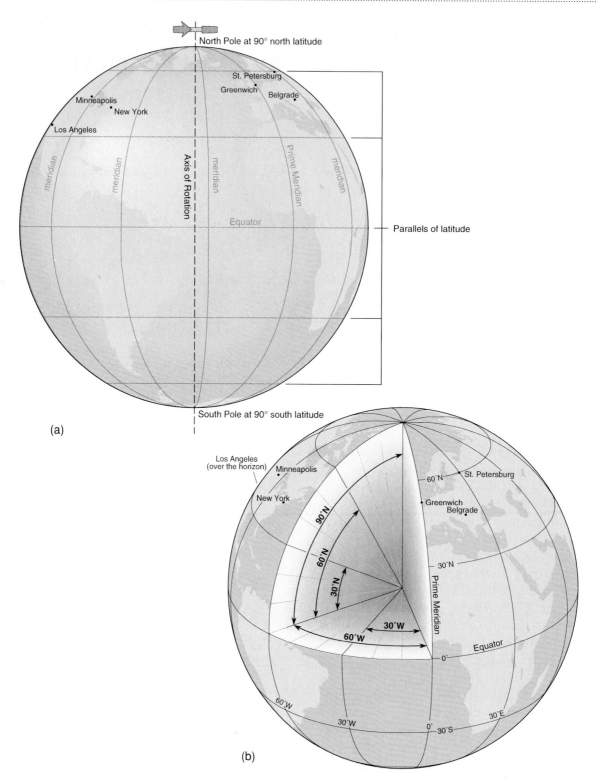

Figure 1-7   The coordinate system used to locate places on Earth's surface. (a) Parallels of latitude and meridians of longitude. (b) The basis of numbering latitude and longitude.

plane of the equator at 30° (Figure 1-7). A circle joining places of the same latitude at Earth's surface is called a *parallel* of latitude. **Parallels of latitude** are concentric circles around the globe. They are often referred to as *lines* of latitude because they appear as straight, parallel lines on some maps. The parallel 45° north of the equator is referred to as 45° N; the parallel 45° south of the equator is 45° S. The North Pole is at 90° N; the South Pole is at 90° S. Philadelphia (Pennsylvania) and Beijing (China) are at nearly the same latitude (39° N); Phoenix (Arizona) and Baghdad (Iraq) are at 33° N; Saskatoon (Canada) and Warsaw (Poland) lie at 52° N; and Washington, D.C. and Athens (Greece) are at 38° N.

One degree of latitude covers approximately 110 km (69 miles) on Earth's surface and is divided into 60 *minutes* (written: ′) to pinpoint locations more accurately. Each minute is further divided into 60 *seconds* (written: ″). The precise latitude of Meades Ranch in Kansas, the control point for a survey of the United States, is 39°13′26.686″N. The *Tropic of Cancer* is at 23°30′N, and the *Tropic of Capricorn* is at 23°30′S. The area between these two parallels is often called the *tropics*. The *Arctic* and *Antarctic Circles* lie at 66°32′N and 66°32′S respectively. The significance of these parallels is explained in Chapter 3.

Each place on Earth is given a precise east-west position by its longitude. **Longitude** measures position east or west of a half circle drawn from the north to the south pole and passing through the old Royal Observatory at Greenwich, near London, England. Its position was arbitrarily chosen by a conference in 1884. All places on this prime meridian are at 0° of longitude. A place at 30° of longitude lies where a line from the surface to Earth's center intersects the plane of the prime meridian at 30° (Figure 1-7). Places of the same longitude are joined by half circles from pole to pole, called **meridians.** The meridian 30° west of the prime meridian is written as 30° W; the meridian 45° east of the prime meridian is written as 45° E. Washington, D.C. and Lima (Peru) are both close to 77° W; Budapest (Hungary) and Cape Town (South Africa) to 19°E; and Beijing (China) and Perth (Australia) to 116°E.

Meridians also help timekeeping around the world. Earth rotates at a rate of about 15° of longitude in an hour, so that solar noon (when the sun is directly overhead) differs by an hour in places about 15° of longitude apart. The world is, therefore, divided into 24 time zones, each about 15° of longitude wide (Figure 1-8). The first time zone is centered on the prime meridian and others are centered every fifteenth meridian east and west from there. Clocks within one time zone are set to the same time. The time in the 15° of longitude centered on the prime meridian is called universal time or *Greenwich mean time.* All other time zones are counted as hours forward or hours behind Greenwich mean time. Places to the west of Greenwich are behind Greenwich mean time; places to the east are ahead. The time differences either side of the prime meridian accumulate so that on either side of the globe, areas close to 180°E and 180°W are a whole day apart. The 180° meridian is the *International Date Line.* Travellers crossing it from east to west lose a day; those going from west to east gain a day.

MAP SCALE AND MAP SYMBOLS

Maps are relatively small pieces of paper that represent much larger areas of Earth's surface. Representing Earth's sphere on flat paper has troubled cartographers for generations. The problems involved and some of their solutions are described in the *Reference Section* on map projections.

Horizontal ground distance is related to map distance by a **scale,** usually quoted as a ratio (1:10,000) or a fraction (1/10,000). This scale means that 1 unit on the map represents 10,000 of the same units on the ground. For example, on a map with a 1:10,000 scale, 1 cm represents 10,000 cm, or 100 meters, on the ground.

Map scales vary with the size of the area to be mapped and the purpose of the map. Small-scale maps show areas at a ratio of 1:250,000 and smaller, providing less detail about more extensive areas. The world maps of climate (Figure 1-9a) and vegetation (Figure 1-9b) are examples of small-scale maps. They have a scale of distance along the equator of approximately 1:120 million and show the approximate distributions and positions of generalized features. Not everything can be drawn on the map exactly as it appears on the ground. Roads, railways, rivers, buildings, vegetation types, and other features of the environment are replaced on all maps by symbols. More symbols are used as the map scale gets smaller. The *Reference Section* contains examples of maps prepared by the United States Geological Survey and lists the symbols used in these maps. It also lists the symbols used on weather

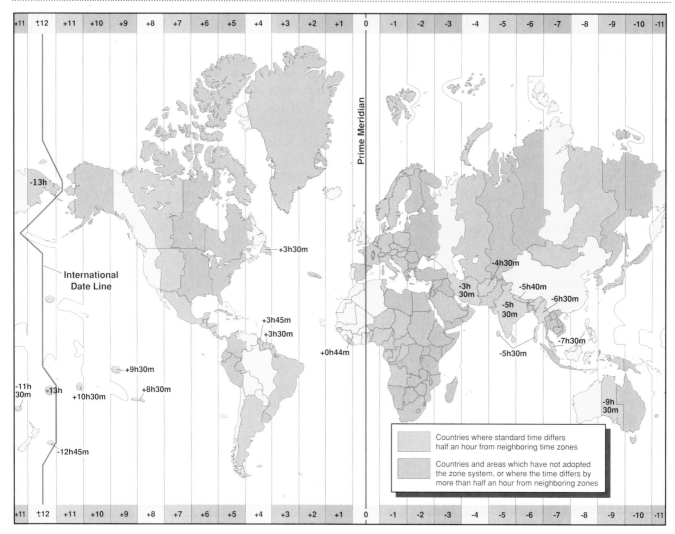

Figure 1-8 World map of time zones and the International Date Line.

maps prepared by the National Oceanic and Atmospheric Administration. Large-scale maps show areas at ratios of 1:10,000 to 1:200,000. They contain greater details about the nature of features and their position. For example, the maps of urban heat islands (Figure 1-9c) show detail that could not be represented on a small-scale global map. On very large scales with ratios of 1:10,000 or better, nearly all the information can be shown at true scale instead of as symbols, and the term *plan* is sometimes preferred to the term *map.*

The distribution of some kinds of geographic data can be mapped by isolines. **Isolines** join together points with the same value. They are frequently used to map temperature when they are called *isotherms* (Figure 1-9d) and atmospheric pressure, when they are called *isobars* (Figure 1-9e). Vertical distance above sea level is shown on some maps by isolines called *contours*. The patterns of contours shows the steepness of slope (the closer the lines, the steeper the slope) and the shape of the landforms (Figure 1-9f). On

large-scale maps isolines can be positioned quite accurately and the increment between neighboring isolines is small. On smaller-scale maps only general positions will be plotted and the increment between neighboring isolines will be greater.

GEOGRAPHIC INFORMATION SYSTEMS

Geographic Information Systems (GIS) are a recent development from computer-based cartography in which standard maps are stored and updated on computer. A GIS is a computer database in which every piece of information is tied to a geographic location. A GIS can collect, store, retrieve, analyze, and display spatial data. The data can be input and output as maps, numbers, graphs, and diagrams. Some modern systems contain mathematical models like those described earlier in this chapter. The GIS principle provides physical geographers with an important tool that makes the dynamic representation, integration, and analysis of many types of spatial data possible.

Figure 1-9 Map varieties. (a) Climate map; (b) Vegetation map; (c) Urban heat island; (d) Isotherm map; (e) Isobar map; (f) Contour map.
Source: part (e) NOAA.

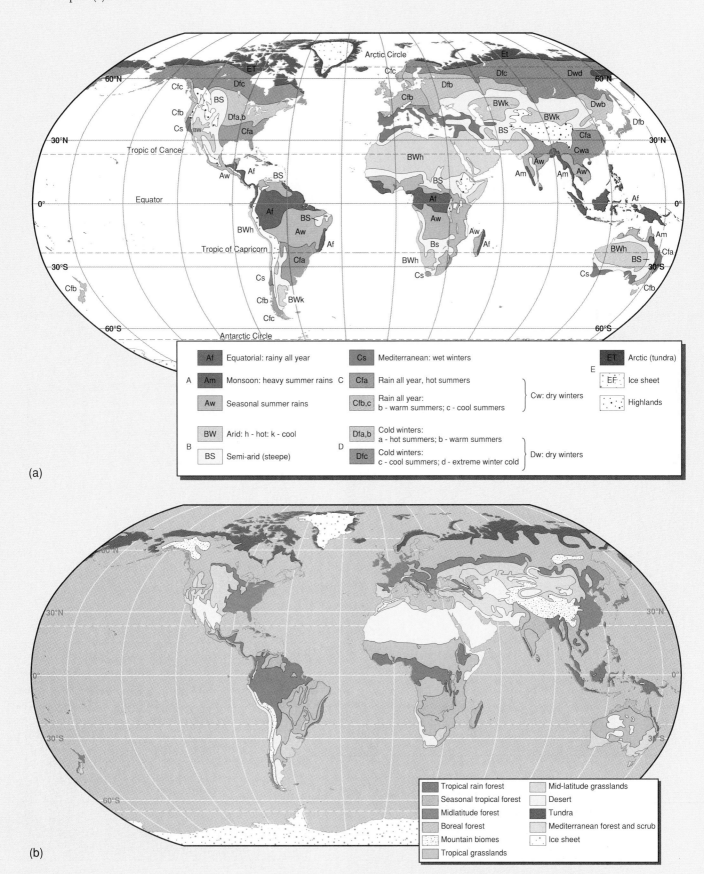

(a)

(b)

Figure 1-9 *(continued)*

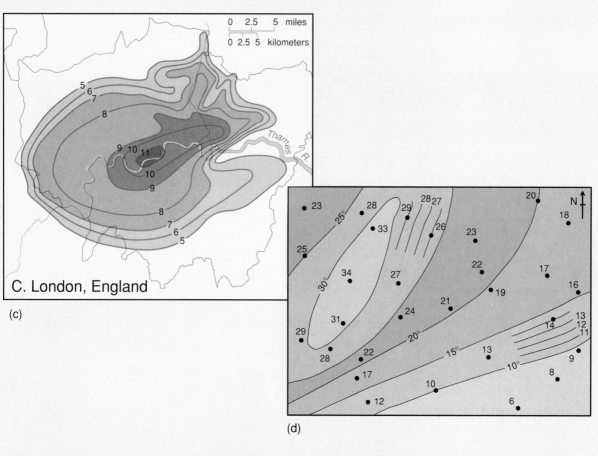

C. London, England

(c)

(d)

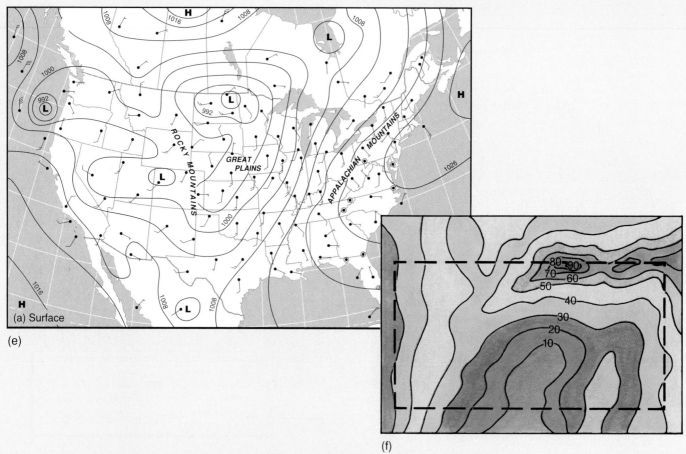

(a) Surface

(e)

(f)

The data in a GIS can be thought of as a series of layers that are linked by a basic grid of spatial position (e.g., latitude and longitude). Figure 1-10 shows the types of data found in a GIS that monitors streamflow into a reservoir. The outputs from this system might include graphs of streamflow against slope angle and a map and statistics about how erosion is related to vegetation change in the local area. Many researchers see the analytic and synthesizing capabilities of GIS as the key to a better understanding of processes in the natural environment.

A GIS allows many individuals or research teams to work on separate elements of a problem and then pool their results. Current research in GIS helps this process by focusing on the problems of integrating data collected by different people with different instruments, at different spatial scales, and at different time intervals.

REMOTE SENSING

Remote sensing is the use of aerial photographs or satellite images to gather information about Earth environments. Remotely sensed data are an increasingly important input to GIS systems and are often the most cost-effective way to survey a large area. In contrast to maps, remotely sensed images give a more complete picture of Earth's surface and, in most cases, can be updated quickly.

Most remote sensing systems collect and record the amount of solar energy reflected toward them from Earth's surface. Different objects reflect energy in distinctive ways that makes it possible to identify and map them from remotely sensed images. Human eyes can detect only a small part of the sun's output. We call this energy *light*. The remote sensors used in physical geography are sensitive to other wavelengths. They literally give human observers a new view of Earth.

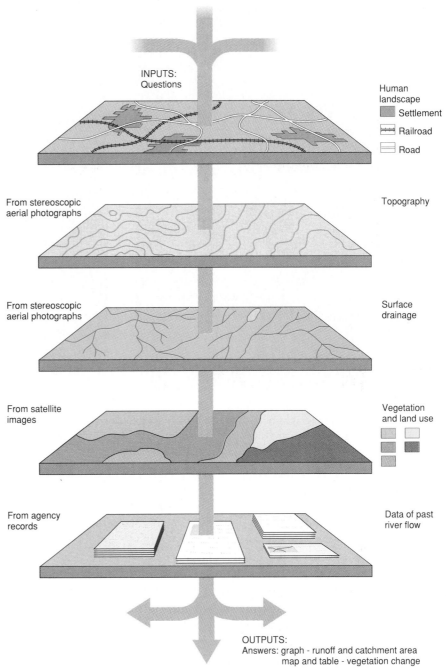

Figure 1-10 A geographic information system. The layers of data in this example are used to monitor a river system that feeds a reservoir. Outputs from the system might include graphs of seasonal streamflow, divided by drainage basin and dominant vegetation-cover type, and maps of changing land use.

Aerial Photographs Aerial photographs can be black and white or color and can record information in the visible and near-infrared wavelengths (Figure 1-11). True color photographs record the landscape in the same colors the pilot sees. False color photographs drop some of the visible wavelengths to include some of the infrared. Figure 1-11b is a false color photograph that shows healthy vegetation as bright red.

(a)

(b)

Figure 1-11 Aerial photographs of (a) Black and white, emphasizing the different light reflectancies of ground surfaces. (b) False color, showing infrared information. A variety of land uses, including farm fields (red color indicates growing plants, crops, etc.), a racetrack, and urban areas, are shown.

Some aerial photographs, particularly black and white ones, are taken with a deliberate 60 percent overlap between adjacent photos. Pairs of overlapping photographs can be placed under an arrangement of mirrors called a *stereoscope* to recreate the three-dimensional landscape seen from the plane. Specially trained technicians measure height and distance from the three-dimensional model the stereoscope produces and use the data in mapmaking. This method has largely replaced older mapmaking techniques based entirely on ground surveys.

Satellite Images   Some satellite images come from cameras operated by astronauts (Figure 1-12). Satellite images are produced most often from information collected by electronic scanners and transmitted to receiving stations on Earth. Earth-observing satellites are normally placed in either a *geostationary* or a *polar orbit*. Geostationary satellites are about 36,000 km (22,500 mi) above the equator. Their orbital speed matches the Earth's rotation, so to an earthbound

observer the satellite appears stationary above a single place. Geostationary satellites view a large area of the globe at a time. Their data yield frequent images that are used primarily for weather forecasting. In contrast, satellites in polar orbits are only about 900 km (550 mi) above the ground. Earth rotates beneath the satellite's orbit so that the sensor records images from different parts of the world on successive days. Images from polar-orbiting satellites show more detail about Earth's surface. They are particularly useful in investigating the surface-relief and the living-organism environments.

Satellite systems must be able to see through the atmosphere to Earth's surface. Their view is often obscured by dust and clouds. This has been a major problem for remote sensing, especially in cloudy regions such as the tropics. Now systems using *radar* technology can penetrate cloud cover and get an uninterrupted view of the surface, day or night. Radar systems send out and collect their own radiation at wavelengths that pass through clouds. They do not depend on the sun's illumination.

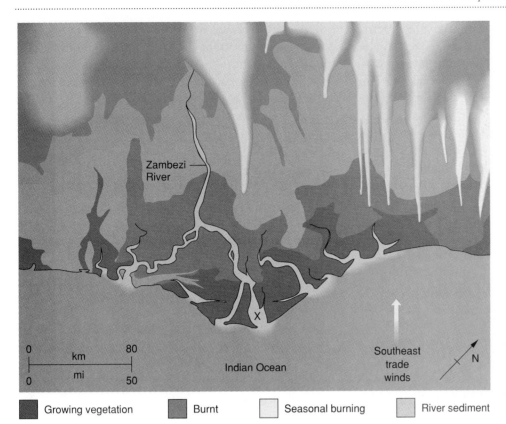

Zambezi
River

X

0   km   80

0   mi   50

Indian Ocean

Southeast
trade
winds

N

■ Growing vegetation   ■ Burnt   □ Seasonal burning   □ River sediment

(a)

(b)

Figure 1-12 The coast of Mozambique, Africa, photographed from the space shuttle. The map (a) provides details and scale. The colors on the photo (b) are false colors from an infrared film.

A close look at a satellite image shows it is actually a grid of little squares (Figure 1-13). Each square is a *pixel*, or *picture element*. The ground area represented by a pixel fixes the amount of spatial detail in the image. The pixel size is designed to match the features being monitored and the needs of those using the data. Oceanic sensors view pixels of several square kilometers while land sensors have pixels tens of meters across. The best pixel size achieved in civilian systems is currently 10 m × 10 m from the French SPOT sensor.

The three images of Plymouth, England, in Figure 1-14 were recorded by a sensor on a polar-orbiting satellite. Each image is recorded using different wavelengths of energy and shows different features of the natural and built environments. Satellite data are often more useful in the combined color format shown in Figure 1-15. Data analysis does not stop at producing a color image; the data can be manipulated further by computer to create more informative images.

Satellite images, although complex, are a more flexible data source than aerial photographs and are of growing importance in physical geography. They are particularly important for studying phenomena at a global scale. Figure 1-16 is an example of a satellite-based study relating ocean-surface temperatures to organic productivity along the western coast of the United States.

Physical geographers collect data that help them describe, explain, and predict patterns and processes in Earth environments. Data can be gathered through field and laboratory work and can be extended by computer-based models.

Maps are important tools for physical geographers. A coordinate system such as latitude and longitude locates places on maps; a scale and symbols represent features of the environment. A geographic information system is a computer database that allows the combination and analysis of many kinds of spatial data.

Remote sensing is the use of aerial photographs or satellite images to gather information about Earth environments. It is a flexible and efficient way of gathering data for physical geography.

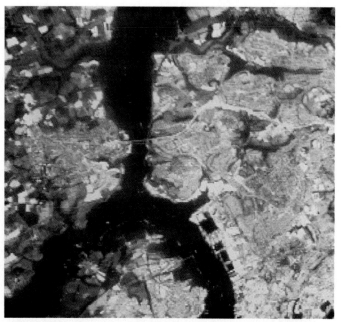

Figure 1-13   A satellite image of part of Plymouth, England, showing the pixel structure of the image. Each pixel represents an area of 30 square meters on the ground.

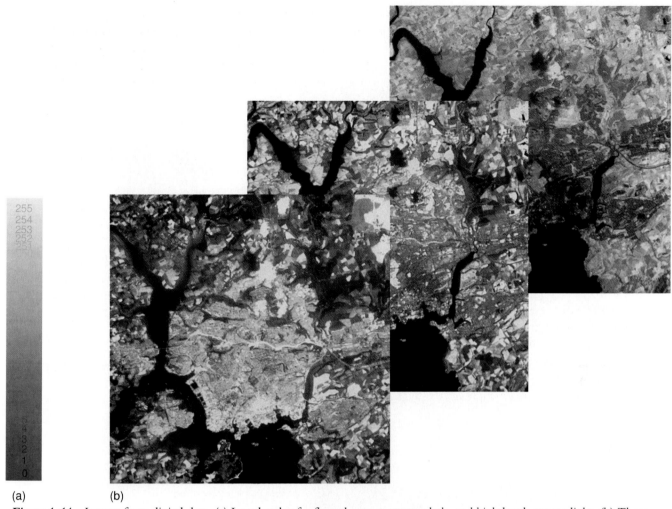

(a)   (b)

Figure 1-14 Images from digital data. (a) Low levels of reflected energy appear dark, and high levels appear light. (b) Three single waveband images of Plymouth, England, showing different information about land use in the area.

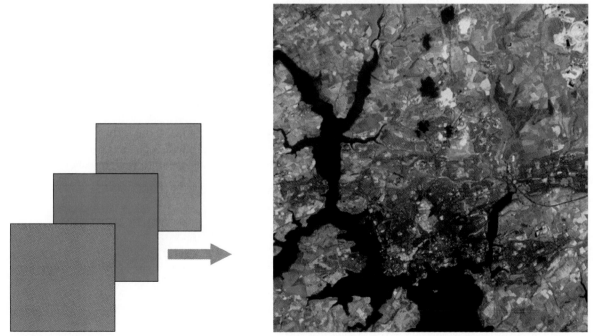

Figure 1-15 Color composite images. Three waveband images are shown in shades of red, green, and blue and are overlaid to produce a composite image in color.

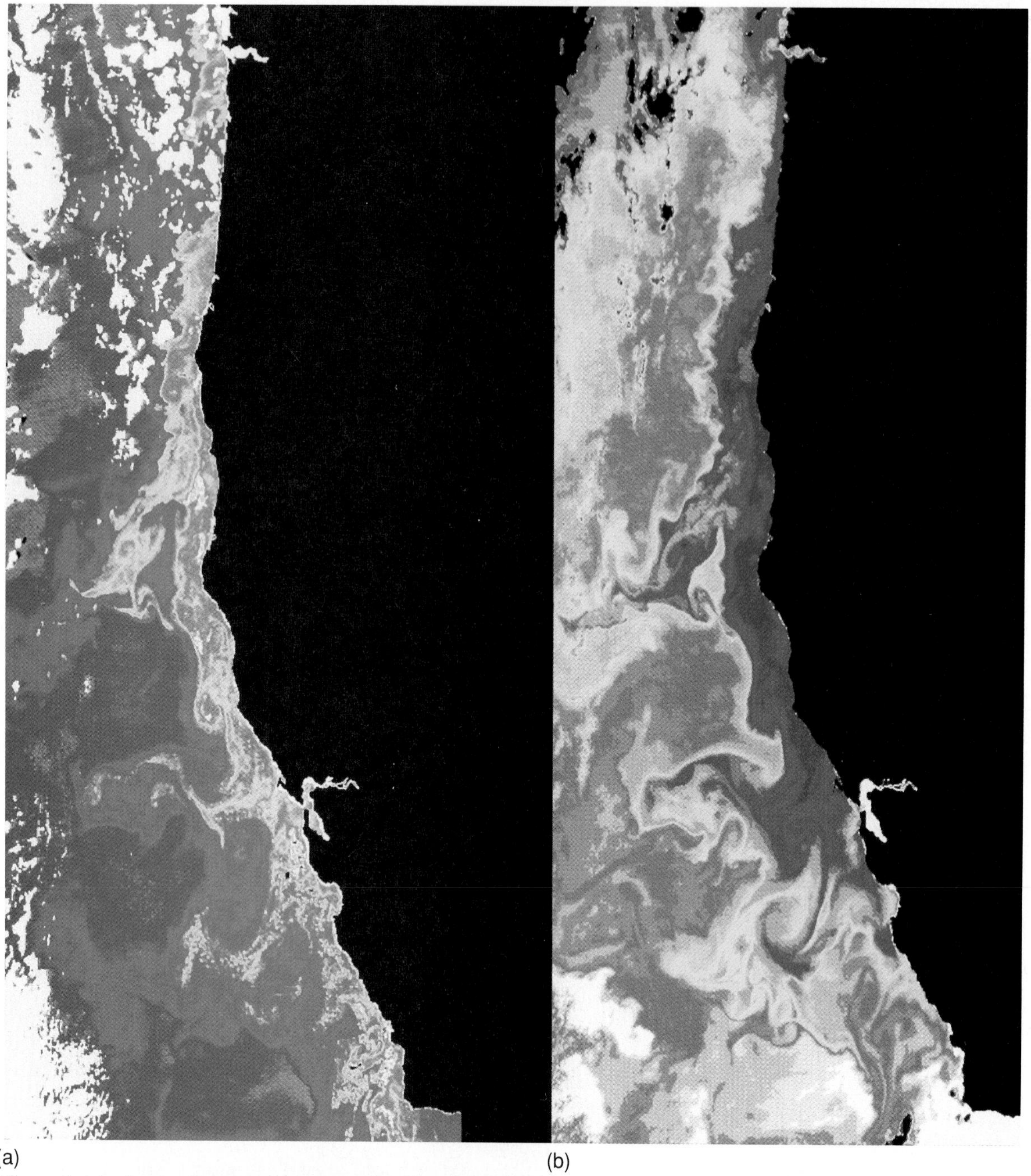

(a)   (b)

Figure 1-16   Sea-surface temperature and living organisms along the coast of California, July, 1981. (a) The coastal waters are rich in nutrients and provide a base for high concentrations of microscopic ocean plants. (b) In the sea-surface temperature image, the California Current (yellow and red) meanders southward, mixing with cooler coastal waters.

CHAPTER SUMMARY

1. Earth can be divided into four dynamic and interacting environments. The atmosphere-ocean environment consists of the atmosphere and all the oceans. The solid-earth environment comprises the rocks, solid and molten, that make up the framework of the planet. The atmosphere-ocean environment and the solid-earth environment interact to form the surface-relief environment. Plants and animals form the living-organism environment, involving interactions among the other three environments.

2. Physical geography is the study of Earth's environments, interactions among them, and changes in conditions over time. Physical geographers also study how Earth environments affect and are affected by human activities.

3. Physical geographers collect data that help them describe, explain, and predict patterns and processes in Earth environments. Data can be gathered through field and laboratory work and can be extended by computer-based models.

4. Maps are important tools for physical geographers. A coordinate system such as latitude and longitude locates places on maps; a scale and symbols represent features of the environment. A geographic information system is a computer database that allows the combination and analysis of many kinds of spatial data.

5. Remote sensing is the use of aerial photographs or satellite images to gather information about Earth environments. It is a flexible and efficient way of gathering data for physical geography.

KEY TERMS

environment   *4*

natural environment   *4*

atmosphere-ocean environment   *6*

solid-earth environment   *7*

surface-relief environment   *7*

living-organism environment   *7*

structure   *7*

process   *7*

physical geography   *10*

latitude   *11*

equator   *11*

parallel of latitude   *13*

longitude   *13*

meridian   *13*

scale   *13*

isoline   *14*

geographic information system   *14*

remote sensing   *17*

CHAPTER REVIEW QUESTIONS

1. What do physical geographers study?

2. Define *environment* and *natural environment.*

3. List and describe the four major Earth environments. Describe how they interact in your local area.

4. Give examples of structures and processes in Earth environments.

5. List three major environmental issues that affect you.

6. How does a physical geographer collect information about Earth environments?

7. What are the main components of maps?

8. Distinguish between latitude and longitude.

9. How are cartography, remote sensing, and geographic information systems related?

SUGGESTED ACTIVITIES

1. Determine how a physical geographer can contribute to a debate over a specific environmental issue.

2. Compare the roles of fieldwork and laboratory analysis in physical geography. What other data sources does the physical geographer use?

3. Discuss how a geographic information system (GIS) can help a geographer in research.

4. Compare the usefulness of maps, aerial photographs, and satellite images to physical geographers.

5. Find a topographic map of the area around your college. What distance on the ground is represented by 1 cm on the map? What vertical distance apart are the contours? What symbol is used to show a train station?

Annual Precipitation
(in millimeters)

above 3,000

2,000-3,000

1,000-2,000

600-1,000

200-600

below 200

CHAPTER

2

ATMOSPHERE AND OCEAN

THIS CHAPTER IS ABOUT:

◆ Atmosphere-ocean environment
◆ The atmosphere
◆ The oceans
◆ Exchanges between the atmosphere and oceans
◆ Evolution of the atmosphere-ocean environment
◆ Environmental Issue: The Ozone Issue

The atmosphere and oceans are closely linked. Water circulates from the oceans into the atmosphere. Winds blow the surface of the oceans into waves that break against coasts—as here in Oregon at the Shore Acres State Park. Both provide living environments for a variety of animals and plants, like the rock pool with its starfish and other creatures.

ATMOSPHERE-OCEAN ENVIRONMENT

Earth is the only planet in the solar system to have both surface water and a relatively dense atmosphere. The ocean waters and atmospheric gases combine in the atmosphere-ocean environment and encircle a solid ball of rock. Increasingly, physical geographers treat the atmosphere and oceans as a single environment because of the many links between them.

Although human beings think of themselves as living *on* Earth, they and the plants and other animals actually live *in* the planet. Living organisms exist and move about in the envelope of gases of the atmosphere and in the water of the oceans.

One of the most noticeable things about the atmosphere is that it is very changeable. Sunlight can reach Earth's surface through clear skies but may be blocked by clouds. Winds blow at different strengths and from different directions. Rain or snow falls for limited periods of time, separated by dry spells. We summarize our experiences of living in the atmosphere in two words—weather and climate. **Weather** is the day-to-day fluctuations of measurable elements such as temperature, atmospheric pressure, cloudiness, winds, and rain. **Climate** describes the longer-term conditions of the atmospheric environment at a place over periods of years, centuries, and millenia. Weather and climate profoundly affect people's lives. Not only do they vary over time, but also from place to place. The number of Americans who travel from the frozen Midwest and Northeast to the warmth of Florida in late winter testifies to a strong understanding of likely climate differences between the two places at that time of year.

In this book, the atmosphere-ocean environment forms the subject matter of Chapters 2 through 6. This chapter focuses on the composition and structure of the atmosphere and oceans—the environmental framework in which the processes of heating, movement, and water circulation take place to cause weather. Chapter 3 explains the heating of the atmosphere-ocean environment and how this causes wind and ocean current movements. Chapter 4 is about the circulation of water through the atmosphere-ocean environment, beginning with the ocean store, continuing in the formation of cloud and rain, and then bringing the water back to the ground and oceans. Chapter 5 combines the results of these processes in a consideration of weather systems such as cyclones, tornadoes, and hurricanes. Chapter 6 considers the longer term view of activity in the atmosphere-ocean environment, providing an account of world climates and changes in their geographic extent over time.

THE ATMOSPHERE

COMPOSITION OF THE ATMOSPHERE

Earth's **atmosphere** forms an envelope around the planet (Figure 2-1). Gases make up most of the atmosphere's volume with dust particles and water droplets suspended in the lower part (Figure 2-2).

The atmospheric gases exist as **molecules**, combinations of atoms held together by electric bonds. An **atom** is the smallest particle of an element that can exist alone; it may combine with atoms of other elements. An **element** is a substance with a distinctive atomic structure containing only one kind of atom (e.g., oxygen, nitrogen, iron). The combined atoms in a molecule may be of the same element, as in oxygen (two atoms of oxygen), or of different elements, as in water (two atoms of hydrogen to one of oxygen), which is a **compound**. An **ion** is an atom or molecule that has a positive or negative electric charge.

Major Atmospheric Gases   The lowest 100 km (65 mi) of the atmosphere contains some 99 percent of its total weight; half of this is in the lowest 5 km (3.5 mi). The atmosphere gradually fades into space several hundred kilometers above the surface. The lowest 80 km (50 mi) of atmosphere has a consistent composition of gases (Figure 2-3). It is made up mainly of **nitrogen** (78.08 percent by volume in dry air), **oxygen** (20.95 percent), and **argon** (0.93 percent).

Minor Atmospheric Gases   The three major gases make up 99.6 percent of the volume of dry air in the lower atmosphere. A "dry" atmosphere is one with little water vapor. The remaining 0.04 percent comprises small amounts of a variety of gases that are often more significant than their proportion suggests. **Carbon dioxide** is the most important member of this group, making up 0.03 percent of the total volume of dry air. Other minor gases include hydrogen, helium, neon, krypton, xenon, ozone, sulphur dioxide, and methane. New gases, such as the chlorofluorocarbons (CFCs) are being added to the atmosphere by human activities. **Water vapor** is a gas in Earth's atmosphere that is more variable than others over short periods of time and from place to place. It ranges from almost zero to 4 percent of the atmospheric volume.

Figure 2–1

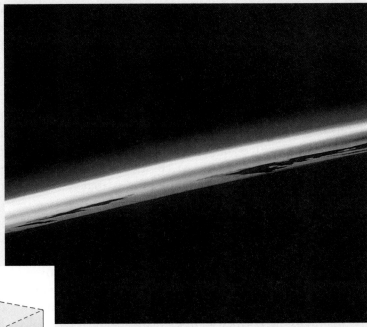

Figure 2–2

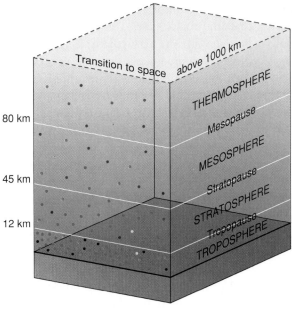

- Hydrogen, helium ions
- Nitrogen molecules
- Oxygen molecules
- Argon molecules
- Water vapor molecules
- Dust particles

Figure 2–1 Earth's atmosphere: a section through layers at sunrise near Hawaii. The darker colors are high clouds (cirrus) and the remains of the Mount Pinatubo ashes erupted in June 1991. The lower atmosphere is red because the sun's rays are scattered by dust, water vapor, and smoke.

Figure 2–2 Earth's atmosphere: composition. The troposphere contains 75 percent of the total weight of the atmosphere, including all the water vapor and dust. What are the main features of the atmosphere above the troposphere?

Figure 2–3 Earth's atmosphere: proportions of gases in the lowest 80 km (50 mi). Compare the major, minor, and variable gases.

Figure 2–3

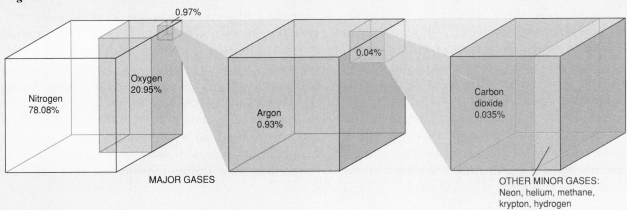

VARIABLE GASES: Water vapor (0–4%)
carbon monoxide, ozone,
sulfur dioxide, nitrogen oxide

Other Atmospheric Constituents   The atmospheric gases produce a buoyant medium that keeps very fine particles in suspension. Fine dust and liquid droplets called *aerosols* may remain in the atmosphere for months or years. They are visible when concentrated in smoke or clouds but are commonly too small to be seen. The fine dust particles come from sea salt, rock dust from volcanoes or exposed soils, fine organic matter, and the products of combustion in furnaces and car engines. The combination of natural and human sources makes them most highly concentrated over cities.

EXCHANGES AND FLUCTUATIONS OF ATMOSPHERIC CONSTITUENTS

Continuous exchanges of molecules and atoms occur among the atmosphere, the oceans, living organisms, soil, and rock minerals. These exchanges maintain the balance of gases that is characteristic of the lowest 80 km (50 mi) of the atmosphere. Fluctuations in the balance occur over long periods of time and depend on the chemical characteristics of the gases. The increasing role of human activities in adding new gases or enhancing the proportions of minor gases is an important environmental issue at present.

Nitrogen makes up over 75 percent of the lower atmosphere but is not very active chemically. It is exchanged between the atmosphere and plants or soil but in a small number of processes. Lightning and some land-based microorganisms "fix" atmospheric nitrogen into soil compounds. Other bacteria release nitrogen from the plants and soil into the air. Chapter 12 describes more fully the nitrogen cycle that results. On average, a nitrogen molecule spends about 42 million years in the atmosphere before being fixed again. Since nitrogen is an important plant nutrient, the fertilizer industry has found chemical processes that can fix nitrogen, increasing the rate of removal of nitrogen from the atmosphere. Although the total impact is relatively small, other effects of increased nitrogen at ground level are noticeable. For instance, rain washes nitrogen-based fertilizer into ponds and lakes, providing extra food for microorganisms and causing them to grow, reproduce, and die so rapidly that their remains foul the waters.

Oxygen is the most chemically active of the major atmospheric gases. During Earth's existence it has become an element in virtually all nonliving matter at the planet's surface. Its main reactions now involve exchanges with living tissue. Plants add oxygen to the atmosphere during the interaction of sunlight with plant leaves—a process called *photosynthesis* that is explained in Chapter 12. Animal and certain types of plant respiration, the decay of organic matter, and the burning of plants and fossil fuels remove oxygen from the atmosphere. Chapters 8 and 12 refer to this cycle of oxygen exchange between the atmosphere, rock minerals, and living organisms. The greater chemical activity of oxygen, compared to nitrogen, means that each oxygen molecule spends approximately 5000 years in the atmosphere.

Argon is the least chemically reactive of the major atmospheric gases and occurs in the smallest quantities. It enters the atmosphere from the breakdown of radioactive potassium in surface rock minerals; any changes in its proportions are very slow.

Many minor gases vary in quantity more rapidly than the major gases. Variations in the amount of carbon dioxide are particularly significant because carbon dioxide complements the role of oxygen in living organisms. Plants take carbon dioxide from the atmosphere for photosynthesis. Animal and plant respiration, organic decay, and the combustion of fossil fuels release it back into the atmosphere. The average carbon dioxide molecule is in the atmosphere for only five years before chemical reactions return it to living tissue or rock-forming minerals. This is shorter than the time it takes to spread the changes in carbon dioxide concentration throughout the atmosphere; therefore, local and seasonal variations in carbon dioxide concentration are common. Bursts of photosynthesis in land plants in spring and early summer deplete the carbon dioxide in midlatitudes, but less photosynthesis and more organic decay restore levels in the late summer and fall.

Carbon dioxide is also one of the greenhouse gases that absorb energy and heat the air around them. If the proportion of the greenhouse gases increases, the atmosphere is likely to get warmer (see Chapters 3 and 6). Other minor gases that contain carbon are also greenhouse gases. They include methane (carbon bonded to four atoms of hydrogen), carbon monoxide (carbon bonded to a single atom of oxygen), and the chlorofluorocarbons. Methane comes from burning vegetation, the decomposition of plant matter, and cattle digestion. Car exhaust gases and burning vegetation add carbon monoxide to the atmosphere. As human populations and their food and energy needs increase, some carbon-based gases in the atmosphere also increase (Figure 2-4). Research shows that carbon dioxide rose from 260 parts per million (0.026 percent) in the eighteenth century

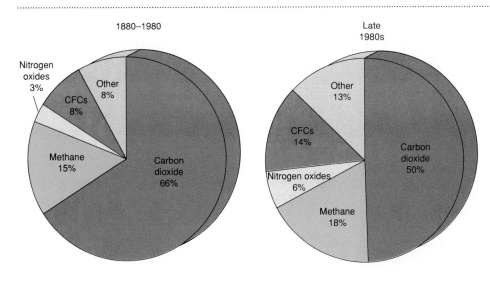

1880–1980

Nitrogen oxides 3%

CFCs 8%

Other 8%

Methane 15%

Carbon dioxide 66%

Late 1980s

Other 13%

CFCs 14%

Nitrogen oxides 6%

Methane 18%

Carbon dioxide 50%

Figure 2-4 Earth's atmosphere: minor gases that absorb long-wave rays and help to heat the surrounding air. The pie charts show the changes in relative proportions of the minor gases during the 1980s. What proportion of Earth's atmosphere do these gases constitute?

Source: Data from V. Ramanathan, et al., *Journal of Geophysical Research* 90:5557–5566, 1985; and J. Hansen, et al., *Journal of Geophysical Research* 93:9241–9364, the American Geophysical Union.

to nearly 350 parts per million (0.035 percent) today, while methane rose from 700 parts per billion (0.00007 percent) to 1700 parts per billion (.00017 percent), over the same period. Water vapor also acts as a greenhouse gas. The atmosphere's content of water vapor is subject to fluctuations produced by variations in the availability of moisture and in the temperature of the atmosphere which are discussed further in Chapter 4.

Sulfur dioxide and nitrogen oxides are two other gases that increase as a result of human activities. Burning fossil fuels releases much more sulfur into the atmosphere than natural processes. Sulfur compounds react with oxygen to form the noxious gas, sulfur dioxide. Sulfur compounds in the air above the United States increased by 50 percent from 1940 to 1970 before antipollution measures caused a slight fall. At the time of rising sulfur levels, sulfur compounds damaged plant and animal tissues and the surfaces of building stones. Nitrogen oxides increase in the atmosphere as part of the nitrogen cycle—for example, forest burning, car exhaust fumes, and the use of nitrogen-based fertilizers. Further reactions in the atmosphere between sulfur dioxide, nitrogen oxides, the sun, and water vapor convert the gases to sulfuric and nitric acids that fall to the ground as acid rain (see Environmental Issue: Acid Rain—Who Is to Blame? p. 94).

Ozone is a compound of three oxygen molecules produced by chemical actions in the presence of sunlight. It is concentrated in a layer 20–40 km (12–25 mi) above the surface, where it plays an important role in intercepting incoming ultraviolet rays from the sun that would damage living organisms at Earth's surface. There is a current international concern over the growing "ozone hole" over the South Pole and the thinning of the ozone layer around the globe (see Environmental Issue: The Ozone Issue, p. 35).

The increase of carbon-based gases, sulfur dioxide, nitrogen oxides, and ozone, cause air **pollution**—the lowering of air quality by adding materials to the atmosphere. Air pollution increased markedly from the early nineteenth century, particularly since manufacturing became concentrated in large urban-industrial areas. By the mid-twentieth century, pollution from burning coal was causing increased death tolls especially in the Pittsburgh region and in many parts of northern Europe. In the 1950s, deaths from air pollution led the United Kingdom government to restrict the burning of coal in open fires and in steam engines. The smoke from burning coal, combined with water droplets in the atmosphere, resulted in smoke fog, or **smog.**

More recently, cars, trucks, and airplanes have added further fossil-fuel pollutants to the atmosphere including nitrogen oxides. These pollutants also reduce visibility, and the term smog is still applied to the effect although it is no longer strictly a smoke fog. Summer haze from these pollutants reduces visibility in the eastern United States to 25 km (11 mi) from the natural 150 km (90 mi) that is still common in parts of the American West. The yellow-brown pollution domes over such western cities as Denver stand out against blue skies. In portions of the Los Angeles basin, visibility on a good day is only 14 km (8 mi), (Figure 2-5). Concern about atmospheric pollution in the 1960s led to legislation that reduced the emissions of coal smoke and other pollutants and resulted in cleaner atmospheres in many U.S. cities by the 1980s.

Figure 2-5   Los Angeles smog: the yellowish haze near the ground is formed by the reaction of sunlight with car exhaust fumes.

The proportions of gases in the lowest 80 km of the atmosphere are relatively constant. Dry air at this level is composed almost entirely of nitrogen, oxygen, and argon. It includes smaller proportions of carbon dioxide, carbon monoxide, methane, hydrogen, helium, neon, krypton, xenon, and ozone. Of the atmospheric gases, water vapor is the most variable in proportion. The concentrations of some gases, notably the carbon-based gases, sulfur dioxide, and nitrogen oxides, increase as a result of human activities. The atmosphere also contains aerosols, water droplets, and ice particles.

STRUCTURE OF THE ATMOSPHERE

The atmosphere is made up of a series of vertical layers defined by their temperature and characterized by the reducing concentrations (or density) of the gases away from Earth's surface. The layering of the atmosphere affects movements of air and is important in controlling surface weather.

Density and Pressure in the Atmosphere   Gases consist of molecules that are not bonded together (as in liquids or solids). They can be compressed into small spaces, or float freely where they are not confined. The downward pull of gravity compresses and concentrates atmospheric gases near the ground. Lighter gases float freely near the top of the atmosphere. The concentration of atmospheric gases, therefore, decreases with height above Earth's surface.

"Concentration" is often measured as **density** (mass per volume). Water, for instance, has a density of

1 gram per cubic centimeter. The density of air near the ground is 0.001293 g/cm^3 (1/800 the density of water). At 7 km (4.5 mi) above the ground, the density falls to 0.00066 g/cm^3 and at 100 km (65 mi) it is 5 ten-billionths of a gram per cubic centimeter.

Pressure is the amount of force per unit of area, and **atmospheric pressure** is the amount of force exerted by atmospheric gases on a specific area, either of Earth's surface or of another level in the atmosphere. Atmospheric pressure at ground level averages 1.034 kilograms per square cm (14.7 lb. per square inch). Meteorologists express atmospheric pressure in units called millibars (mb). The average value at the ground near sea level or ocean surface is 1013.25 mb.

Differences of density result in differences of atmospheric pressure. Where more air molecules fill a given space, they collide more often and exert a greater force on every surface (Figure 2-6). Atmospheric pressure is lower in the less-dense upper atmosphere than just above the ground. At 5.5 km (4 mi) above the surface, the average atmospheric pressure is half that at the surface. People visiting places above 3000 m (10,000 ft) often find their breathing takes time to adjust to the lower oxygen density where atmospheric pressure is less.

Temperature and Atmospheric Layers   The atmosphere is divided into layers by temperature characteristics. Some layers of the atmosphere heat more readily than others. In some layers temperature decreases with height; in others, it increases with height. The relationship between temperature and density of the air affects the amount of heat energy present at each level.

The layer of atmosphere next to the ground is the **troposphere** (Figure 2-7). It is heated from the ground up, and temperatures generally decline with height. The troposphere contains 75 percent of the total weight of the atmosphere, including virtually all the water vapor and dust. The top boundary of the troposphere, the *tropopause*, varies from 8 to 9 km (5 to 6 mi) above the surface over the poles to 16 to 17 km (10 to 11 mi) high over the equator. The height of the troposphere varies because the sun gives more heat to places near the equator and both warms a deeper layer of air and provides for its expansion. The tropopause is an effective "lid" that prevents most atmospheric turbulence generated at Earth's surface from extending upward. The troposphere is the layer in which weather occurs.

The *stratosphere* is the layer above the tropopause. It has a lower atmospheric density than the troposphere. In the lower part of the stratosphere, temperature changes little with height; but in the upper part,

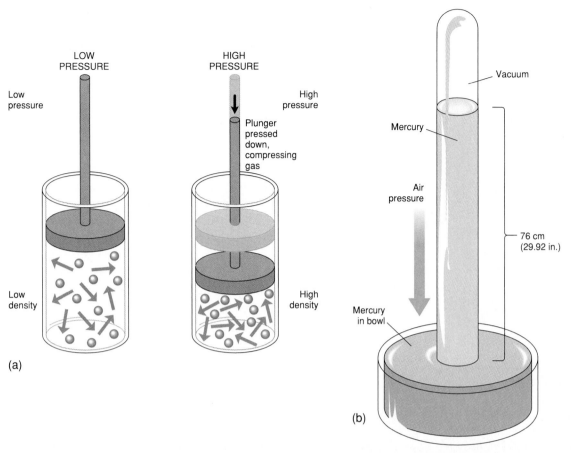

Figure 2-6 Atmospheric pressure. (a) An increase of pressure leads to the compression of gas molecules. This results in greater molecular activity and higher temperatures in the compressed gas. (b) A barometer measures changes in atmospheric pressure based on these principles. The pressure of the weight of air above the barometer forces the mercury downward and up the central tube. Average atmospheric pressure is when a column 76 cm (29.92 in) high is supported; this equals 1013.25 millibars.

temperature rises. Ozone is concentrated in the upper part of the stratosphere where it absorbs ultraviolet rays from the sun and heats the whole layer. The *stratopause* is the upper limit of the stratosphere.

The *mesosphere* lies above the stratopause. Its gases do not absorb the sun's rays and so temperatures fall with height in this layer to –90°C (–80°F) at the upper boundary, the *mesopause*. The mesopause is about 80 km (50 mi) above Earth's surface.

Above the mesopause is the uppermost layer of the atmosphere called the *thermosphere*. The air in this layer is extremely low in density, consisting of ions of the main atmospheric gases and lighter gases such as hydrogen and helium. The tiny ions of the gases in this layer absorb the shortest solar rays (the shortest ultraviolet rays, X rays, and gamma rays) and their temperatures rise to 1200°C (2200°F). The thermosphere may extend out to some 1000 km above the surface, gradually fading to the near-vacuum of space.

A simplified view of this complex layered arrangement treats the atmosphere as a lower zone, the troposphere, in which turbulence produces

surface weather, and an upper zone, all the higher layers, that acts as a filter for the harmful aspects of the sun's rays. As altitude above Earth's surface increases, atmospheric density and pressure decrease. Although the temperatures of individual particles increase in the upper stratosphere and thermosphere, the heat energy available decreases as the density decreases. It is not nice and warm outside a jet at 12,000 m (37,000 ft)!

In the atmosphere, pressure exerted by air is greatest at Earth's surface where most of the mass of the atmosphere is concentrated and air is most dense. Atmospheric pressure is measured in millibars.

The atmosphere can be divided on the basis of temperature into a layered structure of troposphere, stratosphere, mesosphere, and thermosphere. The troposhere is in contact with Earth's surface and is the environment of processes that cause weather. The upper layers act as a filter for the sun's harmful rays.

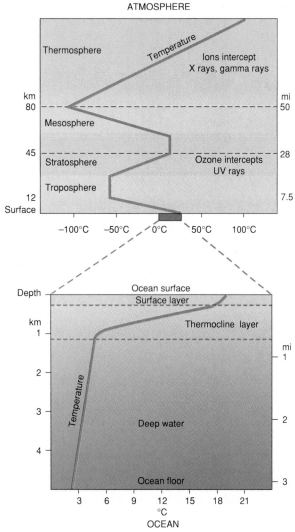

ATMOSPHERE

Figure 2-7 Layers of the atmosphere and ocean. Temperature variations produce distinctive layers. Which layers in the atmosphere are warmer, and which are cooler? How does the ocean compare in depth and heating to the atmosphere?

THE OCEANS

COMPOSITION OF THE OCEANS

The atmosphere is composed largely of gases but also contains some water droplets and dust particles. The **oceans** are made up mostly of liquid water but also contain a variety of salts and some gases, such as oxygen and carbon dioxide, in solution and are covered by masses of ice in the polar regions. The concentration of dissolved salts in seawater is known as **salinity** and is measured in parts per thousand by weight. The average salinity of ocean water is 35 parts per thousand, and ranges from 33 to 38 parts

per thousand. Lesser and greater densities than those in the open oceans occur in almost closed seas, such as the Baltic Sea where dilution by fresh water reduces salinity to 8 parts per thousand and the Red Sea where intense evaporation increases it to 40 parts per thousand. The highest salinities are in large lakes in desert areas, such as the Dead Sea and the Great Salt Lake of Utah (Figure 2-8).

The saltiness of ocean water results from the accumulation of chemicals after the wearing down of continental rocks. The soluble products of chemical reactions with rocks exposed to the atmosphere wash down to the oceans in rivers and glaciers or are blown there by wind. Once in the ocean, further chemical reactions may cause some minerals to come out of solution and form ocean-floor deposits, while organisms remove other elements for use in their own bodies or shells. Salts not removed by these processes remain in the ocean water and cause its salinity. The relative proportions of the major dissolved elements in ocean water—sodium, chlorine, magnesium, calcium, sulfur, and potassium—do not change despite the varying salinities.

The impact of human activities on the oceans receives less media attention than those on the atmosphere. Although oil spills, such as those from huge shipwrecked tankers, get most publicity, there are many other sources of ocean pollution. Coastal towns pump sewage out to sea, and radioactive and other wastes are dumped at sea. People have often treated the oceans as a massive and resilient environment that can cope with such treatment without degradation. The unrestricted development around the coasts of industrial countries, however, can lower the quality of the ocean environment and of the organisms that live in it.

STRUCTURE OF THE OCEANS

Oceans cover 71 percent of Earth's surface. There is an imbalance in the distribution of oceans between the northern and southern hemispheres. Oceans make up a larger proportion of the southern hemisphere (85 percent) than of the northern hemisphere (57 percent).

The water fills huge basins that are produced and rearranged by great movements of the planet's crust (see Chapter 7). The main ocean basins connect with each other and are linked with shallower seas at their margins. Changes in the shapes and sizes of the ocean basins and the withdrawal or additions of water to the oceans affect the level of water in the

Figure 2-8 It is easy to float in the Dead Sea because the greater density of the salty water provides buoyancy.

oceans. For instance, when ice sheets form on land, water does not return to the oceans and sea level falls. When ice sheets melt, the water they contain flows into the oceans and sea level rises.

Water is not as compressible as gas and, in contrast to gravity's effects on the density of atmospheric gases, it has little effect on the density of ocean water. Salinity and temperature control seawater density. Heating of water, as of air, causes it to expand slightly, and this lowers its density. The sun heats surface ocean water to as much as 30°C (86°F). The less dense surface water lies on top of cold, denser water (see Figure 2-7) with temperatures below 5°C (40°F). There is a transitional layer between the warm surface water and the colder deep water, at around 1 km deep, where temperature changes rapidly. It is known as the **thermocline**.

Pressure in ocean water increases with depth as a result of the weight of the overlying water. At 1 km below the surface, the overlying water exerts a pressure that is 100 times the surface atmospheric pressure. The rapid increase of pressure with depth affects the organisms living in the oceans. They have special means of coping with changes of pressure as they change levels by equalizing water pressure inside and outside their bodies. Whales, for instance, can dive from the surface to 500 m below and return within minutes. Humans do not have such capabilities. Deep-sea divers descend and return to the surface slowly so that their bodies can adjust to pressure changes.

EXCHANGES BETWEEN THE ATMOSPHERE AND OCEANS

The atmosphere and oceans continuously exchange gases, water, and solid particles with each other and with other Earth environments. The apparently stable composition of the atmosphere and the oceans results from a balance between these inputs and outputs. Oxygen, carbon dioxide, and water vapor are the most significant gases exchanged. The exchange occurs as a result of photosynthesis and respiration by single-celled plants, algae or phytoplankton, at the ocean surface. Evaporation continuously transfers water from ocean surfaces into the atmosphere, while rain and snow return it to Earth's surface. When waves break, salt particles in solution escape into the atmosphere. These particles are so small that light winds carry them upward. Winds also blow fine particles of dried clay from the land over the oceans where rain brings them down to the ocean surface. Once in the ocean, they sink to the ocean floor.

EVOLUTION OF THE ATMOSPHERE-OCEAN ENVIRONMENT

A speculative scenario of Earth's formation suggests that an accumulation of iron-rich compounds made up the center of the planet and was later covered by somewhat less dense rocky materials that crashed into it around 4 billion years ago. This formed the basis of Earth's inner rocks (core and mantle), from which the surface (crustal) rocks developed. The increased gravity of this growing body attracted the lightest materials—the gases—to produce an envelope of atmosphere around Earth. The gases may have been added by meteorite impacts that also transferred water, rare gases, and even organic chemicals—the basis for living organisms—from the outer parts of the solar system.

In this scenario the atmosphere-ocean environment of today developed out of interactions among the early atmosphere, the surface rocks, and living organisms. Debate continues as to whether the early atmosphere is still with us or whether Earth lost its first atmosphere in a gigantic blast of solar wind. If the latter is true, the present atmosphere

probably formed by the release of gases following the meteorite impacts and volcanic eruptions around 4 billion years ago. At the end of this "outgassing" phase (if it occurred), Earth's atmosphere comprised largely water vapor, carbon dioxide, and nitrogen. Soon after, the oceans were filled with water as rains occurred while the Earth environment cooled (Figure 2-9).

The oxygen content of the atmosphere probably built up slowly, first following the breakup of water molecules reacting with solar radiation and later as the result of photosynthesis (see Chapter 12) by the increasing number of plants in the oceans and on land. The first evidence of marine life appeared over 3 billion years ago, but the oldest fossils of land-based plants and animals occur in rocks only 400 million years old. Not until then did plants provide sufficient atmospheric oxygen to form a shield against the harmful rays. Only then could plants and animals live out of water on the continental surfaces.

Ocean water contains small proportions of dissolved salts and gases. The proportion of dissolved salts to water, known as salinity, averages 35 parts per thousand in the open oceans. The thermocline separates warmer surface water and colder deep ocean water. Materials exchanged between the atmosphere and oceans include oxygen, carbon dioxide, water, salt, and dust. The atmosphere and oceans developed jointly as external and internal conditions changed.

Atmospheric gases interact with other Earth environments, involving chemical reactions and interchanges. The gases are added by the radioactive decay of minerals, by volcanic activity *(solid-earth environment)*, by chemical reactions with surface minerals *(surface-relief environment)*, and by the living processes of plants and animals *(living-organism environment)*. Organic activity and chemical reactions with surface minerals remove them from the atmosphere. Water cycles through ocean, atmosphere, and rocks. Wave action, volcanic activity, desert dust storms, and human activities produce aerosols.

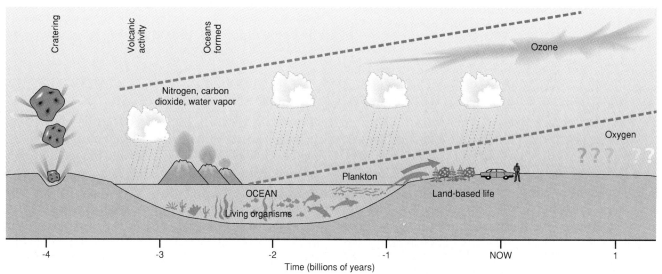

Figure 2-9   Evolution of the atmosphere–ocean environment. Read the diagram from left (over 4 billion years ago) to right to compare the composition of the early atmosphere with today's atmosphere. Explore the link between the ozone layer and the emergence of land plants and animals.

ENVIRONMENTAL ISSUE:

THE OZONE ISSUE

The ozone issue raises several questions:
   What does the ozone layer do?
   Is it thinning, how much, and where?
   What can be done to avoid possible harmful effects?

Ozone is an important component of the stratosphere, where it absorbs many of the incoming ultraviolet rays. Although ozone also occurs near the ground, where it forms a small but dangerous constituent of smog, the stratospheric ozone is the major subject of concern to humans. If the ozone shield in the stratosphere gets thinner, the less intense forms of ultraviolet energy will get through more often, increasing skin cancers and eye cataracts and breaking down the human immune system. The most intense forms of ultraviolet radiation will break apart important molecules in living tissue. Sunscreen products stop the small amounts of ultraviolet rays that penetrate to the ground at present from damaging our skin. The ozone layer in the stratosphere acts as a sunscreen for the planet.

It was not until the mid-1980s that scientists confirmed that a hole the size of the United States develops each spring in the ozone layer over Antarctica. The series of satellite images show how the Antarctic hole has deepened since 1979. Subsequent studies showed less severe but similar depletions in the Arctic and midlatitudes.

Research shows that ozone forms when oxygen molecules combine with additional oxygen atoms in the presence of sunlight, but it is broken apart mainly by reaction with chlorine monoxide. One molecule of chlorine monoxide can destroy many thousands of molecules of ozone. Chlorine monoxide levels rise rapidly over the poles in winter when icy cloud particles form in the main ozone layer and convert chlorine into chlorine monoxide. Sunlight penetrates and warms the polar atmosphere in spring, melting the icy cloud particles, reducing the formation of chlorine monoxide, and encouraging the formation of ozone. The end of winter (October in Antarctica) sees the damage to the ozone layer at its greatest. Damage is less in other parts of the world because exchanges of air take place with adjacent regions; the Antarctic stratosphere is almost isolated from its surroundings.

Chlorine is not a natural constituent of the Antarctic atmosphere. The main source of the added chlorine in the atmosphere is the chlorofluorocarbon (CFC) group of gases. CFCs are manufactured for use in refrigerators, aerosol cans, and the production of plastic insulating foam. They are extremely stable and nontoxic in the troposphere and were originally considered an ideal industrial chemical. They are lightweight gases, however, and when they rise into the stratosphere the intense ultraviolet rays break them down, releasing chlorine. The chlorine is converted to chlorine monoxide, depleting the ozone. CFCs have penetrated the Antarctic atmosphere in small but sufficient quantities to periodically destroy the ozone shield above the continent.

The stability of the CFC gases means that they break down slowly; those in the atmosphere may persist for for up to 100 years. The present problems of ozone depletion in the stratosphere are the result of lower levels of CFC emissions earlier in the twentieth century. High levels of emission over the last few decades threaten further reductions in the ozone layer in the twenty-first century. Concern over the potential effects of the thinning ozone layer in the late 1980s led governments to agree on international standards for CFC emissions. The agreement reached in Montreal in 1986 as part of the United Nations Environmental Program became more rigorous at a London meeting in 1990 when industrialized countries agreed to eliminate all production of CFCs and related chemicals by 2000 A.D. This measure should reduce the destruction of ozone and recreate the ozone layer faster than it is destroyed.

This account of "the hole in the sky" emphasizes the immense complexity of Earth's atmosphere and its processes. It also shows that the best efforts of scientists have resulted in only a partial knowledge of how the atmosphere-ocean environment works. Moreover, the importance now attached to CFCs in the destruction of ozone highlights how humans may unwittingly alter a delicate balance in nature before it is fully understood.

Linkages

The ozone issue demonstrates a number of significant linkages between Earth environments. First, ozone is produced naturally by the reaction of sunlight with oxygen in the atmosphere **(atmosphere-ocean environment).** Second, the ozone layer in the stratosphere provides a protection for living organisms at Earth's surface **(living-organism environment).** Third, human products have unexpected and often negative impacts on the workings of the natural environment.

Questions for Debate

1. Why is the ozone layer over Antarctica thinner than that elsewhere?

2. What happens if the actions agreed to by governments are not carried out? or are not effective?

3. What can be done to reduce the dangers of such unexpected effects from human actions?

ENVIRONMENTAL
ISSUE
(Continued)

(a) Sunbathing will be more dangerous if the ozone layer in the stratosphere continues to diminish. (b) The Antarctic ozone hole as charted from satellites between 1979 and 1992. The hole appears each October at the start of the Antarctic spring. Its intensity fluctuates from year to year.

(a)

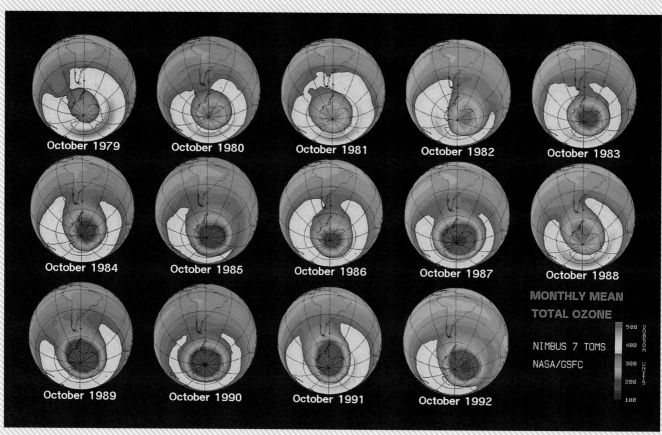

(b)

CHAPTER SUMMARY

1. The proportions of gases in the lowest 80 km of the atmosphere are relatively constant. Dry air at this level is composed almost entirely of nitrogen, oxygen, and argon. It includes smaller proportions of carbon dioxide, carbon monoxide, methane, hydrogen, helium, neon, krypton, xenon, and ozone. Of the atmospheric gases, water vapor is the most variable in proportion. The concentrations of some gases, notably the carbon-based gases, sulfur dioxide, and nitrogen oxides, increase as a result of human activities. The atmosphere also contains aerosols, water droplets, and ice particles.

2. In the atmosphere, pressure exerted by air is greatest at Earth's surface where most of the mass of the atmosphere is concentrated and air is most dense. Atmospheric pressure is measured in millibars.

3. The atmosphere can be divided on the basis of temperature into a layered structure of troposphere, stratosphere, mesosphere, and thermosphere. The troposphere is in contact with Earth's surface and is the environment of processes that cause weather. The upper layers act as a filter for the sun's harmful rays.

4. Ocean water contains small proportions of dissolved salts and gases. The proportion of dissolved salts to water, known as salinity, averages 35 parts per thousand in the open oceans. The thermocline separates warmer surface water and colder deep ocean water. Materials exchanged between the atmosphere and oceans include oxygen, carbon dioxide, water, salt, and dust. The atmosphere and oceans developed jointly as external and internal conditions changed.

5. The ozone issue demonstrates a number of significant linkages between Earth environments. First, ozone is produced naturally by the reaction of sunlight with oxygen in the atmosphere (*atmosphere-ocean environment*). Second, the ozone layer in the stratosphere provides a protection for living organisms at Earth's surface (*living-organism environment*). Third, human products have unexpected and often negative impacts on the workings of the natural environment.

KEY TERMS

weather   *26*

climate   *26*

atmosphere   *26*

molecule   *26*

atom   *26*

element   *26*

compound   *26*

ion   *26*

nitrogen   *26*

oxygen   *26*

argon   *26*

carbon dioxide   *26*

water vapor   *26*

ozone   *29*

pollution   *29*

smog   *29*

density   *30*

atmospheric pressure   *30*

troposphere   *30*

ocean   *32*

salinity   *32*

thermocline   *33*

CHAPTER REVIEW QUESTIONS

1. List the main atmospheric gases and their properties.

2. What is the difference between density and pressure? Illustrate your answer by referring to differences between liquids and gases. Explain the changing density and pressure sequence between the bottom of the ocean and the top of the atmosphere.

3. What factors cause differences in ocean salinity from place to place?

4. Describe the structure of the atmosphere and oceans by reference to differences in temperature.

5. Discuss the factors that affect the level of carbon dioxide in the atmosphere.

6. In what ways are human activities causing changes in the composition of the atmosphere and oceans?

SUGGESTED ACTIVITIES

1. Find out about levels of atmospheric or ocean pollution in your town or region. Have the levels changed in the last 10 years? What regulations are in force and how well do they work?

2. Debate whether the composition of atmosphere and oceans are fixed or in a state of balance that is maintained by continuous exchanges.

CHAPTER

3

HEATING AND MOVEMENT IN THE ATMOSPHERE AND OCEANS

THIS CHAPTER IS ABOUT:

- Heating the atmosphere and oceans
- Temperature differences on Earth
- Heating and winds
- Winds and Earth
- General circulation of the atmosphere
- Ocean currents
- Environmental Issue: Santa Ana Winds and Building Codes

The sun heats Earth's atmosphere and oceans. Different parts of the world have higher or lower temperatures, as the map shows. The heating and cooling of air and water causes movements in the atmosphere and oceans that take heat from near the equator toward the poles. These movements are winds that bend the grass and carry the wind surfer along.

Heat and temperature are basic to the study of weather. Heating of the atmosphere and oceans leads not only to a variety of temperatures in different parts of the world, but also to movements of air and ocean waters and to the circulation of water from the oceans to the continents via clouds and rain or snow. Extreme temperatures affect human comfort, and strong winds lower the effective temperature in wind-chill.

This chapter examines how energy from the sun reaches Earth and interacts with the air and ground to heat the lower atmosphere. It explains the reasons why different places around the world have different temperatures. This understanding provides a basis for demonstrating how differences in heating the atmosphere and oceans cause movements within them such as winds and currents. The geographic patterns of winds and currents, initiated in this way, are seen to be affected by Earth's rotation and by interactions between the winds, land, and water surfaces.

HEATING THE ATMOSPHERE AND OCEANS

Energy is needed to make any system work. Animals, for instance, get their energy from food—a form of chemical energy storage. Virtually all the energy required for the working of the atmosphere-ocean environment comes from the sun. Only 0.1 percent of the energy comes from Earth's hot interior. Energy from the sun is received by Earth and converted to heat; contrasts in heating drive movements in the atmosphere and oceans.

HEAT AND TEMPERATURE

Heat and temperature are closely related but distinct concepts. In the thermosphere, for instance, temperature is high, but heat is low (see Chapter 2). **Heat** is the total energy of molecular movement within a body; the greater the molecular movement, the greater the heat energy. Heat is measured in **calories**: one calorie is the amount of heat needed to raise the temperature of 1 gram of water 1°C at sea-level pressure. (Note: one dietary calorie is 1000 calories.) **Temperature** measures the average energy of the individual molecules in a substance. As heat energy is absorbed, molecules move more rapidly, raising the temperature. The temperature of a substance is measured by a thermometer. Physical geographers normally use either the Celsius or Fahrenheit scales of temperature, in which pure ice melts at 0°C and 32°F. In the thermosphere individual molecules have high

energy and, therefore, high temperatures. Heat is low because of the small number of gas molecules present.

Substances react in different ways to heating. The **specific heat** of a substance is the amount of heat required to change the temperature of 1 g by 1°C. Substances with high specific heats, such as water, require more heat to raise their temperatures than those with lower specific heats, such as metals. For example, one calorie of heat raises the temperature of one gram of water by 1°C and that of one gram of copper by 11°C.

ENERGY TRANSFERS

Energy is transferred between or through substances by three processes—conduction, convection, and radiation (Figure 3-1). Conduction transfers heat energy between adjacent bodies and convection transfers it by movement within a gas or liquid. Radiation transfers a range of energy types across space and through matter.

Conduction   The transfer of heat energy within and between substances in contact with each other, but without movement of the substances, is called **conduction.** The more active molecules in the hotter material transfer their vibrations and heat energy to the colder substance where they enliven the sluggish molecules. Figure 3-1a illustrates a common example of conduction. The metal pan on a burner is heated by conduction from the intense molecular vibrations in the hot burner in contact with the saucepan. The hot metal of the pan then conducts heat to a thin layer of water next to it. Water, however, is a poor conductor and heat is transferred very slowly through it by conduction.

In the natural environment, conduction works best in solids such as rock minerals; air and water are both poor conductors. Conduction is a less important means of heat transfer in the atmosphere-ocean environment than in the solid-earth environment. The poor heat conductivity of air, however, makes it a good insulator. This property is used in placing an air gap between panes in storm windows.

Convection   **Convection** occurs in air and water, causing them to transfer heat by movement. Heating part of an air or water body causes that part to expand and rise as a result of its lower density compared to the rest of the body. Cooling part of an air or water body causes that part to contract, increase in density, and descend. Circulation is established as warmed, lower density matter is replaced by cooler, denser matter, which is then warmed—and so on. In the saucepan example of Figure 3-1a, the lowest section of water is heated by conduction from the heated pan. Since water is a poor conductor, convection plays the main part in spreading heat through the water. The heated water rises toward the surface and is replaced by cooler water which is heated at the base. Over time, the whole body of water is heated.

Figure 3-1b shows how convection circulates heat upward in the atmosphere from heated ground or water surfaces. In the oceans, water cooled at the surface in polar regions descends and flows beneath the warmer surface water.

Radiation   All bodies with temperatures above the point at which molecular movement stops (–273°C, –459°F) radiate energy. **Radiant energy** is transferred in waves that may pass through some substances (e.g., radio waves pass through walls into a room) and be absorbed by others. When a substance absorbs radiant energy, its heat energy increases.

Radiant energy includes forms with different wavelengths or frequencies (Figure 3-1c). The forms range from long wave, low-frequency radio waves through heat (infrared) waves to short-wave, high-frequency visible light waves, ultraviolet waves, X-ray waves, and

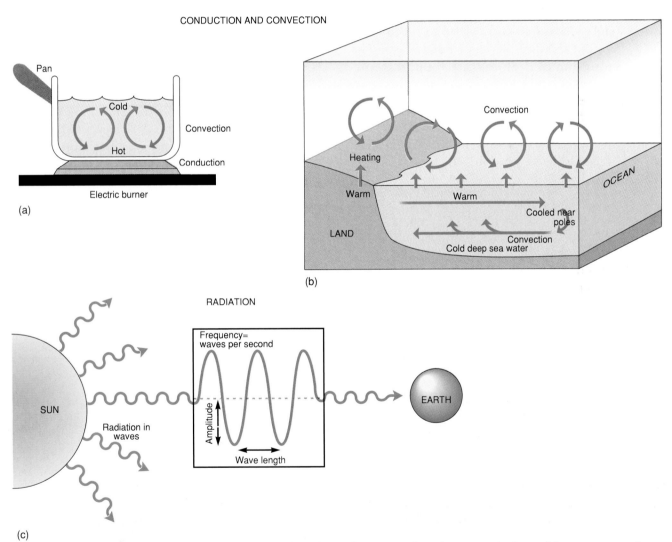

Figure 3-1   Energy transfer. (a) Conduction transfers heat from a hot to a cooler substance at the base of the pan; convection transfers heated water upward, replacing it with cooler water until all the water is boiling. (b) In the atmosphere and ocean, conduction transfers heat into the air a few centimeters above land and ocean; convection transfers heat in moving bodies of air and water. Describe the processes shown. (c) Radiation transfers a range of different forms of energy in waves that pass through some materials and are absorbed by others.

gamma-ray waves (Figure 3-2). The shorter waves have a higher frequency and transfer energy more intensely.

Hotter bodies emit more energy in shorter wavelengths than cooler bodies. For instance, the sun, with a surface temperature of 6000°C, emits a range of rays concentrated in the shorter wavelengths and has its peak in visible light waves. Earth, with an average surface temperature of 12°C, emits less intense, longer wavelength infrared rays.

ENERGY FLOWS THROUGH THE ATMOSPHERE-OCEAN ENVIRONMENT

All three processes—conduction, convection, and radiation—are involved in heating Earth's atmosphere and oceans. Energy radiated from the sun interacts with the atmosphere as it passes through and then with the oceans and land surfaces. Conduction, convection, and radiation all take part in circulating the energy received through the lower atmosphere. The energy returned from Earth to space balances the incoming energy.

Insolation Reaches Earth   The energy radiated from the sun toward Earth is known as **insolation** (INcoming SOLar radiATION). It consists mainly of

short-wave radiation that is transferred across 150 million km (94 million mi) of space without alteration. At such a distance, Earth intercepts only a tiny part of the total solar output, but that is sufficient to provide the energy needed to drive the atmosphere-ocean system. About 9 percent of insolation is in wavelengths shorter than visible light waves, 45 percent is in visible light waves, and 46 percent is in the wide band of infrared rays. The shorter, more intense waves contain nearly all the energy.

The sun's energy output varies little over time, and the insolation arriving at the outside of Earth's atmosphere hardly varies. It can be regarded as unchanging over periods of a few decades. The most recent satellite measurements give an average global insolation value of 1.96 calories per square centimeter per minute.

Places on Earth, however, receive a fluctuating supply of energy because the planet is spherical, rotates on a tilted axis, and has an elliptical orbit around the sun. Earth's *spherical shape* causes insolation to be received at varying angles by different places on the globe (Figure 3-3). Places where the sun is vertically overhead have a greater intensity of

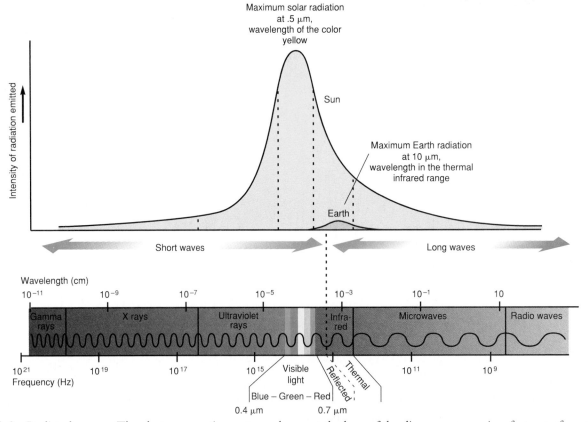

Figure 3-2   Radiated energy. The electromagnetic spectrum shown at the base of the diagram summarizes features of a range of energy forms from those with long wavelengths and low frequencies at the right end (radio waves) to those with short wavelengths and high frequencies at the left end (gamma rays). A micrometer (μm) is one-thousandth of a millimeter. What do the patterns of solar and terrestrial radiation tell about the surface temperatures of the sun and Earth?

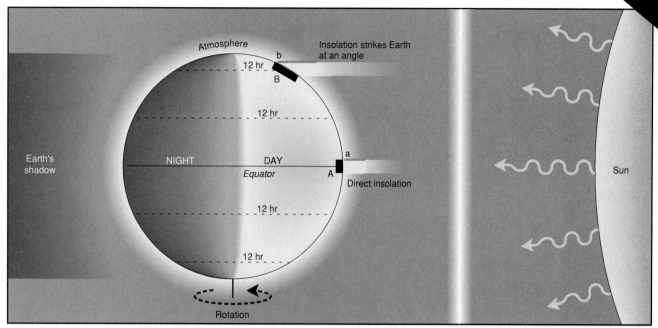

Figure 3-3 Insolation and Earth. At an equinox, the sun is vertically overhead at the equator and everywhere on Earth has 12 hours of day and 12 hours of night. At the equator (a), insolation focuses on a small area and the rays of the vertical sun pass through a short distance of atmosphere (red line). How does the receipt of insolation at the Arctic Circle (b) differ from that at the equator?

insolation than those where the sun is at a low angle since the same amount of radiation is either focused on a small area or spread over a larger area, respectively. The lower-angle rays also take a longer path through the atmosphere, which further reduces their intensity when they reach the ground at high latitudes. Places that face away from the sun receive no insolation.

Earth *rotates* on its axis causing daily variations of insolation around the world. The half of the globe facing the sun receives insolation during daylight hours; the half facing away experiences night. At a given location, the daily cycle of insolation begins at sunrise, increases until the peak at solar noon when the sun is most directly overhead, and decreases through the afternoon.

The *tilt of Earth's axis* of rotation produces seasonal variations in insolation and in the lengths of day and night at different latitudes (Figure 3-4). As Earth revolves around the sun during its annual orbit, the angle of its axis to a line perpendicular to the plane of orbit remains 23.5°. As a result, the sun is vertically overhead at latitude 23.5°N (the Tropic of Cancer) on June 21 and at latitude 23.5°S (the Tropic of Capricorn) on December 21. The two dates are known as the **solstices** because they mark the maximum distance from the equator of places with a vertically overhead sun. When the sun is vertically overhead at the Tropic of Cancer, the northern hemisphere has its

summer season and longer daylight periods. Places within the Arctic Circle (66.5° N) have daylight for twenty-four hours. When the sun is vertically overhead at the Tropic of Capricorn, the northern hemisphere has its winter season with shorter days, and places within the Arctic Circle receive no daylight. On March 21 and September 21 the sun is vertically overhead at the equator and all parts of the world have daylight and night for twelve hours. These occasions are known as **equinoxes** ("equal nights").

Earth's orbit round the sun is elliptical (oval) rather than circular, and the sun is not quite at the center of this oval path. Earth is closer to the sun at one season than the others. At present, Earth is closest to the sun (*perihelion*) on January 3 and receives 7 percent more insolation then than it does on July 4 when it is farthest away (*aphelion*). This difference might cause temperatures to be higher by 4°C in January, but the predominance of oceans in the southern hemisphere (where most insolation is received in January) minimizes the impact.

Insolation Passes Through the Atmosphere The relatively short passage through Earth's atmosphere has major impacts on insolation which travels 150 million km through space with little change. Just about half of the insolation arriving outside of the atmosphere reaches the surface. The gases, dust, and clouds in the atmosphere absorb, scatter, and reflect

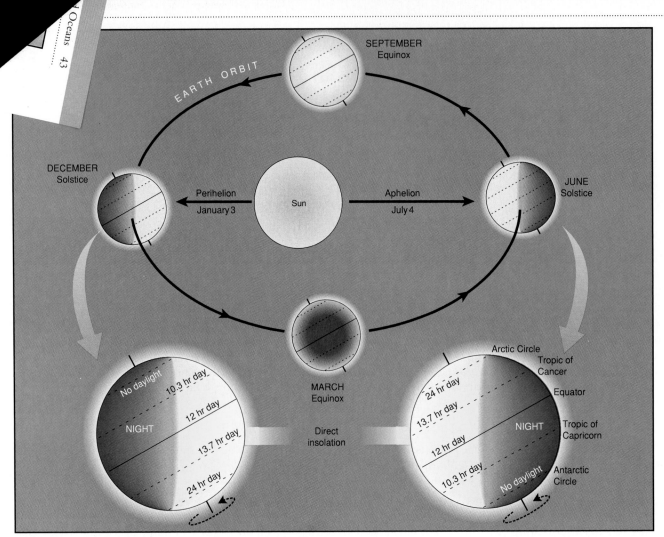

Figure 3-4 Seasonal changes in insolation. As Earth revolves in its orbit around the sun, Earth's tilt causes the sun to be vertically overhead at different positions between 23.5°N (June) to 23.5°S (December). The northern hemisphere summers in June (right globe) have 24 hours of daylight in the Arctic Circle and 24 hours of darkness in the Antarctic Circle. What happens in the northern hemisphere in December and September?

the remainder (Figure 3-5). **Absorption** is the retention of radiant energy by gases, dust, or water droplets in clouds. **Scattering** occurs when radiation waves hit gas molecules, particles of dust, or water droplets and are not absorbed. The radiation bounces off in all directions—upward, sideways, and downward. **Reflection** is scattering in a single direction. Insolation bounces off substances that do not absorb it and returns directly to space. The **albedo** of a surface is the proportion of insolation hitting it that is reflected.

Insolation *absorbed* by atmospheric gases, dust, and water droplets is transformed to heat and raises the temperature of the absorbing substance. The substance then reradiates longer-wave energy that may be absorbed by other materials. Many of the

shortest-wave X rays and gamma rays are absorbed by the gas ions in the thermosphere, heating them to temperatures of 1200°C (2200°F) (see Figure 2-7). Ozone in the upper stratosphere absorbs ultraviolet rays, heating the molecules to temperatures of 20°C (70°F). The combined effect of these two layers absorbing the shortest-wave insolation is to provide a protective shield for living organisms at Earth's surface. In the troposphere, carbon dioxide, water vapor, and other greenhouse gases absorb some of the incoming infrared rays and are heated. The atmosphere, however, absorbs few visible light rays and acts as a transparent window to them.

Some radiation is not absorbed, but is *scattered*. This causes insolation to reach the ground from all parts of the sky, not merely from the direct solar rays. Scattering by atmospheric gas molecules affects the shorter wavelengths of visible light most—the blues and greens. Scattering of blue light in this way produces the blue color of a daytime sky as seen from the ground. Dust and water droplets, which are concentrated over cities, scatter all visible light wavelengths and so produce a much lighter blue-gray sky. The orange and red colors of sunsets (Figure 3-6) are produced when sun passes through the atmosphere at a low angle. All the blue, and much of the green, light are absorbed in the greater thickness of atmosphere. The red and orange colors are particularly vivid where the atmosphere contains dust and smoke particles.

Clouds both absorb and scatter insolation and allow less visible light to penetrate as they become thicker. Below a cloud cover, the visible light is *diffused* (Figure 3-7). During daylight, any point on Earth's surface receives some energy that comes directly from the sun, some that is diffused through clouds, and some that has been scattered by the atmosphere.

Radiation that is not absorbed is *reflected* off substances. Almost one-fifth of total insolation is reflected back to space from the tops of clouds. Dust particles and Earth's surface also reflect insolation back to space. The albedo of Earth as a whole is 0.34, or 34 percent of total insolation received.

Solar Energy and Earth's Surface   Earth's surface reflects or absorbs the insolation that reaches it. The proportions of reflection and absorption depend on the nature of the surface materials and their albedo (Figure 3-8). On land, the main distinction is between light (high reflection) and dark (low reflection) surfaces. The albedo of fresh snow may be over 90 percent but that from a blacktop road may be as low as 5 percent. Over the oceans the albedo depends on the sun angle. If the sun is vertically overhead, nearly all the insolation is absorbed (albedo of 3 percent); if it is only 5 degrees above the horizontal, the ocean albedo may be as much as 50–80 percent.

Insolation that is not reflected is absorbed by Earth's surface and transformed into heat energy. The oceans under a high sun angle and dark surfaces on land absorb high proportions of insolation and the heating process raises the temperature of the surface. A significant proportion of solar energy reaching the ocean surface is taken up by the evaporation of water which demands high energy inputs (see Chapter 4 regarding the impact of solar energy on water surfaces).

About 80 percent of insolation reaching Earth's surface hits the oceans. Although oceans cover 71 percent of Earth's surface, they dominate the tropical zone and so receive a high proportion of the total insolation. The oceans take longer to heat up and are slower to cool down than the continents because of a combination of factors that are summarized in Figure 3-9.

Although the oceans under a high sun angle absorb a higher proportion of insolation than the continents, they have a higher specific heat than the continental rocks and take more energy and a longer time to heat up.

The insolation penetrates ocean waters to several meters, compared to a few centimeters in continental soils or rocks. This factor, combined with the mixing movements in ocean waters, spreads the heating to a deeper layer.

Ocean areas are subject to continuous evaporation, compared to variable and generally lower rates from continental surfaces. More insolation is, therefore, used in evaporation over oceans than it is over land.

Ocean areas tend to be more cloudy than continental areas and so receive less direct insolation.

Overall, the absorption of so much insolation makes the oceans a huge heat store. The slower rate of heating, however, causes ocean temperature highs to lag further behind the period of maximum energy input from the sun than those of the continents.

Radiation From the Ground   Earth's surface temperature averages 12°C (54°F) as a result of absorbing insolation. With this surface temperature, Earth radiates energy at a much lower intensity than the sun. **Terrestrial radiation** is composed of long-wave, low frequency infrared waves (see Figure 3-2).

Only a small proportion of terrestrial radiation escapes directly to space through narrow "windows" in the atmosphere. The rest is absorbed by gases such as water vapor and the carbon gases, and dust particles in the lower atmosphere. The absorbing gases and dust particles are heated and emit long-wave rays in all directions. Some rays from the gases and dust go upward and eventually out to space; some go downward, back to Earth and are known as **counter-radiation.** The counter-radiation helps to reinforce heating of the ground and increases terrestrial radiation. Terrestrial radiation is more effective in heating the lower atmosphere than insolation; heating the lower atmosphere thus begins from the ground and progresses upward. This process and the denser atmosphere at sea level explain why temperatures are higher near sea level than they are at altitudes of several thousand feet.

Figure 3-5   Insolation: reflected or absorbed? When insolation is intercepted by gas molecules, water droplets, or dust particles (all symbolized by the grey box), it is either reflected away (yellow path), absorbed (black arrows), or passes through. The absorbed energy heats the substance, which radiates long waves.

Figure 3-6   Red sky at dusk in the flat area of the Great Plains. What causes the red color?

Figure 3-7   Diffused sunlight through clouds. In this photograph, taken at Florence, Italy, the sun is shining through a layer of cloud. The water droplets in the thin cloud allow a portion of the light through.

Figure 3-5

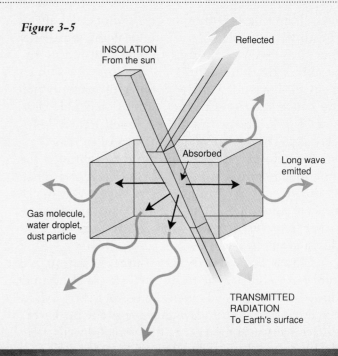

Figure 3-6

Figure 3-7

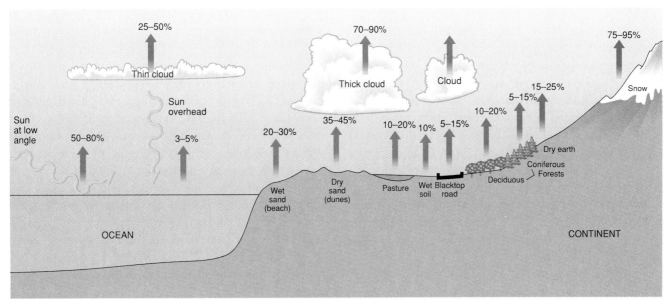

Figure 3-8 Albedos of land and ocean surfaces. The figures illustrate a percentage of insolation that is reflected. Compare different surfaces. How does the albedo affect the heating of the atmosphere immediately above the surface?
Source: Data from R. Barry and R. Chorley, *Weather and Climate,* copyright © 1987 Methuen & Co.

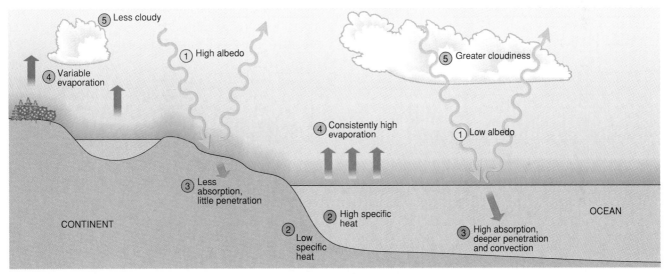

Figure 3-9 Heating of continents and oceans. What is the overall outcome of the five factors shown on the heating of the ocean and continent environments?

The absorption and reradiation of heat by Earth's surface and the dust and gases of the lower atmosphere is complemented by other forms of heat transfer that add to the process of heating the troposphere from the ground up. Some heat is conducted from rocks, soil, and water at the surface into the lowest few centimeters of air. Convection in the atmosphere causes warmed air in contact with the ground to rise.

The process of heating the lower atmosphere from the ground up is also enhanced by the decreasing density of the atmosphere with height. Denser air near the ground contains most of the dust particles, water vapor, and other gases that absorb long-wave energy. Heating is less effective in the less dense air of high altitudes because there are fewer absorbing gases and particles, so temperatures remain low.

The rate at which temperature decreases with height in the atmosphere is known as the **lapse rate.** The lapse rate varies from place to place, depending on the rate of insolation and counter-radiation absorption by the surface and the transfer of this energy into the lower atmosphere. In still air the rate of temperature decrease with altitude at a particular time and place is known as the **environmental lapse rate.** On average, the environmental lapse rate in the troposphere is 6.5°C per km (3.6°F per 1000 feet), but wide variations occur from day to day.

The environmental lapse rate varies from place to place, and over time, because of differences in the heating and cooling of air in contact with the ground. When air near the ground is heated (Figure 3-10a), the difference in temperature between the base of the atmosphere and a point several kilometers above it increases, and this raises the environmental lapse rate. When cooling of air takes place near the ground, the difference in temperature between these two points decreases, and so does the environmental lapse rate (Figure 3-10b). It is common for the environmental lapse rate to increase during daytime heating and to decrease during cooling at night.

In some cases, cooling near the ground is so great in windless and cloudless conditions at night that a layer of air a few tens or hundreds of meters deep becomes cooler than the air above it. This is known as a **temperature inversion** (Figure 3-10c) because it reverses the normal pattern of temperature decreasing with height. Temperature inversions have significant effects on vertical movements of air, as will be seen later in this chapter and in the next one.

Clouds play an important role in these transfers of energy within the lower atmosphere. Clouds absorb, reflect, and scatter insolation from above and also absorb terrestrial radiation from below. They radiate long-wave rays back to Earth, and cloud tops radiate them out to space. The overall effect is to slow the return of heat energy to space and keep more heat in

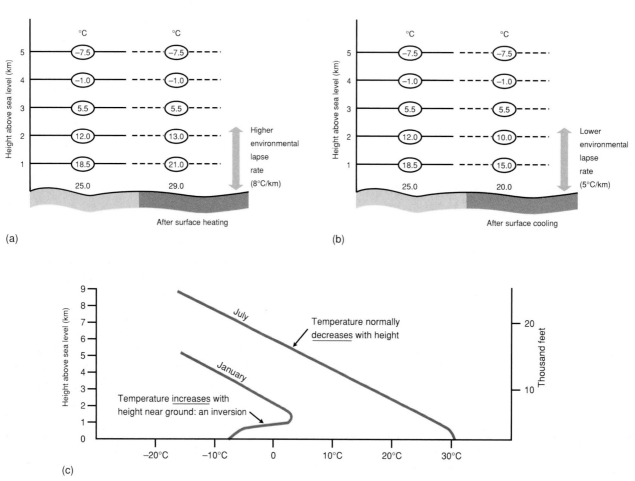

Figure 3-10 Environmental lapse rates: the rate of temperature change in relation to height above the ground. Generally, temperatures decrease with altitude in the troposphere. Heating at the surface (a) causes a rapid rise in temperature of the air immediately above the surface. This increases the lapse rate. (b) Cooling at the surface has the reverse effect, decreasing the lapse rate. (c) Intense cooling at the surface, as shown in the winter graph of the Midwest may cause the air to be colder at the surface than at a kilometer above the surface—an inversion of the lapse rate.

the lower atmosphere. Cloudy nights are warmer than clear nights since the clouds prevent terrestrial radiation escaping directly to space.

Outputs of Energy to Space Energy returns to space as reflected short-wave insolation rays from clouds and Earth's surface and as radiated long-wave rays from the ground, gases in the troposphere, and other heated atmospheric layers. These outputs balance the insolation inputs in total amount, but the type of radiation output to space differs from that which arrives from the sun. Over 70 percent of the output is in infrared wavelengths. Thirty percent of the output is reflected visible light and causes Earth to have a bright appearance when viewed from space (Figure 3-11).

The Time-Lag Between Insolation Arrival and Heating the Atmosphere A time lag occurs between the receipt of insolation at the surface and the rise in temperature of the air above. This delays the time of the highest daily and annual temperatures. The hottest time of day beneath cloudless skies is an hour or so after the time of most intense insolation at noon. It takes the hour or so for the ground to absorb insolation, transfer heat energy to the lower atmosphere by terrestrial radiation, conduction, and convection, and for the lower atmosphere gases to absorb and counter-radiate this energy. In cloudy conditions maximum air temperatures may not be reached until 3 or 4 o'clock in the afternoon.

A time-lag also occurs in seasonal heating and is particularly marked in midlatitudes. Northern hemisphere midlatitude continental interiors have their highest monthly temperatures in July, not June when the sun is highest in the sky. Their coldest month is January, not December. Midlatitude coastal locations experience an even greater lag with highest monthly temperatures (northern hemisphere) in August and lowest in February. This is because coastal areas are affected by radiation from the oceans, which take longer to absorb insolation and release the heat energy into the atmosphere than land surfaces. For instance, the Scilly Isles in the Atlantic Ocean at the southwestern tip of England (at about 50°N) have a maximum summer temperature delayed to August of 16°C (60.8°F), and a minimum winter temperature of 7.6°C (45.7°F) in January. Kiev, Ukraine, in the middle of Eurasia, is also close to 50°N latitude but has a summer maximum of 19.2°C (66.7°F) in July and a winter minimum of –6°C (21.2°F) in January.

Figure 3-11 Blue planet. The view of Earth from space shows the dominance of blue oceans. Much of the atmosphere is transparent, but clouds reflect insolation out to space.

Temperatures in the atmosphere depend on energy from the sun. Most insolation consists of short-wave rays that are subject to filtering during passage through the atmosphere. The upper atmosphere absorbs gamma rays, X rays, and most ultraviolet rays, while visible light rays and some infrared rays reach the ground.

Energy absorbed by the ground from visible light rays heats the surface. Terrestrial radiation from the surface in long-wave rays is absorbed by water vapor, carbon dioxide, and dust heating the lower atmosphere from the ground upward. Air warmed near the surface by conduction transfers heat upward by convection. The heated gases in the lower atmosphere radiate energy downward to the ground (counter-radiation) and upward to space.

BALANCE OF ENERGY IN THE ATMOSPHERE-OCEAN ENVIRONMENT

The heat transfer processes of radiation, conduction, and convection maintain a complex balance of inputs from the sun and outputs from Earth (Figure 3-12). This **heat balance** keeps temperatures in the lower atmosphere at levels that do not fluctuate wildly.

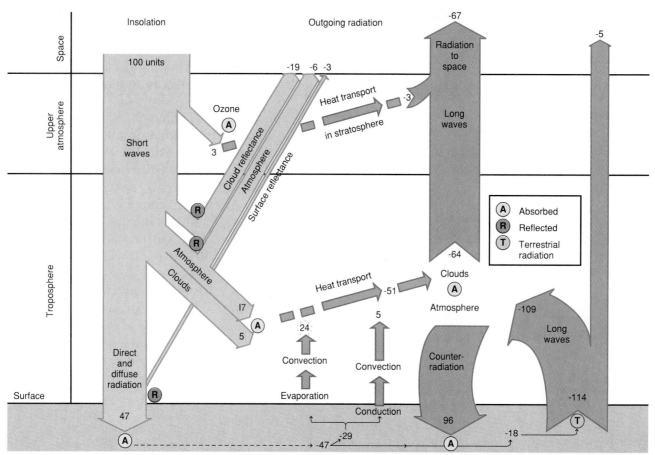

Figure 3-12 Heat balance in Earth's atmosphere. The total insolation is treated as 100 units. Of this, fewer than half of the units get through the atmosphere to the ground (47 units), although 25 of the units that do not reach the ground are absorbed on the way (3 in the upper atmosphere, 17 in the lower atmosphere, and 5 by clouds) and help to heat layers of the atmosphere there. Reflection causes an immediate loss of 28 units to space. The 47 units absorbed by the ground are radiated as terrestrial radiation (18 units, of which 5 go directly to space). Heat is conducted into the atmosphere in contact with the ground (5 units), and 24 units are used in the process of evaporation. Both of these produce heat that is convected upward. The blue arrows show how long-wave energy is circulated in the lower atmosphere by absorption and counter-radiation. What different types of heat transfer contribute to the circulation in the lower atmosphere?

The input of insolation is straightforward. The upper atmosphere absorbs the shortest-wave rays, while the visible light rays that are not reflected away to space by clouds bring energy to the surface.

The output component of the balance is more complex. Most terrestrial radiation is absorbed in the lower atmosphere from where it is counter-radiated back to the ground or radiated out to space. A small proportion of energy is transferred from the ground into the lowest few centimeters of the atmosphere by conduction. Convection in the lower atmosphere carries air heated near the ground upward. The heat convected from the surface joins the heat energy absorbed from incoming solar infrared rays and the heat absorbed from terrestrial radiation. Much energy is concentrated in the lower atmosphere and transferred back and forth.

The possible impact of changes in the balance between insolation and radiation to space stimulates present worries over *global warming* (see Environmental Issue: The Future—Global Warming or Cooling? p. 165). Increases in materials such as carbon dioxide, dust from industrial activity, and water vapor, that absorb terrestrial radiation in the troposphere might result in rising temperatures in the lower atmosphere. Attention has been drawn to this concern through the popular model of the **greenhouse effect.**

The atmosphere acts in some ways like a giant greenhouse. In a greenhouse (Figure 3-13) the glass lets in visible light rays but prevents most heat from escaping. Once inside the greenhouse, the visible light is either reflected out or absorbed by plants, soil, and atmospheric gases. These materials radiate heat waves that the glass absorbs and reradiates—mainly back inside. Since the heated air is trapped inside the glass, temperatures in the greenhouse rise higher than those outside.

In the atmosphere, visible light rays pass through the atmosphere. After the ground absorbs this energy, terrestrial radiation takes place but is intercepted by gases such as carbon dioxide and water vapor that are known collectively as the *greenhouse gases.* The lower atmosphere is heated and the greenhouse factor is thus a positive feature of Earth's atmosphere. If the atmosphere did not contain greenhouse gases, surface air temperatures would be 33°C lower (-18°C, 0°F, instead of 15°C, 59°F).

The term *greenhouse effect* has mistakenly been used to describe an *enhanced greenhouse effect.* Human activities, such as the burning of fossil fuels, add greenhouse gases and dust to the atmosphere. Additional greenhouse gases absorb more heat and raise temperatures near the surface. The atmosphere-ocean environment, however, works in very complex ways that include exchanges and feedback mechanisms that may mute or add to the enhanced greenhouse effect. As more and more human activities contribute to the greenhouse gas store, however, governments are showing serious signs of concern about the issue and are discussing possible actions. The U.S. Global Change Research Program monitors changes in the atmosphere-ocean environment to understand their possible impacts so that governments may be advised on the amount of change to be expected.

Finally, it is worth pointing out that the atmosphere does not always behave like a greenhouse. In particular, in a greenhouse the glass acts as a physical barrier to the upward convected transfer of heated air and the influence of winds bringing colder or warmer air. The glass ensures that the air inside the greenhouse is isolated; there are no such barriers in the atmosphere. The greenhouse exploits only one part of the workings of the atmosphere. It is also the case that a greenhouse environment can be controlled by letting out some of the heat or adding heat from burners. The atmosphere cannot be controlled by humans because the scope and complexity of processes in the atmosphere are too great.

GLOBAL DISTRIBUTION OF SOLAR ENERGY

Insolation is not distributed evenly over Earth's surface because of the planet's spherical shape, its rotation on a tilted axis, and its fluctuating distance from the sun. The tropical and polar regions contrast markedly with each other in the total amount of insolation they receive and in its seasonal distribution. In the tropics more energy arrives than is radiated back to space and there is a net gain, or surplus, of radiation (Figure 3-14). Polar regions radiate out more energy than they receive and have a net loss, or deficit.

Tropical regions do not, however, get progressively hotter and polar regions do not grow continuously colder. A balancing mechanism operates to maintain the temperatures in different parts of the world. Heating contrasts in the atmosphere and oceans initiate convective movements within the atmosphere-ocean environment (Figure 3-15). The atmosphere is heated at its base in the tropics and midlatitudes and cooled higher in the troposphere and at higher latitudes. Oceans are heated at the surface in the tropics and cooled at the surface toward the polar regions and at depth.

The movements through the atmosphere and oceans involve huge horizontal transfers of energy primarily

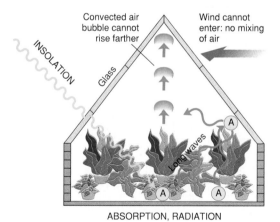

Figure 3-13 Greenhouse effect. Insolation passes through the glass and is absorbed by interior surfaces. All the surfaces radiate long-wave energy that does not escape through the glass but is absorbed there and reradiated inside the greenhouse. The greenhouse atmosphere warms more rapidly than the outside. What are the similarities between the heating of a greenhouse and the lower atmosphere?

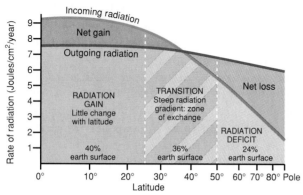

Figure 3-14 Heat balance for different parts of the globe. The net gain of radiation in the lower latitudes is offset by a net loss of radiation in higher latitudes. Spacing on the horizontal axis is proportional to the areas within each 10° of latitude.

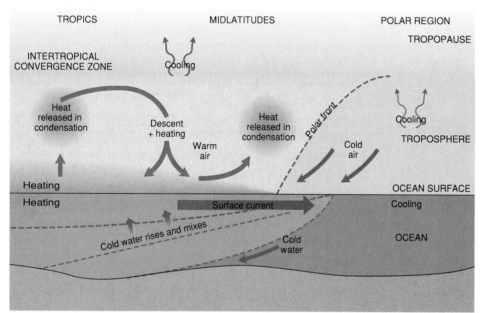

Figure 3-15 Heating and cooling surfaces in the atmosphere-ocean environment. Heated air and water expands, becoming less dense; cooled air and water contracts, becoming more dense. How do these differences affect movements of air and water?

through convection (Figure 3-16). Air heated in the tropics moves poleward bringing heat to the midlatitudes. In the tropics, ocean waters absorb much more energy than the atmosphere, and water movements transport this heat poleward. The ocean transport of heat energy makes up nearly 75 percent of total heat transfers in the atmosphere-ocean environment at 20°N and half at 30–35°N. During winter in midlatitudes, the ocean often contributes more heat to the lower atmosphere than do the sun's rays. The local

contribution of such transferred heat compared to direct insolation increases toward the poles. The transfers of energy involve movements of the atmosphere (winds) and ocean (currents) that are studied later in this chapter and the circulation of water through the atmosphere studied in Chapter 4.

TEMPERATURE DIFFERENCES ON EARTH

Differences in insolation and the transfers of energy around the globe are reflected in differences of surface temperature from one part of the world to another. Temperatures measured at different locations are recorded and plotted on maps by **isotherms**—lines joining places of equal temperature (Figure 3-17).

The world maps of temperature at Earth's surface in January and July (Figure 3-18) depict average conditions and are based on temperature records over approximately thirty years. The main differences of temperature are between the high temperatures of the tropics and the low temperatures of the polar regions. Furthermore, the summer and winter hemispheres reveal a seasonal contrast in the number and pattern of isotherms between the equator and poles. On the January map (Figure 3-18a), the isotherms are closer together during the northern hemisphere winter than the southern hemisphere summer. The greater number of isotherms in winter reflects a more rapid change of temperature during a journey away from the equator. In other words, the **temperature gradient** (the rate of temperature change with distance) is greater.

The pattern of temperature distribution shown on world maps is determined by interactions among the factors involved in heating the atmosphere-ocean environment. The isotherms generally run along the parallels of latitude, indicating temperature decreases from equator to poles. This reflects the fact that the greatest solar heating is in the tropics.

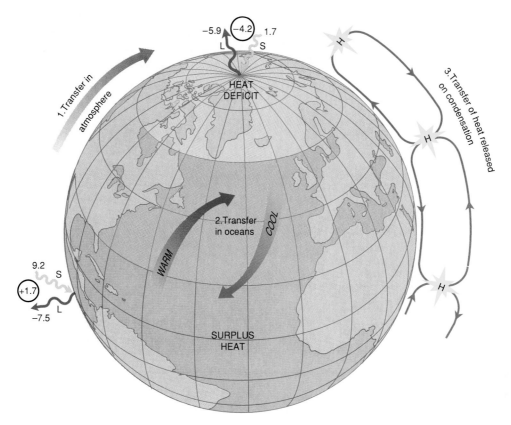

Figure 3-16 Global heat transfers. There are three main mechanisms of heat transfer from low to high latitudes: by movements of the atmosphere and oceans and by transfer of heat in humid air. Flows of warm air and water toward the poles are balanced by flows of cold air and water toward the equator. Longwave terrestrial radiation (L) at the equator is less than shortwave insolation (S). At the poles the terrestrial radiation exceeds insolation.

If Earth had a uniform surface, the isotherms would run precisely east-west along parallels of latitude. The ocean and continental surfaces, however, react differently to insolation and cause the isotherms to depart from east-west lines. On the January map, the northern hemisphere winter isotherms bend toward the poles over the warmer oceans and toward the equator over the colder continents. On the July map, the northern hemisphere summer isotherms bend toward the equator over the cooler oceans and toward the poles over the warmer continents. These patterns result from the greater seasonal temperature contrasts over the continents than over the oceans. Another result is that the northern hemisphere, with its greater proportion of continental surfaces, has warmer summers and cooler winters than the southern hemisphere. Isotherms are closer to the parallels of latitude in the southern hemisphere because the midlatitude area is almost all ocean.

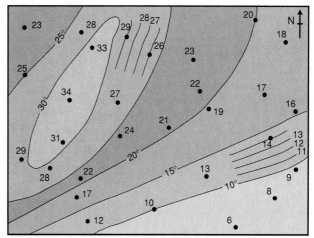

Figure 3-17 Isotherms. Point measurements of temperature and the drawing of isotherms at 5° intervals. Isotherms are drawn at the correct ratio between the measurement points: for instance, the 20° isotherm is drawn two-fifths of the way between measurements of 22° and 17°.

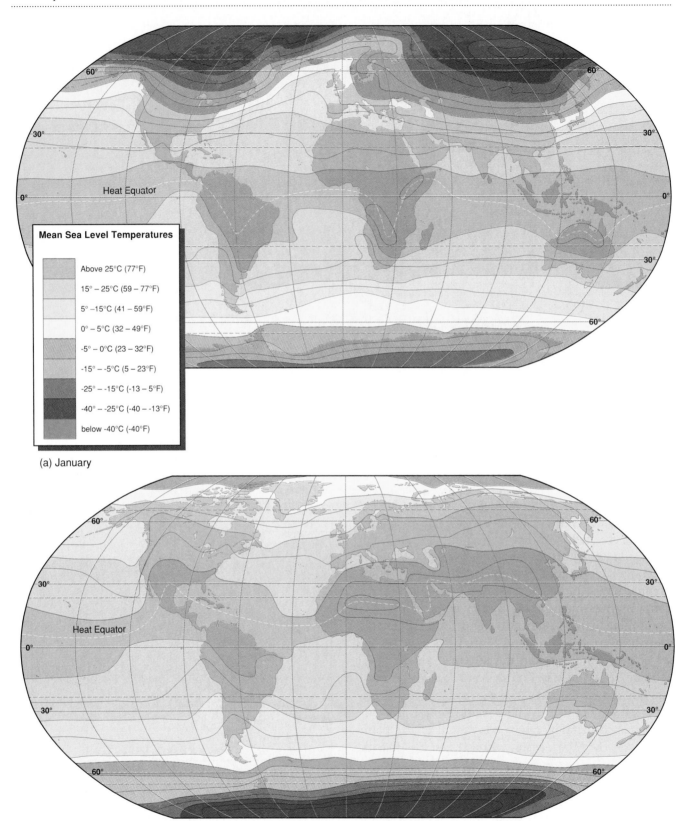

(a) January

Mean Sea Level Temperatures

Above 25°C (77°F)

15° – 25°C (59 – 77°F)

5° –15°C (41 – 59°F)

0° – 5°C (32 – 49°F)

-5° – 0°C (23 – 32°F)

-15° – -5°C (5 – 23°F)

-25° – -15°C (-13 – 5°F)

-40° – -25°C (-40 – -13°F)

below -40°C (-40°F)

(b) July

Figure 3-18 World isotherm maps: January (a) and July (b). The isotherms are based on average temperatures over 30 years. The heat equator connects the points of highest temperature at each meridian of longitude. How do the northern and southern hemispheres compare in terms of summer and winter temperatures?

The *continentality* of a place is recognizable by its high annual range of temperature compared to other places on the same latitude. Anchorage, on the southern coast of Alaska (62° N), has a warmest-month average temperature of 12°C (58°F) and a coldest-month temperature of -12°C (13°F), an annual range of 24°C (45°F). Verkhoyansk, in central Siberia (66° N), has a warmest-month average temperature of 15°C (59°F) and a coldest-month average temperature of -45°C (-49°F), an annual range of 60°C (108°F). Anchorage has a low continentality and Verkhoyansk one of the highest in the world.

Temperatures in air over the oceans are also affected by the greater level of *cloudiness.* The higher percentage of cloud cover over oceans lowers insolation and reduces radiation losses directly to space. This factor enhances the oceanic characteristic of warmer winters and cooler summers compared to continents on the same latitude.

Although their effects do not show up clearly on world maps, *mountain ranges* also affect the distribution of surface temperatures. Temperature decreases with **altitude,** or height above sea level, because the atmosphere is less dense. The decrease in density is accompanied by a decrease in the quantity of greenhouse gases and hence the heat absorbed.

The interactions between insolation and surface conditions at a place are not the only factors that determine surface temperatures. The atmosphere and oceans are constantly in turbulent motion. The movements of air and water transfer heat between oceanic and continental areas as well as from tropics toward polar areas. Warm and cold ocean currents affect the atmospheric temperatures of adjacent coasts. Warm currents occur off east coasts and cold currents off west coasts between approximately 10 and 40 degrees North and South of the equator. The effects of these currents can be seen on the world temperature maps for January and July, where, for instance, temperatures are higher on the east coast than on the west coast of the United States. Air in winds from the ocean, for instance, is warmer in winter and cooler in summer than the air over land. Islands and places on coasts have cooler summers and warmer winters than places in the interiors of continents.

Human activities have modest and localized impacts on the world isotherm map since their contribution to total atmospheric heating is tiny when compared to that from solar energy. The image constructed from satellite data (Figure 3-19) shows the heat generated from urban settlements (white), grass and forest fires (red), and oil flares and burns (yellow) in Europe, Africa, and Asia. In the depth of winter, the buildings of densely populated western European cities contribute more heat to the atmosphere around them than direct insolation does.

Earth's atmosphere has a balance of incoming and outgoing energy. The lower atmosphere traps heat in a manner that resembles the workings of a greenhouse. An enhanced greenhouse effect will occur if more heat is trapped in the lower atmosphere than is returned to space.

The surplus of heat received in the tropics is distributed to cooler parts of the globe by movements in the atmosphere and oceans. Temperature at a place on Earth's surface is controlled by latitude, the distribution of continents and oceans, ocean currents, wind directions, and (in a small way) human activities.

HEATING AND WINDS

Winds are the atmosphere in motion. They may come from any point of the compass and their strength varies from destructive hurricane-force winds to light breezes. Gusty conditions bring moment-to-moment variations in wind strength and direction on a local scale. Air can also be still.

Winds are horizontal movements of air relative to Earth's surface. Winds have a speed (kilometers or miles per hour) and a direction (e.g., westerly or easterly). The speed of winds is determined by the forces that set the air moving and those that may slow it down. Horizontal flows of air result from differences in air density. The flows occur from higher toward lower density and tend to even up the differences.

Solar heating is the main cause of density differences in the atmosphere. If a parcel of air is warmed above land or sea, it expands (i.e., its volume increases) and becomes less dense than the colder air surrounding it. It therefore rises by convection. At the same time, air also descends toward the ground under gravity to replace the rising air. The ascent of lower-density air reduces atmospheric pressure on the surface, while the descent of air increases surface atmospheric pressure (Figure 3-20).

Horizontal pressure contrasts in the atmosphere just above the ground are shown on weather maps by **isobars**—lines that join places of the same

Figure 3-19   Human activities and heat output. This view of half the world—including Africa, Europe, and Asia—is compiled from satellite images that record surface heat patterns. White represents city lights, red shows forest and agricultural fires, and yellow shows gas flares.

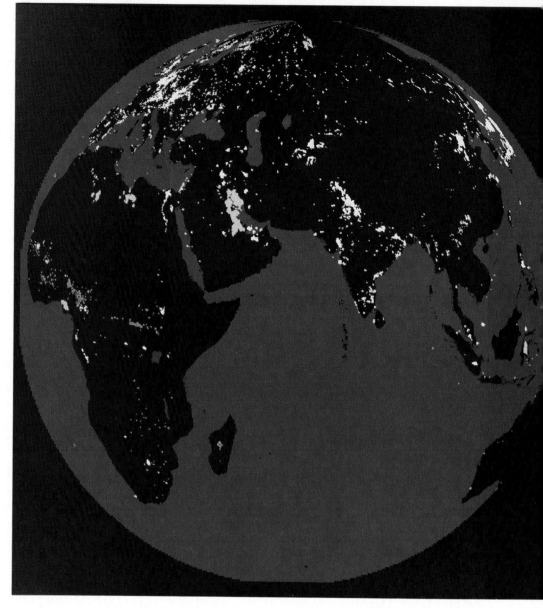

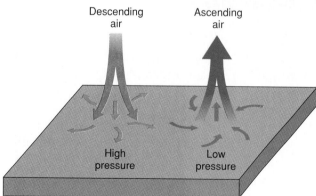

Figure 3-20   High pressure areas occur where air is still or descending, but not rising. Winds blow outward at the surface. Low pressure areas have air ascending and winds blow inward at the surface.

atmospheric pressure. Winds are flows of air over the surface, across the isobars, from areas of high pressure to areas of low pressure. The difference in pressure per unit distance between the high and low pressure forms a **pressure gradient.** It creates a force (the pressure gradient force) that acts on air between centers of high and low pressure. The pressure gradient force determines the strength and initial direction of the wind. A steep pressure gradient—a large difference in pressure over a small distance—causes strong winds. When isobars on weather maps are close together, they indicate a steep pressure gradient and strong winds. When isobars are farther apart, the pressure gradient is low and the winds are light.

The effect of surface pressure differences on air movement is seen at the local scale, when the heating contrast between land and sea produces daytime breezes blowing from the sea to the land and a reversal of this pattern at night (Figure 3-21). During the day, air over land heats more rapidly than that over the sea. The heated air expands and rises, lowering atmospheric pressure over the land (Figure 3-21a). Cooler air from the higher pressure over the sea flows in to replace the rising air, causing a sea breeze. Descending air a few kilometers offshore replaces the air blowing toward the land, and a convection cell is established. The strength of the sea breeze grows as the difference in heating between land and sea increases during the day; it is often greatest in mid-afternoon. The circulation reverses at night when air over the land cools more rapidly than that over the sea (Figure 3-21b). Surface air flows toward the sea. The night contrast in temperature between land and sea is not as great as that during the day, however, and the winds are usually weaker.

WINDS AND EARTH

Wind direction, initially determined by differences in surface air pressure, is modified by relationships with the solid planet. Earth's rotation, friction near the ground, and the barrier and channelling effects of mountains and valleys are of major significance in determining wind directions.

WIND DIRECTION AND EARTH'S ROTATION

People standing on the ground have no impression that Earth is rotating since they move with it. Movements in the atmosphere, however, are deflected when tracked against the rotating Earth beneath.

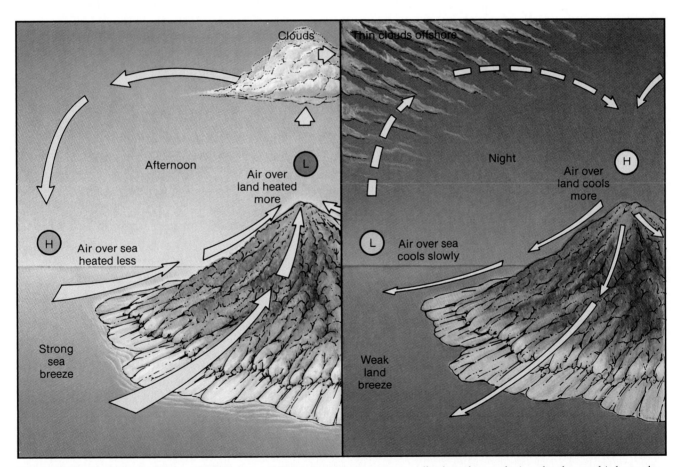

Figure 3-21   Sea breezes (day) and land breezes (night). Land heats more rapidly than the sea during the day, and it heats the air above first. Air above the land expands and rises, lowering surface pressure and establishing a pressure gradient from sea to land. Air flows along this pressure gradient as a sea breeze. Air moving over the sea toward the land is replaced by air descending, and a convectional circulation forms. At night the flow reverses as air over land cools more rapidly than air over the sea. The dense cooled air flows out to sea, but the pressure gradient at night is not as strong as that in the day and the winds are weaker.

Winds in the northern hemisphere bend to the right of the direction set by the pressure gradient, while those in the southern hemisphere bend to the left (Figure 3-22). This deflection of wind direction can be seen as the result of the rotating Earth imparting a spin to the atmosphere above—like eddies in a rotating pan of water. In the absence of the factors that are discussed below, the deflection results in air turning to flow parallel to the isobars, instead of across them at right angles. Winds blowing parallel to isobars are known as **geostrophic winds.**

In 1835 Gaspard C. De Coriolis proposed a "force" termed the **Coriolis effect** to explain such observations. This force acts equally and in the opposite direction to that determined by the pressure gradient. Earth's spin causes the Coriolis effect and it has the greatest effect at higher latitudes where Earth's surface is almost parallel to the plane of rotation and no effect at the equator where Earth's surface is almost vertical to the plane of rotation.

The way in which the Coriolis effect works can be visualized by imagining a wind caused by a pressure difference between two places in the northern hemisphere (Figure 3-23). The air begins to flow in response to the pressure gradient and travels at right angles across the isobars. As the wind accelerates toward the speed dictated by the pressure gradient, the Coriolis effect increases too and turns the wind fully to the right in a direction parallel to the isobars instead of perpendicular to them.

WIND DIRECTION AND FRICTION WITH THE GROUND

Winds in contact with the ground are slowed by friction, which is greatest in mountainous terrain and least over the oceans or flat ice sheet surfaces. The depth of atmosphere affected by friction with Earth's surface averages 1 km, but it is deeper over mountains and shallower over oceans and plains. Above this friction layer, the winds tend to be geostrophic.

The friction also reduces the Coriolis effect so that winds cross isobars at angles between the direct pressure-gradient route from high to low pressure and the geostrophic pattern parallel to the isobars. Winds crossing rugged mountains have a high angle with the isobars because of the increased friction factor, but those blowing over the sea or flat plains have directions that are almost parallel with the isobars.

WIND DIRECTION, MOUNTAINS, AND VALLEYS

Earth's surface features affect wind directions in different ways at different geographic scales. On the continental scale, mountain ranges act as barriers to winds. The best example is the world's highest mountains, the Himalayan-Tibet massif, much of which is over 8 km (5 mi) high. In winter these mountains exclude cold airflows from high-pressure centers in Siberia toward lower pressure areas to their south, allowing higher temperatures to be maintained across the Indian subcontinent to the south.

At the regional scale, valleys channel winds through lower routeways between mountains. The cold *mistral* wind blows down the Rhône valley in southern France when a high-pressure area over northern Europe coincides with a low-pressure area over the Mediterranean Sea. In spring, cold air passes southward through the valley, which concentrates it into strong winds that can harm early flowering fruit trees. The *Santa Ana* winds affect southern California when high pressure over the interior desert causes hot, dry winds, often carrying

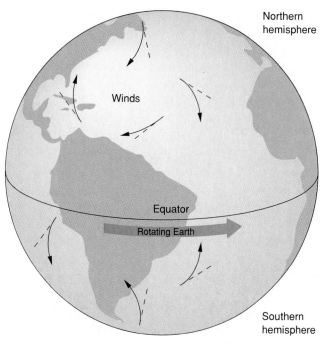

Figure 3-22 Earth rotation and wind deflection. Winds in the northern hemisphere are deflected to the right by Earth's rotation; those in the southern hemisphere are deflected to the left.

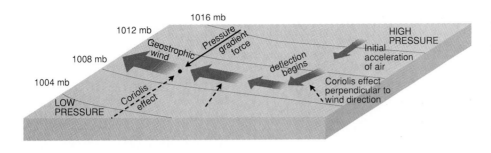

Figure 3-23 Geostrophic wind. The change in direction that results from Earth's rotation may lead to a geostrophic wind that blows parallel to isobars instead of across them. Earth's rotation acts as a force of equal and opposite strength to the pressure gradient force. This diagram suggests a sequence as the pressure gradient force gives rise to a wind blowing from high to low pressure. The wind is deflected by Earth's rotation. This sequence occurs instantaneously.

dust, to be concentrated as they blow through the narrow valleys of the San Bernardino Mountains toward the Pacific coast (see Environmental Issue: Santa Ana Winds and Building Codes, p. 70).

At the local scale, upland areas commonly have winds blowing upslope in the afternoon and downslope at night (Figure 3-24). Daytime heating of the upper slopes causes the air above to expand and rise by convection. Lower air moves upward to replace the rising air. At night rapid radiation from the upper slopes cools the air and this dense air drains downward to the valley floor. Where the valley floor is confined, cold air may accumulate, rapidly lowering the temperature. This may produce a temperature inversion. Cold air draining into a valley floor can be disastrous for fruit crops. The effect of cold night winds on the grapes growing in the Napa Valley of central California was combated by heaters and smoky smudge pots until environmental legislation forced their replacement by windmills that circulate and mix the cold air with higher, warmer air (Figure 3-25).

Winds result from heating in the atmosphere-ocean environment. Vertical movements of air triggered by differences in solar heating produce variations in surface atmospheric pressure. Wind is the movement of air from high to low pressure.

Earth's rotation deflects the apparent wind path according to the Coriolis effect. The resultant geostrophic winds blow parallel to isobars. Friction at Earth's surface causes winds to slow and flow across isobars. The major relief features affect the paths of airflows by blocking them or channelling them through valleys.

WINDS AT DIFFERENT LEVELS IN THE TROPOSPHERE

Figure 3-26 shows the surface (a) and upper troposphere (b) winds on maps of winds and pressure at one moment over North America. The map of the higher level (b) shows geostrophic winds blowing parallel to the closely spaced isobars at high speeds. An airplane flying in the same direction as these high-level winds has high pressure on the right and low pressure on the left. The effect of Earth's rotation induces the geostrophic wind directions at this level.

The surface winds, however, are slower and few are near geostrophic in direction (Figure 3-26a). They blow away from high-pressure areas and toward low-pressure areas but are neither parallel to the isobars nor do they cross them at right angles. The light winds blowing across the Appalachian and Rocky Mountains cross the isobars at high angles; strong winds blow almost parallel to the isobars over the North Pacific Ocean and the Great Plains. Such patterns demonstrate the importance of friction in slowing wind speeds and deflecting their paths.

The strong, high-level winds at the upper level (Figure 3-26b) are known as **jetstreams.** The most continuous jetstreams occur in midlatitudes, where tropical and polar air interact and set up extreme differences in upper-air temperature and pressure over small distances. The *polar-front jetstream* (Figure 3-27), a jetstream blowing from the west, has wind speeds up to 500 kph (300 mph) in its core although average wind speeds are half this extreme. Jetstream velocities are higher in winter than in summer in response to greater temperature and pressure gradients. Aircraft save time and fuel by flying eastward along

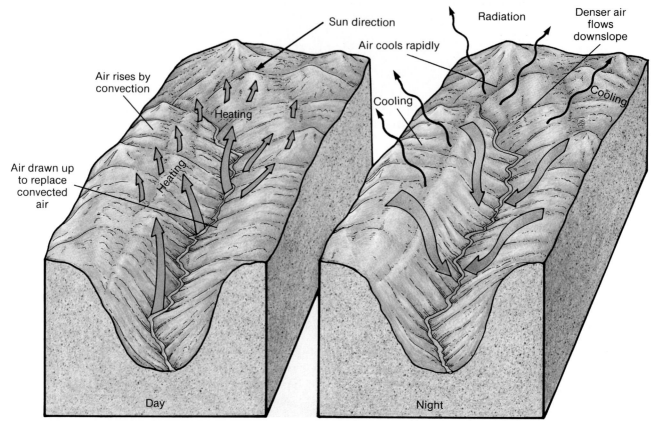

Figure 3-24 Mountain winds. During the day, air on the mountainsides is heated, expands, and rises. Air from the valley floor is drawn upward to replace the rising air, giving rise to upslope winds. Night cooling affects air above the upper slopes first, and dense air drains down to the valley floor.

Figure 3-25 Napa Valley, central California: The geographic center of the Californian wine industry. Cold air draining from the hills can kill the early flowers. The heaters that once kept temperatures above freezing have been replaced by fans that circulate the air, mixing cold and warm air to prevent frost.

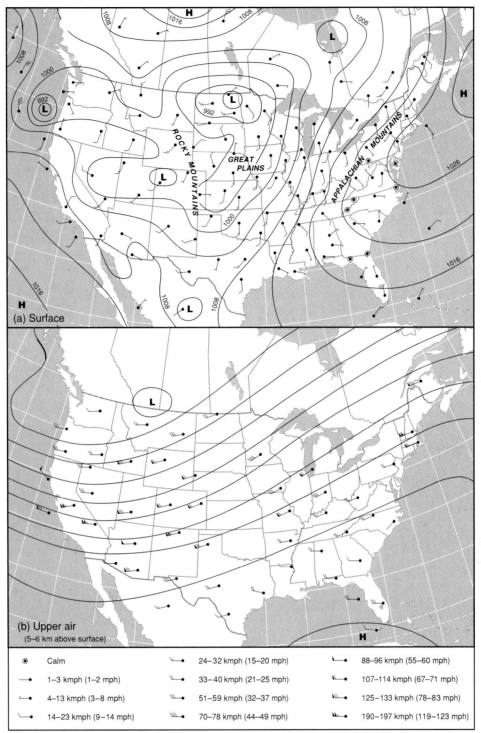

Figure 3-26   Atmospheric pressure and winds over North America. The surface map (a) has atmospheric pressures approximately twice those at the upper troposphere level (b). These patterns occurred on a day in October and are taken from the Daily Weather Report forecast maps. The key shows wind directions and strength in symbols; arrows indicate direction, and the number of "feathers" the wind speed. Wind speed is given in knots, or nautical miles per hour: a knot is approximately 1.8 kph (1.2 mph). What are the differences between surface and upper troposphere winds?
Source: NOAA.

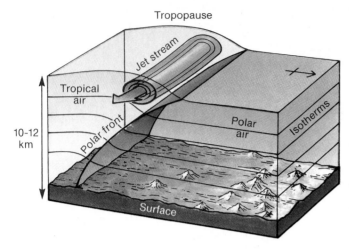

Figure 3-27   Jetstream. The polar-front jetstream is a powerful wind, blowing up to 300 kph in its core. This jetstream occurs above the polar front, a major zone of temperature and pressure change between tropical and polar air in midlatitude regions.

the margins of westerly jetstreams and avoiding them in westward flights. Other jetstreams occur where steep seasonal temperature and pressure gradients exist over the subtropics (a jetstream blowing from the west) and over the equator (a jetstream blowing from the east).

The path of the polar-front jetstream forms wave patterns known as **Rossby waves** as it circles the globe in midlatitudes. There are generally four to six of these waves around the globe (Figure 3-28). The wave forms have a cyclic pattern of development lasting for several weeks that includes a phase of small, open waves with the jetstream blowing mainly west-to-east (Figure 3-28a), followed by increasingly meandering (more wavy) phases in which the jetstream becomes more north-south in flow direction (Figure 3-28b,c). Lobes of cold air, or troughs, develop on the polar side of the jetstream; and lobes of warm air, or ridges, develop on the tropical side. Finally, the

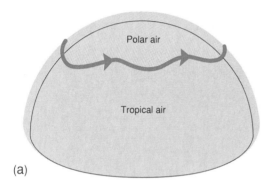

(a)

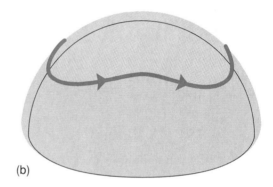

(b)

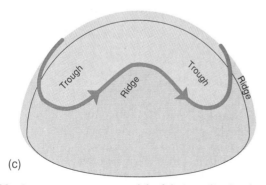

(c)

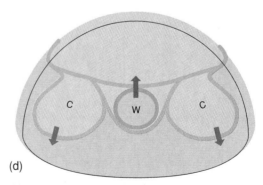

(d)

Figure 3-28   Jetstream waves: a model of their cyclic development. (a) A strong westerly flow, with little mixing of tropical and polar air aloft. (b) The waves grow larger. (c) Ridges and troughs become pronounced, and the airflows become increasingly oriented to north-south and south-north directions. (d) The waves break down, transferring large pools of polar and tropical air across the polar front. The initial stage resumes.

west-to-east jetstream resumes, cutting off troughs of cold air south of the jetstream and ridges of warm air to the north. This is part of the heat transfer processes between tropical and polar atmospheres.

The polar-front jetstream is the strongest wind on Earth and is sufficiently powerful to control mid-latitude surface wind patterns by exchanges of air between the surface and upper sections of the troposphere. Ascent of air is characteristic of low-pressure areas where surface flows come together, or **converge.** Air descends over high-pressure areas, where surface flows spread out, or **diverge.** The wave patterns in the jetstream produce zones of internal convergence and divergence of air. Convergence in a jetstream occurs where it enters a bend, and the pileup of air there forces some downward, raising atmospheric pressure at ground level (Figure 3-29). Divergence in a jetstream occurs where it comes out of a bend and the spreading airflow pulls air upward from surface low-pressure centers. Figure 3-29 shows the northern hemisphere case where winds circulate around high-pressure centers in a clockwise direction and around low-pressure centers in a counterclockwise direction. In the southern hemisphere, winds blow in a clockwise pattern around low-pressure areas and in a counterclockwise pattern around high-pressure centers (Figure 3-30).

GENERAL CIRCULATION OF THE ATMOSPHERE

Understanding the factors that cause wind movements makes it possible to construct a general model of atmospheric circulation that links the observed wind patterns at the ground and in the upper troposphere. Figure 3-31 shows the geographic distribution of surface pressure and the main wind patterns linked to the pressure patterns.

Low pressures prevail around the equatorial region throughout the year. Winds blow toward the equator from large **subtropical high-pressure systems** centered at approximately 30° N and 30° S. These winds blow from the northeast in the northern hemisphere and from the southeast in the southern hemisphere and are known as the **trade winds.** This name recalls the time when the northeasterly winds in the tropical Atlantic Ocean made it possible for early European explorers and traders to sail westward to the Americas.

Near the equator where the trade winds meet, horizontal air movement is often very weak, and the zone was termed the *doldrums* by sailors becalmed there. The light winds interrupted by occasional squalls earned this name because of the listlessness and despondency the conditions induced in sailors.

On the poleward flanks of the subtropical high-pressure zones, the winds blow from the west around and into the southern flanks of low-pressure areas and are known as **westerlies.** The explorers of the Americas returned to their European base on a northern route that exploited the westerly winds.

As early as 1735, George Hadley showed that air circulated in the tropical atmosphere in a single convection cell. The **Hadley cell** comprises surface trade winds, vertical ascent of air over the equator, poleward flow aloft from the equator to approximately 30°N and 30°S, and descent at those latitudes. In 1856, William Ferrel suggested a three-cell model of global circulation dominated by vertical exchanges of air along meridians and based on the distribution of high- and low-pressure centers at the surface as shown in Figure 3-31. Although the Hadley cell is still a satisfactory basis for describing the tropical atmospheric circulation, air movements in midlatitude and polar regions follow different patterns (Figure 3-32).

The *tropical wind circulation* between the Tropic of Cancer and the Tropic of Capricorn affects 40 percent of Earth's surface. Hadley cell circulation dominates the wind patterns in this zone. Surface trade winds bring warm, moist air to the equatorial zone from the northeast and southeast. This zone is known as the **Inter-Tropical Convergence Zone** (ITCZ). On convergence, the warm, moist air rises producing clouds and rain along the ITCZ. When the rising air reaches the top of the troposphere, it spreads out and flows poleward. At approximately 25° to 30° North and South of the equator, the subtropical jetstream forces some of the air downward to the surface high-pressure zones. The descent warms and dries the air, which flows outward at the surface, rapidly gathering evaporated water vapor when it reaches the ocean. This surface airflow forms the trade winds blowing toward the equator and completes the Hadley cell circulation. The strength and persistence of the subtropical high-pressure zones, plus the weakness of the effect of Earth's rotation on the atmosphere near the equator, makes the trade winds the world's most consistent winds in direction and strength.

The *midlatitude wind circulation* affects a further 40 percent of Earth's surface, extending from the tropics

Figure 3-29 Jetstream and surface winds. Air in the jetstream converges at bends in the wavelike flow, forcing air downward above surface high-pressure centers. Air in the jetstream diverges as it moves out of a bend and draws air upward from surface low-pressure centers. What are the surface wind patterns? What is the southern hemisphere pattern?

Figure 3-30 Pressure centers and wind patterns, northern and southern hemispheres.

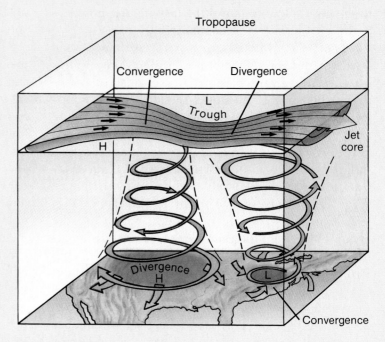

Figure 3-29

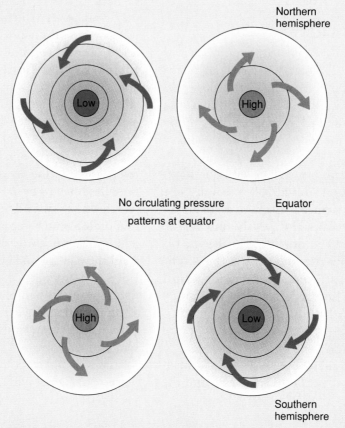

Figure 3-30

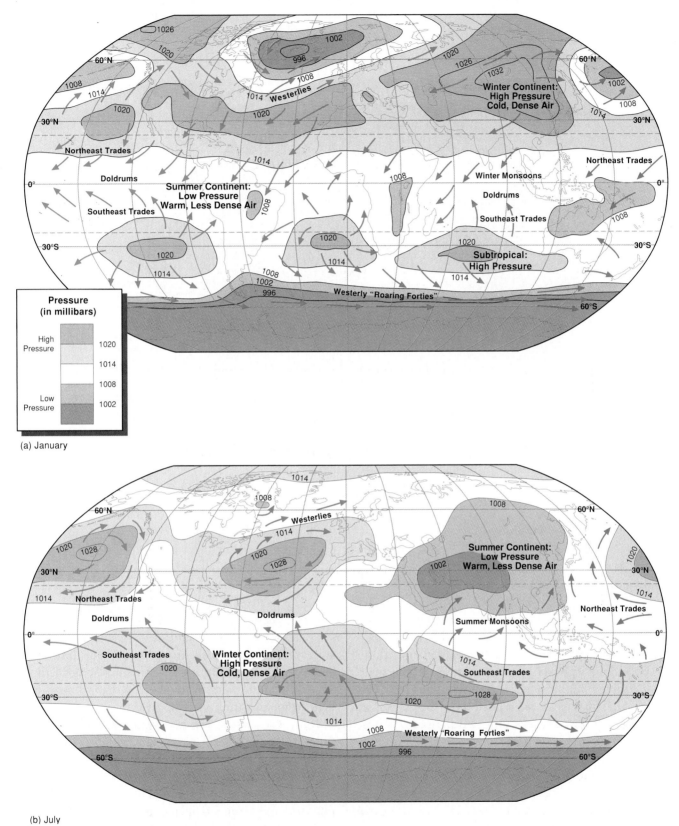

(a) January

(b) July

Figure 3-31   Atmospheric pressure (isobars) and wind directions: world maps for (a) January and (b) July.

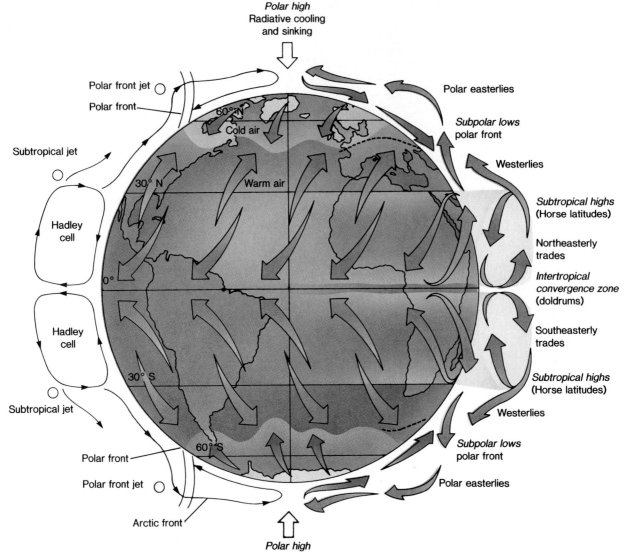

Figure 3-32 Global wind circulation. Compare the surface pattern with vertical movements of air in the troposphere (left side). What are the differences between circulation in the tropics, midlatitudes, and polar regions?

to 60° N and S. Conflict occurs between converging tropical and polar air at the surface and causes steep horizontal temperature and pressure gradients along the polar front. The polar-front jetstream is the main driving force for atmospheric activity in this zone. The surface westerly winds are named after the dominant component of wind direction. Midlatitude winds, however, vary much more in direction than those in the tropics since Earth's rotation produces circulating wind systems around centers of high and low pressure. These pressure centers move toward the east. The strongest and most consistent westerly winds occur over the southern oceans between 40°S and 60°S where there is little land to slow them down by friction. These strong winds are still called the "Roaring Forties" after the name given by early navigators.

The *polar atmospheric circulation* affects places poleward of 60 degrees of latitude. Circulation is weak, since these regions lack strong motivating forces such as solar heating or high-level jetstreams. High latitude areas lose more heat by outward radiation than they gain from insolation, and the deep cooling of the atmosphere leads to a general sinking of air around the poles, forming polar high-pressure zones. Dense, cold air builds up and flows out at the surface in bursts of **polar easterlies.**

The arrangement of continents also affects the extent and influence of polar winds in the two hemispheres. Antarctica acts throughout the year as a massive and single cold core for the southern hemisphere. The surrounding ocean waters are warmer and prevent Antarctica's influence from spreading northward.

The contrast between the Antarctic cold and the oceanic atmosphere to the north, however, produces a strong temperature gradient along the polar front and a very strong westerly wind circulation. In the northern hemisphere, the Arctic Ocean core of high pressure and cold air expands in winter as intense cooling lowers temperatures over northern Siberia and northern Canada. Cold air flows out from this enlarged cold-air source to affect extensive areas on the midlatitude continents of the northern hemisphere.

Global wind patterns include both surface and upper troposphere components. Strong jetstream winds in the upper troposphere affect the surface circulation, especially in midlatitudes. The jetstream moves in Rossby waves whose amplitude increases and decreases in a repeated cycle of events. Vertical movements dominate the general circulation of the tropical atmosphere; horizontal conflicts between polar and tropical air control circulation of midlatitudes; and loss of heat, cooling, and sinking control that of polar regions.

OCEAN CURRENTS

The world map of surface ocean currents (Figure 3-33) shows that they form circular flows, termed **gyres,** apart from the direct west-to-east flow in the southern oceans around Antarctica. Although seawater density and the Coriolis effect influence ocean water movement, surface winds and the shapes of the ocean basins have greater effects on the patterns of ocean current flow.

WINDS AND CURRENTS

Winds blowing across the ocean surface are the most important factor in determining the direction and strength of ocean current movements. The prevalence of westerly winds in midlatitudes causes the dominant eastward flow of ocean currents. In the tropics the northeast and southeast trade winds blow water westward.

The wind-induced flows of water cause the most rapid movements to occur close to the ocean surface. An ocean current typically flows at 2 percent of the wind speed. The speed of water flow in a current falls off with depth, and below 100 m the water flow is unaffected by

Figure 3-33   Ocean surface currents, January. Devise a rule to describe the distribution of warm and cold currents. Compare this map with the January map of pressure and winds (Figure 3-31).

winds. As water depth increases, the effect of Earth's rotation increases relative to the effect of surface winds. It deflects water flow just below the surface to the right in the northern hemisphere and to the left in the southern hemisphere. The overall water movement is at approximately right angles to the wind direction. In California and Peru where wind direction is often parallel to the coast and toward the equator, water movement in the surface layers is away from the coast. Cooler waters rise, bringing nutrients to the surface that have accumulated in the deeper waters. Such nutrients provide a basis for some of the world's richest fisheries.

OCEAN BASIN SHAPES

Continents are barriers to ocean currents, obstructing and diverting flow. Figure 3-33 shows that continents turn the flows and that the gyres are made up of northward flows on the western sides of oceans and southward flows on eastern sides. Ocean-margin currents are known as **boundary currents.** There are two main types. Narrow, warm currents up to several hundred meters deep flow rapidly poleward along the western margins of oceans. The Gulf Stream is an example of

such a current, concentrating tropical waters in the Caribbean and flowing out through the constricted gap between Cuba and Florida. Warm water flows in a stream some 100 km (63 mi) wide at speeds of up to nearly 3 m (10 ft) per second. After flowing parallel to the coast of the United States, Cape Hatteras deflects the current eastward, and the water flows toward northern Europe as the slower North Atlantic Drift beneath the westerly winds. The eastern margins of oceans have shallower, wider, and slower cold currents that flow toward the equator (Figure 3-34).

SURFACE AND DEEP OCEAN FLOWS

Exchanges of warmer surface water and cold deep water occur at few places in the oceans, unlike the many vertical exchanges in the atmosphere. There appear to be just two major locations where dense surface water dives downward—at the northern and southern ends of the Atlantic Ocean (Figure 3-35). Ocean water off the southern tip of Greenland comes from cooled, saline, former Gulf Stream water that meets cold water from the Arctic Ocean. At the southern location, the Antarctic ice cools water in contact with it.

Figure 3-34   Effect of a cold ocean current. San Francisco beach on a fine day in August. The only bathers are hardy people in wet suits.

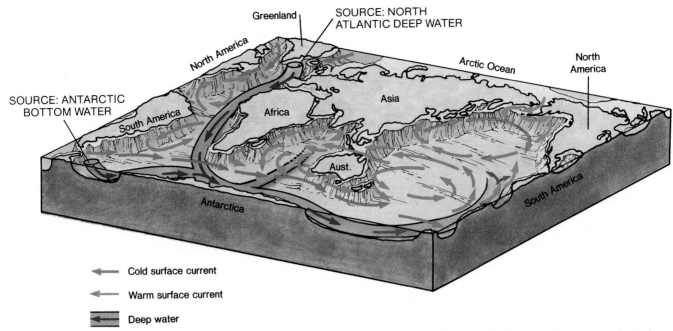

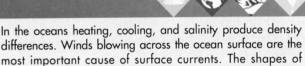

- ← Cold surface current
- ← Warm surface current
- ⇐ Deep water

Figure 3-35   Deep ocean circulation. Cold water descends at the two source areas in the Atlantic Ocean and continues through the Indian and Pacific Oceans, as determined by continental barriers.

North Atlantic Deep Water and Antarctic Bottom Water join in the southern Atlantic Ocean. Once the deep water descends to the Atlantic Ocean floor, lower density surface water heated by the sun prevents the cold water from rising. The bottom water commonly remains at depth for several centuries before gradual mixing with overlying layers and upwelling along western American coasts bring it back to the surface. The deep water flows through the Indian and Pacific Oceans around the southern tips of Africa and Australia. Remaining at depth, it flows across the Pacific Ocean until it comes to the coasts of Peru and California. There it rises as winds blow surface waters away from the land, causing the upwelling of cold, deep water.

In the oceans heating, cooling, and salinity produce density differences. Winds blowing across the ocean surface are the most important cause of surface currents. The shapes of ocean basins turn the surface currents northward and southward in boundary currents. Dense cold and saline water flows down to the ocean floor in the Atlantic and then flows through the deep oceans, returning to the surface along coasts where winds blow offshore and cause upwelling.

Heating of Earth's atmosphere and oceans, and the setting up of winds and ocean currents, mainly affect the **atmosphere-ocean environment.** Other major environments are also involved. The **solid-earth environment** and **living-organism environment** characteristics cause surface differences of albedo and hence of heating in the air above. The **surface relief-environment** and the patterns of ocean basins and continents affect the paths of winds and ocean currents.

ENVIRONMENTAL ISSUE:

Santa Ana Winds and Building Codes

Some of the issues raised include:
What causes Santa Ana winds and when are they most common?
Why are the winds often associated with extensive and uncontrollable fires?
Should building codes be changed to prevent the use of materials that burn easily?

In October, 1993, parts of the Los Angeles area of California were struck again by fires that raged out of control for several days and destroyed many homes. Some twenty-six fires swept southern California in the fall of 1993, killing four people and causing damage estimated at over $1 billion. The United States focused attention on the plight of people living in the area. It was not the first occurrence of such fires, but it was one of the worst. It became clear that although most of the fires may have been started by arsonists, their strength depended on the Santa Ana winds, and the end effects were bound up with whether the materials used in building houses were fire resistant.

Santa Ana winds occur regularly in the Los Angeles basin when high pressure is established over the desert inland of the San Bernadino Mountains and air flows toward a low-pressure center offshore, as the diagram shows. Very dry air funnels through valleys in the mountains, and the wind force increases due to the channelling. When the winds blow into the Los Angeles basin, they are hot, dry, and often very strong.

After long periods of drought, the vegetation around the large homes on the outskirts of Los Angeles is dry as tinder. Lightning or human actions may start a fire. If the Santa Ana wind continues, the fire is fanned into an uncontrollable disaster sweeping down valleys and across hillsides. There is little the fire services can do until the winds die down or humid winds from the ocean push the Santa Ana back. Several times in October, 1993, the winds died down for a little and the fires were brought under control. But the winds became stronger and the fires caused further damage.

At one stage the fires reached Malibu on the coast. The Santa Ana winds reached San Nicholas Island, 160 km (100 mi) offshore. It took several days for the humid ocean air to blow the dry air back inland and end the fire hazard for the time being.

As property owners searched among the scorched wreckage of their expensive homes, questions were asked about the need for more rigorous planning regulations. Homes with wooden shingle roofs caught fire easily, whereas many with stone walls and roofs were preserved. Homes were often built in positions where they were in the path of the strongest winds and so liable to be caught by a firestorm.

Questions were also asked about the presence of homes in areas subject to the Santa Ana hazard. The building of roads and houses affects the vegetation, the amount of water in the soil, and the stability of slopes. If provision is made to avoid bad effects on these aspects of the natural environment, and buildings are fire-proofed or placed in better positions, the costs of development will rise.

Linkages
The winds blowing from the inland high-pressure area are channelled through the San Bernadino Mountains, a feature produced by the **solid-earth environment** and sculpted by rivers in the **surface-relief environment.** The dry vegetation (**living-organism environment**) of the Los Angeles basin is a response to the arid climatic conditions of the area. Human beings have placed increasing pressures on the fragile ecology of this region.

Questions for Debate

1. Can people be expected to understand the operation of natural events that may affect them and their homes?

2. To what extent should building codes relate to the natural environment? [Note: This region has been subject to landslides as well as fires.]

3. Discuss the proposition: "Human beings prefer to compromise with nature rather than accept that natural phenomena should be left alone."

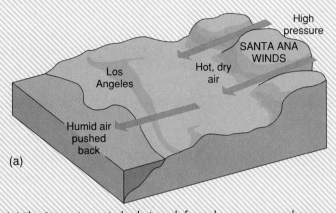

(a)

(a) The Santa Ana winds: their path from desert to ocean shore through the San Bernardino Mountains. (b) The charred remains of many homes and a few survivors dot the hillside at Las Flores Cyn Malibu near Los Angeles, 1993.

(b)

CHAPTER SUMMARY

1. Temperatures in the atmosphere depend on energy from the sun. Most insolation consists of short-wave rays that are subject to filtering during passage through the atmosphere. The upper atmosphere absorbs gamma rays, X rays, and most ultraviolet rays, while visible light rays and some infrared rays reach the ground.

2. Energy absorbed by the ground from visible light rays heats the surface. Terrestrial radiation from the surface in long-wave rays is absorbed by water vapor, carbon dioxide, and dust heating the lower atmosphere from the ground upward. Air warmed near the surface by conduction transfers heat upward by convection. The heated gases in the lower atmosphere radiate energy downward to the ground (counter-radiation) and upward to space.

3. Earth's atmosphere has a balance of incoming and outgoing energy. The lower atmosphere traps heat in a manner that resembles the workings of a greenhouse. An enhanced greenhouse effect will occur if more heat is trapped in the lower atmosphere than is returned to space.

4. The surplus of heat received in the tropics is distributed to cooler parts of the globe by movements in the atmosphere and oceans. Temperature at a place on Earth's surface is controlled by latitude, the distribution of continents and oceans, ocean currents, wind directions, and (in a small way) human activities.

5. Winds result from heating in the atmosphere-ocean environment. Vertical movements of air triggered by differences in solar heating produce variations in surface atmospheric pressure. Wind is the movement of air from high to low pressure.

6. Earth's rotation deflects the apparent wind path according to the Coriolis effect. The resultant geostrophic winds blow parallel to isobars. Friction at Earth's surface causes winds to slow and flow across isobars. The major relief features affect the paths of airflows by blocking them or channelling them through valleys.

7. Global wind patterns include both surface and upper troposphere components. Strong jetstream winds in the upper troposphere affect the surface circulation, especially in midlatitudes. The jetstream moves in Rossby waves whose amplitude increases and decreases in a repeated cycle of events. Vertical movements dominate the general circulation of the tropical atmosphere; horizontal conflicts between polar and tropical air control circulation in midlatitudes; and loss of heat, cooling, and sinking control circulation of polar regions.

8. In the oceans heating, cooling, and salinity produce density differences. Winds blowing across the ocean surface are the most important cause of surface currents. The shapes of ocean basins turn the surface currents northward and southward in boundary currents. Dense cold and saline water flows down to the ocean floor in the Atlantic and then flows through the deep oceans, returning to the surface along coasts where winds blow offshore and cause upwelling.

9. Heating of Earth's atmosphere and oceans, and the setting up of winds and ocean currents, mainly affect the **atmosphere-ocean environment.** Other major environments are also involved. The **solid-earth environment** and **living-organism environment** characteristics cause surface differences of albedo and hence of heating in the air above. The **surface relief-environment** and the patterns of ocean basins and continents affect the paths of winds and ocean currents.

KEY TERMS

heat   *40*

calorie   *40*

temperature   *40*

specific heat   *40*

conduction   *40*

convection   *40*

radiation   *41*

insolation   *42*

solstice   *43*

equinox   *43*

absorption   *44*

scattering   *44*

reflection   *44*

albedo   *44*

terrestrial radiation   *45*

counter-radiation   *45*

lapse rate   *47*

environmental lapse rate   *48*

temperature inversion   *48*

heat balance   *49*

greenhouse effect   *50*

CHAPTER REVIEW QUESTIONS

1. Explain how the sun and Earth radiate energy at different wavelengths.

2. Give examples of atmospheric processes that demonstrate heat transfer by radiation, by conduction, and by convection.

3. How does the composition of Earth's atmosphere affect the passage of insolation and terrestrial radiation through it?

4. Describe the different albedos of wet and dry soils, trees, ice, fresh snow, and blacktop roads

How do these different surfaces affect the temperature of overlying air on a still, sunny day?

5. What are the differences between the greenhouse effect and an enhanced greenhouse effect?

6. The polar regions have a heat deficit; the tropics have a heat surplus. Why do temperatures in these regions vary little from year to year?

7. Account for the differences in temperature in the lower atmosphere of oceanic and continental areas.

8. Compare the atmosphere with the oceans in terms of the types of movement that affect them and the causes of the movements.

9. Why does air circulate in counterclockwise directions around low-pressure centers in the northern hemisphere and in clockwise directions around low-pressure centers in the southern hemisphere?

10. Account for the fact that winds at the surface cross isobars, but those at altitudes of several kilometers flow parallel to the isobars.

SUGGESTED ACTIVITIES

1. Measure pressure gradients (millibars per 100 km) on weather maps for areas of high and low pressure. Relate the gradients measured to wind speeds shown.

2. Discuss how the world pattern of pressure and winds is related to the distribution of temperature studied in Chapter 2.

3. Compare local weather records of temperature and wind direction or strength to weather maps for the same period of time. How do the maps show temperatures, atmospheric pressure, and winds?

CHAPTER 4

WATER IN THE ATMOSPHERE

THIS CHAPTER IS ABOUT:

- Surface water storage
- From surface water to atmospheric gas
- Water vapor in the atmosphere
- Condensation
- Precipitation
- Water budgets
- Environmental Issue: Watering California: Is There Enough to Go Around?
- Environmental Issue: Acid Rain—Who Is to Blame?

Water circulates from oceans, lakes, soil, and plants into the air. Clouds form from this moisture and rain and snow bring the water back to Earth's surface, as in this rainshower in Monument Valley, Arizona (inset). The amount of water available at a place affects what can be grown there.

One of the most important features of planet Earth as a home for living things is the presence of water on its surface. Water circulates from surface storage areas into the atmosphere, moves through the atmosphere, and then returns to Earth's surface as part of the hydrologic cycle (Figure 4-1). The processes that drive the hydrologic cycle include the transformations and transfer of solar energy in the atmosphere-ocean environment and the patterns of global winds. Thus, an understanding of the hydrologic cycle builds on the previous chapter. This chapter follows the sequence of the hydrologic cycle, starting with ocean water changing to atmospheric gas and ending with the return from the atmosphere to land or ocean surfaces.

SURFACE WATER STORAGE

The oceans cover 71 percent of Earth's surface and contain 97.2 percent of Earth's water (Figure 4-2). The remaining 2.8 percent exists as fresh water or ice on the continents and as water vapor in the atmosphere. The ice sheets of Antarctica and Greenland and the glaciers of high mountains store nearly three-fourths of the continental water, and rocks contain the rest below ground.

Only small amounts of Earth's total water circulate through the hydrologic cycle at any one moment, as the third section of Figure 4-2 shows. The average amount of water in Earth's atmosphere would be sufficient for only ten days of rain or snow if no water

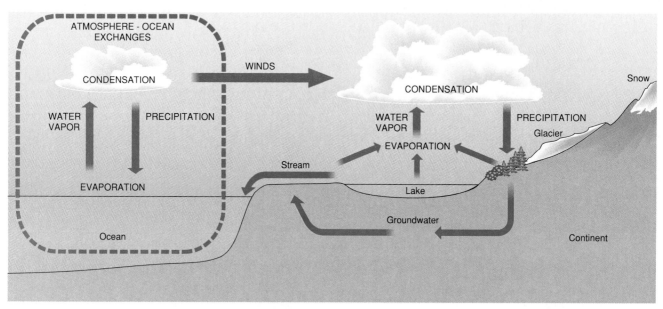

Figure 4-1 The hydrologic cycle. What are the main sources of evaporation into the atmosphere? Note the exchanges between liquid water, gaseous water vapor, and ice.

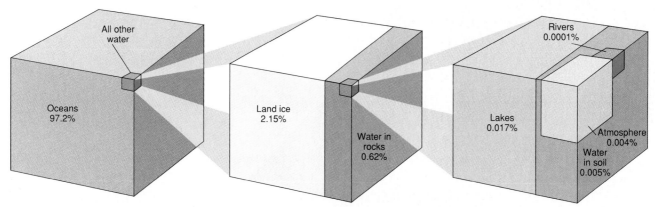

Figure 4-2 Water storage at Earth's surface. Percentages shown are by weight. Note the tiny proportion involved in movement through the atmosphere at any one moment. (Lakes include inland seas.)

was added from the ocean surface. Most of Earth's water is stored, often for several thousand years, before moving through the hydrologic cycle. The water moving through the hydrologic cycle greatly influences weather and climate and is also important in carving surface landforms and sustaining living organisms. See Environmental Issue: Watering California, p. 92, for an example of the importance of water in human activities.

FROM SURFACE WATER TO ATMOSPHERIC GAS

Water begins its journey through the atmosphere when evaporation or transpiration converts it from liquid to gas. **Evaporation** is the change from a liquid (e.g., water) to the gaseous form (e.g., water vapor). **Transpiration** is the process by which plants pass water vapor into the atmosphere through the pores in their leaves.

Evaporation occurs where water surfaces are exposed to the air. It requires a lot of energy to break apart the bonds linking the water molecules and so release each molecule into the atmosphere as a free gas molecule. For instance, the evaporation of 1 g (0.05 oz) of water requires the same amount of energy as the heating of 6 g (0.30 oz) of water from freezing to boiling point (5 g, or 0.25 oz, fill a teaspoon). There is usually sufficient heat energy in the atmosphere for evaporation to occur continuously from water surfaces.

The energy that powers evaporation is not "used up" but exists in the gaseous water as latent heat. **Latent heat** is the amount of heat absorbed or released when a body changes its state (e.g., from solid to liquid or liquid to gas) *without any temperature change* in the body. In water, heat is absorbed as latent heat during the melting of ice and the evaporation of water; it is released during condensation and freezing. The immediate and local effect of evaporation is to cool air in contact with the water surface. By contrast, **sensible heat** is heat energy that *changes the temperature* of a body when it is transferred by convection or conduction.

Transpiration occurs in plants. Plants draw water through their roots and use it to manufacture and transport their own foods (see Chapters 12 and 13). The water passes up the stem and into the leaves, which transpire it as vapor into the atmosphere. A plant may transpire up to 98 percent of the water absorbed by its roots. The amount transpired depends partly on the size of the plant. For instance, a corn plant transpires 2.4 l (.74 gal) of water vapor per day, and an oak tree up to 675 l (185 gal). In a forest, transpiration takes place from multiple layers of leaves, and it may equal evaporation from an open water body covering the same area.

The proportion transpired affects the water demands of crops. Sorghum, for instance, transpires 275 parts of water for every part of dry plant matter produced. The figure for wheat is 507 and for alfalfa it is 1068. That means sorghum can grow in drier conditions, since it requires less water than alfalfa.

Evaporation and transpiration over land are difficult to measure separately and are commonly referred to collectively as **evapotranspiration.** Rates of evaporation and transpiration (or their combined total) vary according to the presence of surface water, the availability of heat energy, and the capacity of the air to accept more water vapor. Evaporation over oceans or lakes is uninterrupted. Over land, surface water is not always available since the soil surface may dry out and reduce the rate of evaporation.

Evapotranspiration is most intense in the tropics and on hot days elsewhere. The highest rates of evaporation occur over tropical oceans, especially above the warm ocean currents on the western margins of the oceans. Up to 2.5 m (8 ft) of water evaporate each year from the surface of the Gulf Stream, compared to less than 0.2 m (8 in) per year from oceans in higher latitudes. On average, 1.2 m (just under 4 ft) of water evaporates from the oceans each year; they are replenished by precipitation and streamflow. There is virtually no evapotranspiration in the coldest parts of the world—over the ice sheets, the frozen Arctic Ocean, or the northern interiors of North America and Siberia in winter.

Water vapor entering the air adds to the molecules of other gases already there but acts separately. The amount of water vapor that can coexist with the other atmospheric gases depends on the temperature of the atmosphere. As temperature rises molecules become increasingly excited until water molecules escape from water surfaces and become water vapor in the atmosphere (see Chapter 3). The capacity for evaporating water increases with the temperature of air. The highest rates of evaporation occur at sea level in the tropics where the pressure exerted by water vapor may be around 40 mb of the average 1013 mb atmospheric pressure—i.e., almost 4 percent of total atmospheric pressure.

The atmospheric temperature limits the amount of water vapor the air can hold (Figure 4-3). When as much water vapor is changing back to water droplets as is being evaporated, the air is said to be **saturated.** Air that is not saturated, in which evaporation

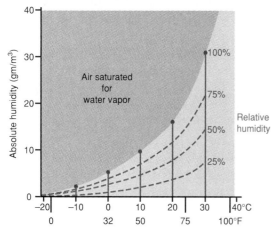

Figure 4-3 Water vapor in the atmosphere: the effect of temperature. Warmer air can hold more water vapor. The curved blue graph line shows the maximum amount of water vapor that air can hold at each temperature—the *dew point.* Below the line the percentage of the maximum water vapor possible at each temperature gives the *relative humidity.*

exceeds the formation of water droplets, is known as "dry" air. Winds help to maintain evaporation rates by moving saturated air away from a water or vegetated surface.

WATER VAPOR IN THE ATMOSPHERE

Although water vapor cannot be sensed easily by people—it is invisible and has no taste—it plays an important part in the weather and affects personal comfort. The amount of water vapor in the atmosphere is known as the **humidity.** Water vapor is an important greenhouse gas; humid air traps more heat near the ground than dry air. High humidity also slows evaporation from the skin and when combined with high temperatures may result in heat exhaustion.

Humidity is measured in several ways. *Absolute humidity* is the mass of water vapor per volume of air (grams per cubic meter). Figure 4-3 shows quantitatively how the amount of water vapor that a body of air can hold increases with temperature. At 10° C (50° F) air can hold 10 g (.4 oz) of water vapor per cubic meter, at 20° C (68° F) it can hold 18 g (.7 oz), and at 30° C (85° F) it can hold 31 g (1.25 oz). The rate of increase in vapor-holding capacity is greater at higher temperatures.

Relative humidity is the amount of water vapor in the atmosphere, expressed as a percentage of the total water vapor the air can hold at a particular temperature and pressure. The maximum amount of water vapor that the atmosphere can hold depends on pressure as well as temperature, but temperature variations are generally more significant to weather

processes. At 20° C and sea level pressures, 100 percent relative humidity is 18 g/m³; 50 percent relative humidity is 9 g/m³ (Figure 4-3). At higher altitudes atmospheric pressure and temperature both decrease, and the amounts of water that can be held by the atmosphere also decrease.

Relative humidity is frequently referred to in weather reports. It is sometimes measured by comparing the temperatures of two thermometers: one has its bulb covered by a moistened wick (wet bulb); the other is a conventional thermometer with no cover (dry bulb). The temperature reading is generally lower on the wet-bulb thermometer because water evaporating from the wick cools the thermometer. The difference in temperatures between the two thermometers is greatest when the relative humidity is lowest. When relative humidity is high and evaporation almost balances the formation of water droplets in the atmosphere, the dry-bulb reading may equal that of the wet bulb. In measuring relative humidity, it is important to make sure that air passes over the wet bulb, since the water evaporated from the wick will increase humidity around the thermometer. Some systems involve whirling the thermometers before taking a reading.

The hydrologic cycle circulates water from the oceans and continental surfaces into the atmosphere and back to the oceans. Evaporation and plant transpiration are the main processes by which liquid water changes to water vapor in the atmosphere.

The humidity of the atmosphere is the variable amount of water vapor that it contains at a given place and time. Higher atmospheric temperatures enable more water to evaporate into a volume of air before condensation balances evaporation.

CONDENSATION

CONDENSATION PROCESSES

The Basic Mechanism Water vapor changes to liquid water by **condensation,** which converts the invisible gas to masses of water droplets that form dew, frost, fog, and clouds (Figure 4-4). Condensation releases the latent heat trapped in water vapor at the time it evaporated into the air.

Condensation and evaporation occur side by side in the atmosphere, and there is a constant interchange between water vapor and liquid water droplets. Evaporation exceeds condensation in air until the point of saturation. The temperature at

Figure 4-4 Fog at Golden Gate Bridge, San Francisco. Such fogs are common along the Pacific coast near San Francisco. They result from warm, moist air flowing over the cold ocean current offshore.

which saturation occurs in air of a given absolute humidity is known as the **dew point** and is shown as the curved blue line in Figure 4-3. Condensation exceeds evaporation in saturated air. The relative importance of condensation and evaporation depends on the supply of water for evaporation, the amount of water vapor already in the air, and the temperature and pressure of the air.

If water vapor enters the atmosphere after saturation, some of the water vapor condenses. Saturation is reached earlier if the temperature or pressure of the air falls. In a cooling air body, condensation exceeds evaporation when the air temperature falls below dew point. Cooling air is the main way that large quantities of water vapor condense in the atmosphere.

Condensation in the atmosphere occurs around dust, smoke, or salt particles, known as *condensation nuclei.* These particles are plentiful in the atmosphere and provide solid surfaces on which water condenses. The geographic distribution of condensation nuclei varies, and they are ten times more numerous over land than over the oceans.

Large-Scale Condensation Forms   Clouds and fog result from three main types of cooling and condensation processes that occur in the atmosphere. One mechanism is the movement of a large body of humid air horizontally from a warm to a cold sur-face. A second process is air cooling just above the ground by prolonged radiation on cloudless nights. A third process occurs when a body of air lifts into air that is both cooler and of lower pressure. The rising air body expands as pressure in the surrounding atmosphere falls (Figure 4-5). As expansion occurs the heat energy in the rising body does not change but spreads through a greater volume of air and so the temperature of the rising body drops. This mechanism is known as **adiabatic expansion.** *Adiabatos* is Greek for "impassable," indicating that heat energy does not escape from the rising air body.

Once water droplets begin to form, they continue to grow by further condensation. The maximum size reached by fog and cloud droplets is tiny compared to raindrops (Figure 4-6). The smallness of cloud droplets means that they are light enough for clouds to be supported above the ground by the upward air currents feeding them with moist air.

DEW, FROST, AND FOG

Dew, frost, and fog are forms of condensation that result from the atmosphere cooling through contact with the ground. The amount of air movement is an important factor in determining which form will occur.

Dew (liquid droplets) and frost (frozen crystals) form on the ground at night when air is still. **Dew**

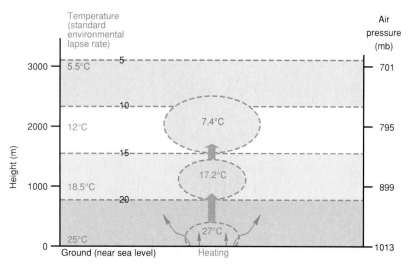

Figure 4-5   Adiabatic expansion and cooling of air. What happens to the body of air that is heated at the ground and rises? What caused its temperature to be different from that of the surrounding air?

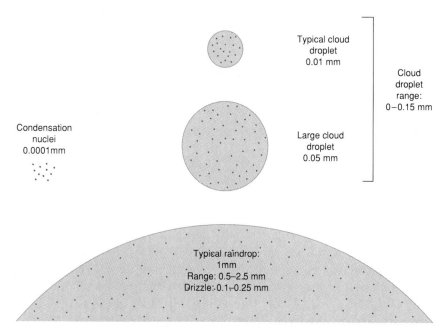

Figure 4-6   Raindrops, cloud or fog droplets, and condensation nuclei: a comparison of sizes. Note the difference between the raindrop and cloud droplets.

forms on surfaces cooled below dew point. Condensation on walls and windows is a form of dew. **Frost** forms on surfaces cooled below freezing point. It forms feathery crystal patterns on windows or fern-like hoar frost on grass and trees. Frost is different from the freezes feared by Florida citrus growers, as the latter cause water inside the plants to freeze, often killing them.

A slight movement in the air above the ground causes cooling to extend upward for several tens of meters. Condensation spreads through this layer of air in contact with the ground to produce fog. The official meteorologic definition of **fog** is a reduction in visibility to under 1 km (.5 mi). Dense fogs may be a hazard to motorists, sailors, and airplane pilots.

Fogs form in several ways, including the movement of air horizontally, or *advection,* and cooling air in contact with the ground at night. *Advection fog* forms when warm, humid air passes over a cold surface that has a temperature below the dew point of the moving air or when cold air mixes with warm air and lowers the overall air temperature. The wind speeds with advection fog are up to 30 kph (20 mph). Greater wind speeds cause sufficient turbulence to lift the mass of condensed moisture off the ground, forming low clouds.

Examples of advection fogs include the coastal fogs of San Francisco, California, which occur when warmer air crosses the cold offshore current (Figure 4-4), and the fogs over the Grand Banks of Newfoundland, which occur where warm air off the Gulf Stream meets cold water from Greenland. Advection fogs are common over land in winter if moist air moves over a snow-covered area. Another type of advection fog forms in upland areas when air cools on the upper slopes, becomes denser than its surroundings, and flows into the valley. The air in the valley may be moist because of evaporation from a stream or lake or transpiration from vegetation. The mixture of the cold air and the moist air causes condensation.

An *upslope fog* is similar to advection fog. It occurs when moist air is blown upslope, as from the lowlands of Manitoba, Canada, to the higher Alberta Plains. If adiabatic cooling is sufficiently strong, saturation occurs and a moving fog forms.

Radiation fogs form during calm atmospheric conditions with light breezes up to 10 kph (6 mph). Such fogs are common when longer fall and winter nights with clear skies follow warm days. The daytime warmth increases evaporation into air just above the ground. The clear night skies provide a long period of radiation that cools the ground surface. Conduction then lowers the temperature of air in contact with the ground, causing condensation that gentle air movements spread through the lowest 15 to 100 m (50–300 ft). Stronger winds mix the cooled air with warmer air to a greater depth, making condensation less likely. The radiation mechanism works slowly and may require several hours before any condensation occurs. Once begun, condensation spreads rapidly and visibility decreases suddenly. The following morning the sun quickly warms the atmosphere and the fog evaporates.

When fog-forming conditions in urban areas coincide with a buildup of pollution, the large numbers of condensation nuclei produce a rapid loss of visibility. The combination, known as smog, is common in Los Angeles. Warm air from the interior desert moves into the Los Angeles basin above cooler, moist air from the ocean. This establishes a temperature inversion that lasts for several days preventing the mixing of air and trapping pollutants and moisture near the ground. This type of advection fog is more likely to produce smog than overnight radiation fog.

CLOUDS

Clouds are masses of water droplets, ice particles, or both suspended in the atmosphere. Nearly all clouds form following adiabatic expansion in rising bodies of air. Advection produces shallow, low clouds only if wind turbulence is strong enough to lift fog.

When viewed from the ground, clouds appear to be solid, stable masses, but a time-lapse film of clouds moving across a landscape reveals that they are in constant motion. Water vapor in air rising beneath a cloud condenses as it enters the cloud. If such condensation balances or is exceeded by evaporation into the drier surrounding air at the tops and sides of a cloud, the cloud will not grow bigger. Clouds are dynamic features and continue to exist as long as condensation exceeds evaporation. It is important to understand the mechanics of cloud formation in order to interpret the significance of the different types of clouds.

Atmospheric Stability and Instability   Several mechanisms cause bodies of air to rise. Surface heating can cause a bubble of air to rise by convection. Air blowing toward mountains has to rise over them—an *orographic* effect. A slow-moving or stationary mass of dense, cold air acts as a wedge and forces faster-moving warm air to climb over it. These mechanisms often act in concert, reinforcing the amount and speed of uplift. Convectional uplift, or uplift over a mountain range, may act together with uplift over a wedge of cold air.

The rising body of air either rises so far and then stops, even falling back toward the ground, or keeps on rising. The relationship between the rising air and surrounding air determines what happens next. If the rising air becomes cooler and denser than the surrounding air, it stops rising—a condition known as atmospheric **stability.** Any cloud developing in a stable atmosphere will be shallow, forming a layer or isolated heap cloud. Alternatively, if the rising air remains warmer and less dense than the surrounding air, it continues rising. This condition is known as atmospheric **instability.** Clouds developing in an unstable atmosphere are likely to be tall and active. The conditions of stability and instability control the nature and shape of clouds.

Chapter 3 discussed the decrease of temperature with height in the troposphere—the environmental lapse rate. This lapse rate varies according to heating or cooling of the surface (see Figure 3-10). A rising body of air must negotiate its way through the atmospheric environment, and at each level its progress depends on the relationship between its temperature and that of the surrounding air. The rate at which temperature decreases *within* a rising body of air experiencing adiabatic expansion differs from the rate in the *surrounding* atmosphere. The **adiabatic lapse rate** is the rate at which temperature decreases in a rising body of air. It is constant at $9.8°$ C per kilometer ($5.4°$ F per 1000 ft) for dry air. The environmental lapse rate is more variable but averages $6.5°$ C per kilometer. This means that, in general, the temperature in the rising body of air falls more rapidly than that in the surrounding air and soon becomes colder and denser. It stops rising and falls back toward earth (Figure 4-7a) without forming large clouds. The atmosphere is therefore normally in a stable condition.

When the temperature of the rising air does not fall as rapidly as that in the surrounding air, it remains warmer than the air around it and the atmosphere is unstable (Figure 4-7b). Air keeps rising, cooling, and, therefore, condensing the water vapor it contains. Tall clouds develop.

Along with the variability of the environmental lapse rate, the presence of water vapor in the atmosphere encourages instability. A rising body of air that is not saturated with water vapor experiences a lapse rate of $9.8°$ C/km. This is known as the *dry adiabatic lapse rate* (DALR). Once the rising body of air cools to dew point and becomes saturated, condensation begins and releases latent heat. The released heat slows the rate of temperature decrease within the rising body of saturated air. The new rate of cooling is known as the *wet adiabatic lapse rate* (WALR) and is more variable but always less than the dry adiabatic lapse rate.

The wet adiabatic lapse rate depends on the temperature of the air as it left the ground. Warmer air at

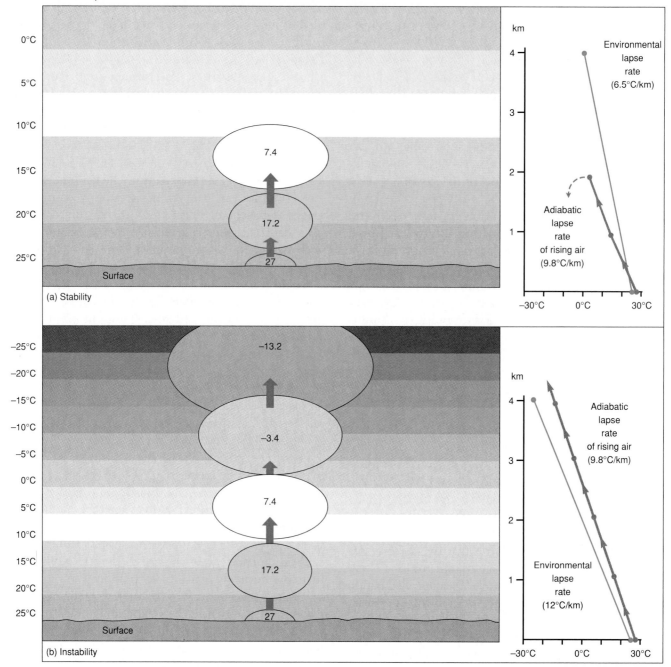

Figure 4-7   Stability and instability in the atmosphere. What are the differences between a stable and unstable atmosphere, and how do they affect a rising body of air in the atmosphere?

the ground takes up more moisture by evapotranspiration than colder air. When it reaches dew point, the warmer air has more water vapor to condense and more latent heat to release. For example, air that is at 0° C (32° F) before rising contains little water vapor and has a wet adiabatic lapse rate of 6° C/km (3.3° F/1000 ft); air that is at 30° C (86° F) before rising contains greater quantities of water vapor and condensation reduces the wet adiabatic lapse rate to 3.4° C/km (1.8° F/1000 ft). In general, the wet adiabatic lapse rate in tropical regions, where water is abundant and surface temperatures are high, is lower than that in higher latitudes.

Frequently, the saturation of an air body approaching stability may change it to an unstable condition. The lower wet adiabatic lapse rate in the newly saturated air body may alter its relationship with the surrounding air to one of instability in which the air body remains warmer and continues to rise instead of falling back. Such situations are common and are necessary to overcome the basic stability of the atmosphere. They are particularly common in the tropics where they produce many tall clouds.

The process of adiabatic expansion in ascending air bodies is complemented by adiabatic compression in descending air bodies. In many parts of the world, air flows up wind-facing slopes of mountain ranges and down a lee slope. The Sierra Nevada Mountains of California provide an example of the results (Figure 4-8). As air rises up the windward slope, it cools at the dry adiabatic lapse rate until it reaches saturation; then it cools at the wet adiabatic lapse rate. The rate of temperature fall is therefore much less over the upper slopes. On descending the lee slopes, however, the air is compressed as it enters air of increased density and pressure. As it descends, this air "dries out" and its temperature rises at the dry adiabatic lapse rate. This mechanism results in lower relative humidities and higher temperatures on the lee slopes for a given altitude. Similar dry and relatively warm downslope winds form on the eastern side of the Rocky Mountains (chinook winds) and in the valleys of the Alps (foehn winds).

Types of Clouds   Clouds are classified by their shape, height, and water and/or ice content. The shapes are described by terms such as *cirrus* (wispy, hairlike), *stratus* (layer), and *cumulus* (heap). There are groups of clouds at high, middle, and low altitudes. Another descriptive name used is *nimbo-*, or *-nimbus,* denoting that rain is likely to fall. These characteristics combine in various ways to produce a range of types (Figure 4-9). For instance, cumulus clouds formed of water occur at low and middle altitudes, altocumulus clouds formed of water and ice occur at middle levels, cirrocumulus clouds formed entirely of ice occur at high levels, and cumulonimbus clouds rise through all levels.

Cumuliform clouds are separate heap-like masses that grow vertically during the convectional lifting of bubbles of air (Figure 4-10) or during the uplift of air over mountains or over denser air. Such clouds have a distinctive flat base at the dew-point level, where condensation begins. The height reached by the cloud tops depends on the amount of moisture available and the conditions of stability or instability in the atmosphere. Smaller *cumulus* clouds form in stable atmospheric conditions. Each cumulus cloud exists for five to thirty minutes. They disperse as the supply of moist air from below is cut off when winds blow the clouds away from their source or when surface heating dies down later in the day.

Cumulonimbus clouds are very tall and form in unstable conditions where air rises and water vapor condenses. The water droplets and vapor turn to ice when temperatures in the tops of such clouds fall below freezing. In warm midlatitude weather, all droplets freeze at heights greater than 4 to 6 km (2.5 to 4 mi) above the ground and cloud tops may be 12 km (8 mi) high; in winter the midlatitude freezing level occurs less than 3 km (2 mi) above the ground. Near the equator, cloud tops may tower 18 km (12 mi) above the ground (Figure 4-11). Ice crystals give the tops of cumulonimbus clouds a diffuse and fibrous appearance in contrast to the clean-cut outlines of the lower parts composed of water droplets. Upward air currents inside cumulonimbus clouds are very strong, and speeds exceed 50 kph (30 mph). High-level winds often cause the ice particles in the top of a cumulonimbus cloud to spread downwind in an anvil-like shape. When convection ceases, these huge clouds decay and leave behind a complex series of layer clouds at various levels.

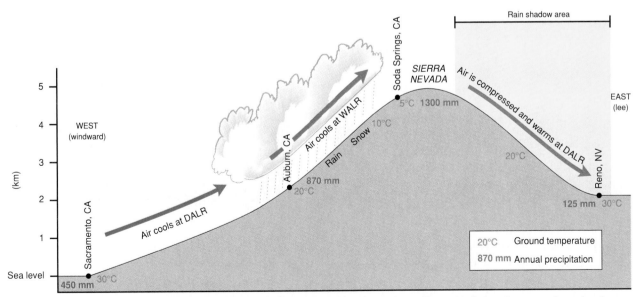

Figure 4-8   Air flowing across the Sierra Nevada, California. Air forced to rise up the west-facing slopes cools at the dry adiabatic lapse until saturation and then at the wet adiabatic lapse rate. Air descending the lee slopes to the east of the Sierra warms at the dry adiabatic lapse rate. How does this pattern affect temperatures on either side of the mountain?

(a)

(b)

(c)

(d)

Figure 4-9 Some cloud types. (a) Cirrus cloud. (b) Fair weather cumulus cloud. (c) Altocumulus cloud. (d) Cumulonimbus cloud with ice-crystal, anvil-like top. Under what atmospheric conditions does each form?

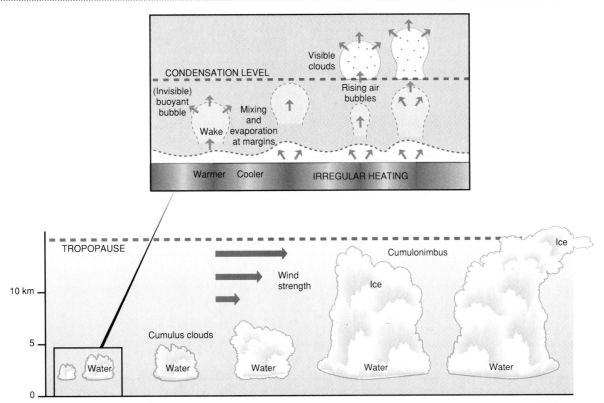

Figure 4-10   The development of cumuliform clouds. Air bubbles rise above heated ground (inset). Clouds form above the condensation level. As the clouds get taller, the precipitation produced increases. Ice particles form at the tops of the tallest clouds.

Figure 4-11   Cumulonimbus clouds, Mekong Delta, Vietnam. This space shuttle view links strong surface heating and plentiful surface water to the formation of tall clouds. The clouds grow very tall and have flat anvil heads just below the tropopause.

Stratiform clouds are layer clouds that form gray, structureless sheets rather than identifiable single clouds. They form during the widespread uplift of a large air mass as it crosses a mountain range or rises over a mass of denser air. Stable atmospheric conditions often prevent a great vertical development of *stratus* clouds by stopping the air from rising further (Figure 4-12). Winds creating sufficient turbulence to lift off the ground the mass of condensed water droplets formed at low levels by advection also produce stratus clouds.

When stratiform clouds break into distinct forms within the general cover, they are called *stratocumulus* at lower levels, *altostratus* between 2 and 8 km above the ground, and *cirrostratus* at high levels. The dark gray *nimbostratus* clouds are thicker. They form in unstable air and often produce continuous heavy rain or drizzle. Several layers of stratiform clouds may occur in a vertical sequence at one time, separated by clear sections of dry atmosphere; these layers are often seen from an airplane as it ascends through them after takeoff.

Cirriform clouds consist entirely of ice and occur at extremely high levels in the troposphere. Their base height is at least 8 to 11 km (5 to 7 mi) above the ground, and most form near the top of the troposphere. The low temperatures, small amounts of water vapor present, and high winds at these altitudes produce clouds consisting of ice-particle trails that are several kilometers long and descend from the level of formation. The ice crystals fall under gravity until they sublime, or change directly from ice to water vapor.

Decreasing atmospheric temperature leads to increasing relative humidity until saturation occurs at dew point and condensation begins to exceed evaporation. Fogs and clouds form as the result of large-scale condensation following cooling of air as it moves over a cold surface, as radiation takes place, or as adiabatic cooling occurs in a rising body of air.

The relationships between the temperatures of a rising body of air and the surrounding air determine whether the body stops rising (stability) or continues to rise (instability) and produces tall clouds.

Cloud types are differentiated by the processes of condensation involved in their formation. Stratus clouds form following widespread uplift of air. Cumulus clouds form following local convection and may grow upward in unstable conditions. Cirrus clouds are high-level, wispy ice clouds formed by cooling below the tropopause.

PRECIPITATION

TYPES OF PRECIPITATION

Precipitation is the deposition of atmospheric water on Earth's surface. Over most of the globe, rain is the main form of precipitation (Figure 4-13). In cold regions, snow is dominant.

Water drops fall as **rain** or *drizzle,* the difference being drop size (see Figure 4-6). Rain freezes when it hits very cold surfaces to give *freezing rain.* **Snow** falls as ice crystals that are grouped into flakes. The size of a snowflake depends on the water vapor available during crystal growth. Small flakes are more common in very cold conditions, whereas large wet snowflakes stick together in warmer conditions and may have diameters up to 5 cm (2 in).

Other forms of precipitation include hail and sleet. **Hail** is rounded or jagged lumps of ice that may have an internal structure of concentric layers like an onion (see Figure 5-7). *Sleet* consists of raindrops that freeze as they fall through very cold air. Sleet and hail bounce, freezing rain does not.

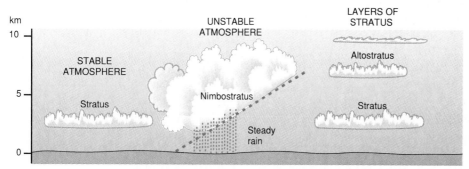

Figure 4-12 Stratiform clouds. Layer clouds form as warm, moist air ascends of cold air. When a cumulonimbus cloud has precipitated most of its moisture, it often leaves several layers at different heights.

Figure 4-13   Rain falling from a cumulonimbus cloud over Nag's Head, North Carolina. Other cumulus clouds in the background are arranged in rows that are formed by strong winds.

The size and weight of raindrops, hail, and droplets in drizzle determine their rate of fall and the impact precipitation has on the ground. Large drops of rain falling at rates of several hundred millimeters (a few feet) a second for a few minutes wash away surface soil. Drizzle comprises small drops that may have no such effect after falling for several hours.

Different clouds produce different types of precipitation. Drizzle generally falls from stratus clouds. Rain and snow fall from nimbostratus and cumulonimbus clouds. Hail falls exclusively from cumulonimbus clouds that have the strong convective updrafts needed to carry the hailstones up and down. Cirrus clouds do not produce precipitation because they are too thin and because the ice crystals convert directly to water vapor. Acid rain occurs when the chemistry of rainwater is made acidic by reactions in the atmosphere (see Environmental Issue: Acid Rain—Who Is to Blame? p. 94).

MECHANISMS OF PRECIPITATION

Cloud droplets stay suspended above the ground because upward air currents, caused by convection or rising over mountains and denser air, prevent them from falling. Growing cloud droplets to a size that will result in precipitation requires different mechanisms from those that produce cloud droplets. Two complementary mechanisms cause precipitation.

The first process requires the formation of *ice crystals in the tops of clouds*. They form particles that are large enough to fall to the ground. This mechanism occurs in clouds in which the upper layers are below freezing point. Freezing occurs in the tops of most cumulonimbus clouds outside the tropics and in the tallest cumulonimbus clouds in the tropics.

Water vapor, liquid droplets, and ice occur together in clouds with temperatures between 0° C (32° F) and −20° C (−4° F). Ice forms at temperatures below −9° C (16° F) on special nuclei that include tiny particles of clay. Once ice particles begin to form in the top of a cloud at around −10° C (14° F), they grow rapidly by attracting water droplets and water vapor. All the remaining cloud droplets freeze when the temperature falls below −20° C (−4° F). The ice crystals may splinter to form additional particles, which then grow until they are heavy enough to fall. On falling, ice crystals often combine in loose aggregates (snowflakes), which melt to raindrops at the 0° C (32° F) level in the atmosphere.

The second means of forming precipitation requires the *collision and coalescence* of water droplets in clouds. This process occurs in clouds where the lowest temperatures do not permit the formation of ice, as well as in those containing ice.

Droplet collision and coalescence is particularly common in clouds with prolonged updrafts of warm, moist air, such as the clouds of moderate height above the tropical oceans. It also occurs in stratus clouds in the tropics and midlatitudes. Water droplets in these clouds range up to about 0.05 mm in radius, the size of the drop being related to the

size of the condensation nucleus. Falling larger droplets collide and coalesce with slower-moving smaller droplets carried by updrafts or downdrafts in the cloud. Larger droplets draw smaller droplets into their wake.

Most raindrops produced by collision and coalescence are fine drizzle, with a maximum diameter of 2 mm, but larger drops form in tall clouds when ice falling from the top collides and coalesces with cloud droplets lower down. When snowflakes or hailstones fall, they often melt and then grow by collision and coalescence to form large raindrops. Very large drops become unstable and break up as they fall. In exceptional circumstances, such as have been observed in strong updrafts over Hilo, Hawaii, warm clouds may produce raindrops 4 to 5 mm in diameter.

WORLD DISTRIBUTION OF PRECIPITATION

The world map (Figure 4-14) shows that the global distribution of precipitation is uneven from place to place. Since sufficient precipitation is essential to the survival of living organisms on the continents, the

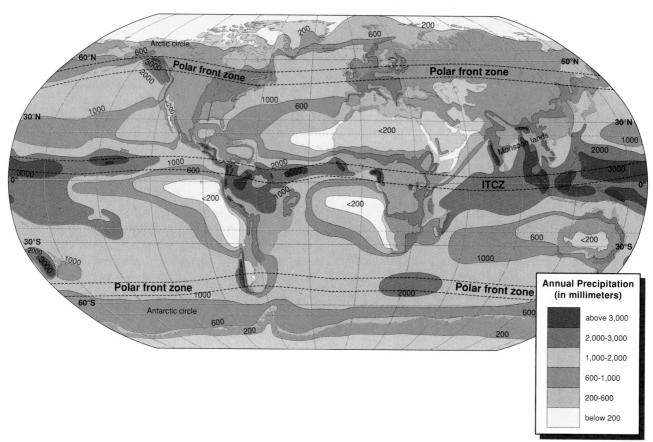

Figure 4-14   World distribution of precipitation: annual average totals. Notice the link between the Inter-Tropical Convergence Zone (ITCZ), the polar front zones, and the areas of heaviest rainfall.

geographic distribution of areas of high and low precipitation is of vital significance. The two key ideas in understanding the distribution of wet and dry areas are the availability of air containing water vapor and the location of zones of atmospheric uplift that produce condensation and precipitation.

Air containing large quantities of water vapor originates over the oceans and the tropical rainforest areas. Air with little water vapor originates over the arid areas situated beneath the sub-tropical high-pressure zones, midlatitude continental areas at great distances from the ocean, and the polar regions covered by ice and snow. **Aridity** is a lack of available water and occurs where there is an excess of evapotranspiration over precipitation and where other water sources are not present.

Atmospheric uplift, cooling, condensation, and precipitation occur mainly in the two major zones of air convergence identified in Chapter 3, the ITCZ and the polar front. Both zones are marked by a high proportion of cloudiness (Figure 4-15) and precipitation (see Figure 4-14). Mountain ranges force airstreams to rise, and some of the highest precipitation totals occur where air convergence zones and mountain ranges coincide to force moist air upward. Over the Sahara, western Australia, southwestern United States, southwestern Africa, the west coast of Peru, and northern Chile, air descends and diverges, producing dry regions. Descent of air on the lee sides of mountain ranges results in the dry conditions of rainshadow areas.

WATER BUDGETS

The processes that make up the hydrologic cycle influence the natural availability of water at a place. Inputs of precipitation and outputs of evaporation, transpiration, flow through soils, and flow over the land determine whether a given place will have a surplus

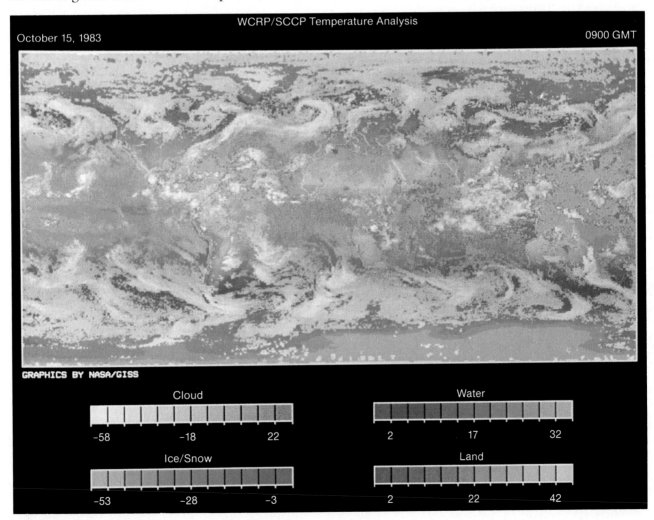

Figure 4-15 World cloud patterns. This map, compiled from satellite images, shows apparently disorganized groups of clouds along the equator and swirls of cloud in midlatitude regions. The cloud belts occur along zones of uplift. What other information does this map provide?

or deficit of water for one season or for the whole year. Calculations of these inputs and outputs are, therefore, important as a basis for estimating domestic and industrial water supply, river flow, liability to flooding, and deciding whether crops require irrigation.

Over Earth as a whole, precipitation and evaporation balance at around 880 mm (35 in) of each per year. There is an excess of evaporation over precipitation in ocean areas, however, and of precipitation over evaporation on the continents. The inputs and outputs of water at a particular location on Earth's surface constitute a **water budget** (Figure 4-16).

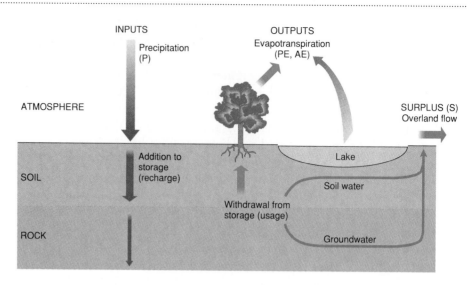

Figure 4-16   Water budget. Inputs of precipitation are balanced against outputs of evapotranspiration. If there is a surplus of input, water is added to ground storage in soils and rocks by recharge; if ground storage is full, water runs off in overland flow. If evapotranspiration exceeds precipitation, water is drawn from ground storage until it is emptied.

In a local water budget, precipitation provides the inputs, or gains. The first set of outputs, or losses to the local budget, are by evapotranspiration. *Potential evapotranspiration* is the amount of evapotranspiration that would occur if water were always unlimited at the surface and at plant-root level. *Actual evapotranspiration* is what actually occurs. The actual may be equal to the potential when water is available, but it will be less in a dry season.

If precipitation exceeds evapotranspiration, there is a *surplus* of water at ground level and the extra water goes into soil and rock storage underground. If these storage zones are full, the surplus water runs off over the surface into streams. The process of replacing water in soil and rock is known as *recharge.*

If evapotranspiration exceeds precipitation, plant roots extract water from the soil and rock stores faster than these stores are replenished. Such *usage* of water can continue only as long as water is available in the stores. When they become empty, there is a water *deficit* and usage and actual evapotranspiration cease.

Figure 4-17 shows water budget diagrams for three places across the United States. The diagrams illustrate different relationships between precipitation (P) and potential evaporation (PE). San Francisco (Figure 4-17a) receives most precipitation in the winter 6 months, October through March, when PE is low and recharge occurs. A surplus contributes to surface runoff in February and March. In summer low precipitation and high PE combine to create a major deficit period when farmers in central California need to irrigate their crops. In Nebraska (Figure 4-17b), the main period of precipitation coincides with the highest PE. There is a small late-summer deficit and a short period of water recharge but little surplus water for runoff in winter (the period of low precipitation). In Boston (Figure 4-17c), precipitation occurs throughout the year, and there is a very short period of deficit in late summer. There is a long period of water surplus and surface runoff in winter.

Precipitation requires special mechanisms for increasing the size of cloud water droplets so that they are large enough to overcome the force of upward currents and fall to the ground. The formation of ice particles in the upper parts of clouds and collision-coalescence are the two main ways in which this occurs.

Most precipitation falls in global zones where humid air rises and the highest totals are where mountain ranges facing into the winds force further uplift. The water budget of a place is the balance between inputs and outputs of moisture. It may vary from year to year or from season to season.

The exchanges of the hydrologic cycle link the atmosphere and ocean in the **atmosphere-ocean environment** with the **surface-relief environment** and the **living-organism environment.** Evaporation from water surfaces combines with transpiration from plants to provide water vapor in the atmosphere. On cooling, water vapor condenses on particles that enter the atmosphere following volcanic activity or are blown there from broken rock materials by the wind. Precipitation as rain or snow brings water to the continents, where it fashions landforms and sustains living organisms.

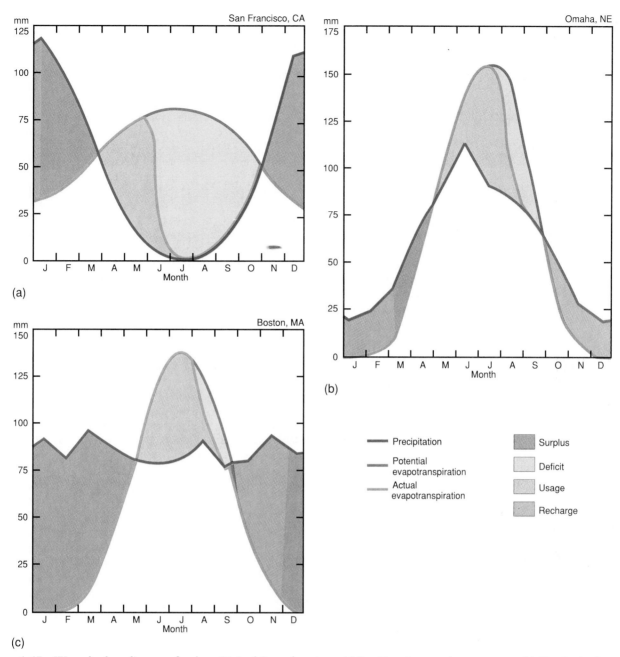

Figure 4-17 Water budget diagrams for three United States locations. (a) San Francisco on the west coast. (b) Omaha in the center. (c) Boston, on the east coast. Compare their features. How will the water budget at each place affect the growing season for plants?

ENVIRONMENTAL ISSUE:

Watering California—Is There Enough to Go Around?

Some of the issues in this controversy include:

What is the source of California's water?
Are other supplies of water available to support further expansion of the economy?
How are forecasts made, and how useful are they?

California has the largest agricultural business in the United States (about $7.5 billion in 1991). It also has large urban-industrial complexes and some of the world's most lavish lifestyles. The wealth of land and people depends partly on a rich natural bounty of good soils, almost constant sunshine, and, until recently, ample supplies of water. The water supplies are now in question, and it is increasingly important to know how much is available. Over 80 percent of the water is used by only 2 percent of the population, the farmers. Expanding cities are increasingly competing for the limited water available.

Water comes to the Central Valley, the biggest agricultural area, from the Sacramento and San Joaquin Rivers, which draw their water from snowmelt in the Sierra Nevada. A network of dams stores the water. The largest is the Shasta Dam, built where the Sacramento River reaches the Central Valley (see map). More than 800 km (500 mi) of aqueducts and canals distribute water to fields in central and southern California. Federal and state government bear the cost of bringing water to the valley. Eventually water users must repay the government investments, but they pay no interest and the water remains inexpensive, especially to farmers.

The southern half of California has an arid climate. During the twentieth century, massive engineering projects delivered water from the wetter northern half of the state and from the Colorado River to the farms and cities of the south. Most of this water goes to agriculture, but increasing

amounts are used on lawns, golf courses, and areas reclaimed from the desert such as Palm Springs.

In the southernmost part of California, the Imperial Valley uses water brought 125 km (80 mi) by the All-American Canal from the Colorado River. California takes about 20 percent of the total flow of the Colorado. The source of 80 percent of the water in the Colorado River is in the snowpack of mountain ranges in its headwater areas. A compact signed in 1922 allocates use of the Colorado's water among the seven states through which the river runs; California has used water not being used by other states. A later treaty guarantees a supply to Mexico at the river's mouth. The average annual flow of water in the river is now less than when the compact was signed, and there is often insufficient water to meet the agreed allocations. With the completion of the Central Arizona Project in 1990 taking Colorado water to the areas around Phoenix and Tucson, California can no longer ask to use the water unused by other states. California is reaching the limits of possible water supplies.

The National Weather Service issues reservoir inflow forecasts and water supply outlooks to the dry western states. It also warns of spring floods from snowmelt and provides general forecasts of river flow. Given the importance of water to agriculture in California and of the produce to the nation, it is crucial that forecasts for the state are as accurate as possible. The National Operational Hydrologic Remote Sensing Center, based in Minneapolis, gathers, checks, and disseminates data for forecasts across the United States. Measurements of rainfall, river flow, and evaporation are compared to airborne and satellite surveys of the extent of and water equivalent of snow cover. A computer network distributes the data as tables of figures and images, such as the two maps included here. They provide a comprehensive picture of water supplies for federal, state, and private agencies, including the California farmers.

Weather satellite images are classified into "snow," "no snow," and "cloud" and are added to a Geographic Information System that contains data on altitude and

watershed boundaries. Percentage snow cover is calculated in five elevation zones for each watershed. Airborne sensors make it possible to calculate the snow water equivalent. The Office of Hydrology then issues longer-term planning documents for a region. In 1991, the analysts noted that storms in late February and early March increased the water content of the Sierra snowpack from about 15 percent of normal to about 50 percent of normal in the northern Sierra Nevada and from 70 or 80 percent in the southern areas. Water levels in three major reservoirs, however, confirmed that despite high precipitation from the storms California was still a long way from recovering from the drought of the later 1980s. Statewide, reservoir storage in California in March 1991 was only 55 percent of normal and the outlook for the rest of the year was grim.

To owners of parched fields and lawns in California, this was bad news. Knowing the worst at the beginning of the summer dry season, however, gives both individuals and authorities the chance to conserve an increasingly precious resource.

Linkages

Flow in Californian rivers depends on the processes that make the atmosphere humid and on the presence of high mountains that force air upward. Precipitation over the mountains in winter produces snowpack storage of water—the main source for the rivers' water when they flow through desert areas in their lower courses.

Questions for Debate

1. Should more money be spent to store and distribute water to every part of California, whatever the demands from growing population and economic development?

2. Should the water that costs so much for the nation remain inexpensive to local users who use it for producing cattle feed, crops such as rice that use lots of water, and more "frivolous" uses such as watering lawns and golf courses?

(a)

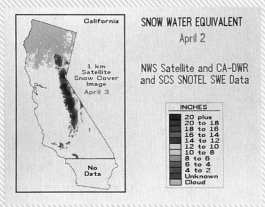
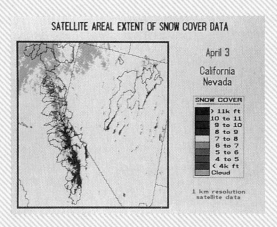

(b)

(a) California's water supply. Central Valley draws its water from the headwaters of the Sacramento and San Joaquin rivers. In the south, Imperial Valley and Los Angeles rely on water from the Colorado River. (b) Snow cover and snow water-equivalent on maps produced from satellite and airplane images. The first map shows the depth of snowpack and the second assesses the amount of water in the snowpack.

ENVIRONMENTAL
ISSUE:

Acid Rain—Who Is to Blame?

Some of the issues involved include:
    What causes acid rain?
    What are its effects?
    Can it be stopped?

Acid precipitation is rain or snow that has an enhanced acidity caused by atmospheric pollution. The increasing acidity of precipitation in the northeastern United States is shown on the series of maps. Similar increases over the same period of time occurred in northwestern Europe. Acid fog occurs around industrial areas.

The acidity of water governs its ability to react chemically with substances in soils and rocks. Acidity is usually measured on the pH scale, which registers the proportion of hydrogen ions (components of acids). The scale goes from 1—highly acid, with many hydrogen ions—to 14—strongly alkaline, the opposite of acid. Normal rainwater usually has a pH of 5.5, which is slightly acid because it contains dissolved carbon dioxide (weak carbonic acid). Acid rain has a pH between 5.5 and 2.5.

Acid rain is caused by sulfur and nitrogen compounds being emitted into the atmosphere following the burning of coal in power plants and of gasoline in automobiles. The sulfur and nitrogen unite with oxygen in the presence of sunlight, forming oxides that may become dry deposition near the source, as the diagram shows. If they travel farther from the source, further chemical reactions convert the oxides of sulfur and nitrogen into sulfuric and nitric acids. These acids fall to the ground in rain or snow several hundred kilometers downwind of their source. The building of high chimneys at coal-fired power stations since the early 1970s reduced dry deposition but has increased wet deposition at a distance.

Acid rain is blamed for fish kills in upland lakes, loss of tree leaves, destruction of building materials, and hazards to human health. The effects on upland lakes in the Adirondack Mountains of upper New York State, parts of eastern Canada, and Scandinavia are most obvious. The thin soils of these areas, which have an acidic chemical environment themselves, rapidly pass the acidity in rain and snow to the lakes. The greatest effect of acid rain may be on crop yields, but little data are available.

While the urban-industrial atmospheric sources of acidity have received most blame for the declining quality of soils and lakes, other human activities also cause or enhance the condition. Adding lime to poor soils makes them less acid, but farm management practices that do not replace the lime after crops have extracted lead to the impoverishment of soils by increasing acidity. In 1993 it was shown that another source of acidity in the atmosphere comes from nitrate fertilizer used on farm fields. This washes into rivers and the sea. The tiny organisms living near the ocean surface around northern Europe grow and reproduce at increased rates because of the added nitrate, and they release more sulfur compounds into the atmosphere. It is estimated that between one-third and three-quarters of acid rain falling on less industrialized countries, such as Ireland and Norway, could be due to this source of sulfur.

Acid rain has become a public issue at an international level between the United States (source) and Canada (recipient), and between other European countries, such as Great Britain and Germany, (source) and Scandinavia (recipient). Plans are now under way to reduce emissions of sulfur and nitrogen in the source countries by using low-sulfur fuels, installing filters at power plants, or by installing catalytic devices to cars. The more developed countries bear the additional expense, but the expansion of industrialization in the developing world may spread acid rain sources to countries that cannot afford the costs of control.

Linkages
Acid rain demonstrates the linkages among Earth environments. Human activities alter the composition of the atmosphere and precipitation brings particles and acids down to the soil and surface waters. The effect of increased acidity is greatest in those soils and waters overlying rocks with a high natural acidity.

Questions for Debate
1. Should all sources of sulfur and nitrogen oxides in the atmosphere be banned?
2. Do we need to find more conclusive evidence before taking further (costly) action to reduce sulfur and nitrogen emissions?
3. What should be done about such emissions in developing world countries?

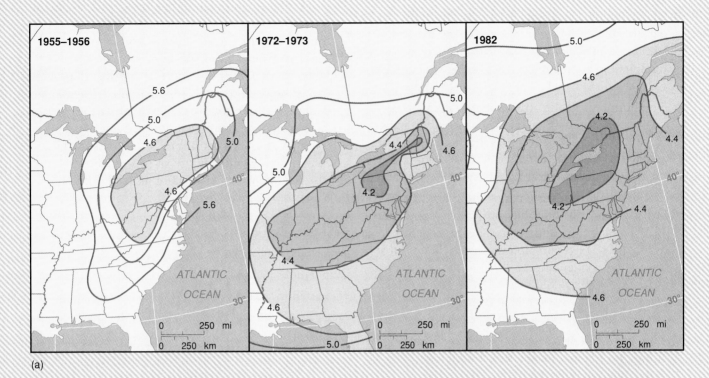

(a)

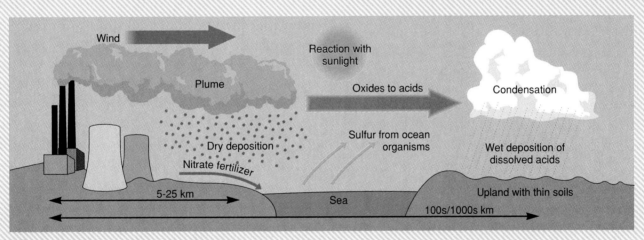

(b)

(a) The distribution of acid precipitation (pH values) in eastern North America, 1955–82, showing the increase in intensity and geographic coverage. (b) The processes that give rise to acid rain.

Source: (a) Top left and center: Park after Likens; top right, U.S. National Atmospheric Deposition Project.

CHAPTER SUMMARY

1. The hydrologic cycle circulates water from the oceans and continental surfaces into the atmosphere and back to the oceans. Evaporation and plant transpiration are the main processes by which liquid water changes to water vapor in the atmosphere.

2. The humidity of the atmosphere is the variable amount of water vapor that it contains at a given place and time. Higher atmospheric temperatures enable more water to evaporate into a volume of air before condensation balances evaporation.

3. Decreasing atmospheric temperature leads to increasing relative humidity until saturation occurs at dew point and condensation begins to exceed evaporation. Fogs and clouds form as the result of large-scale condensation following cooling of air as it moves over a cold surface, as radiation takes place, or as adiabatic cooling occurs in a rising body of air.

4. The relationships between the temperatures of a rising body of air and the surrounding air determine whether the body stops rising (stability) or continues to rise (instability) and produces tall clouds.

5. Cloud types are differentiated by the processes of condensation involved in their formation. Stratus clouds form following widespread uplift of stable air. Cumulus clouds form following local convection and may grow upward in unstable conditions. Cirrus clouds are high-level, wispy ice clouds formed by cooling below the tropopause.

6. Precipitation requires special mechanisms for increasing the size of cloud water droplets so that they are large enough to overcome the force of upward currents and fall to the ground. The formation of ice particles in the upper parts of clouds and collision-coalescence are the two main ways in which this occurs.

7. Most precipitation falls in global zones where humid air rises, and the highest totals are where mountain ranges facing into the winds force further uplift. The water budget of a place is the balance between inputs and outputs of moisture. It may vary from year to year or from season to season.

8. The exchanges of the hydrologic cycle link the atmosphere and ocean in the **atmosphere-ocean environment** with the **surface-relief environment** and the **living-organism environment.** Evaporation from water surfaces combines with transpiration from plants to provide water vapor in the atmosphere. On cooling, water vapor condenses on particles that enter the atmosphere following volcanic activity or are blown there from broken rock materials by the wind. Precipitation as rain or snow brings water to the continents, where it fashions landforms and sustains living organisms.

KEY TERMS

hydrologic cycle   76

evaporation   77

transpiration   77

latent heat   77

sensible heat   77

evapotranspiration   77

saturation   77

humidity   78

condensation   78

dew point   79

adiabatic expansion   79

dew   79

frost   80

fog   80

cloud   81

stability   81

instability   81

adiabatic lapse rate   81

cumuliform cloud   83

stratiform cloud   86

cirriform cloud   86

rain   86

snow   86

hail   86

aridity   89

water budget   90

CHAPTER REVIEW QUESTIONS

1. What factors cause and affect evaporation and transpiration? Focus on the rates of these processes and their global distributions.

2. Define *condensation* and *precipitation*. How are these two processes linked and in what ways do they differ?

3. Compare the two main mechanisms that produce precipitation.

4. Why do some parts of the world have more precipitation than others?

5. Attempt to explain why some parts of North America have more precipitation than others.

6. Describe the main features of the hydrologic cycle and water budget diagrams (see Figures 4-1 and 4-16). Point out similarities and differences between the two diagrams.

SUGGESTED ACTIVITIES

1. Explain why a person measuring relative humidity should ensure that air is passing over the wet bulb thermometer.

2. Use the different types of lapse rate to explain:
   (a) the formation of cumulonimbus, rather than cumulus, clouds;
   (b) the fact that Reno, Nevada, often receives warmer weather than Auburn or Sacramento, California (see Figure 4-8).

3. Study cloud types and the associated weather for a week. In each case, describe the temperature, humidity, and wind speed conditions and assess whether the atmosphere is stable or unstable.

-51° C

INITIAL STAGE

Growth

Snow-
flakes

-8° C

Water
droplets

0° C

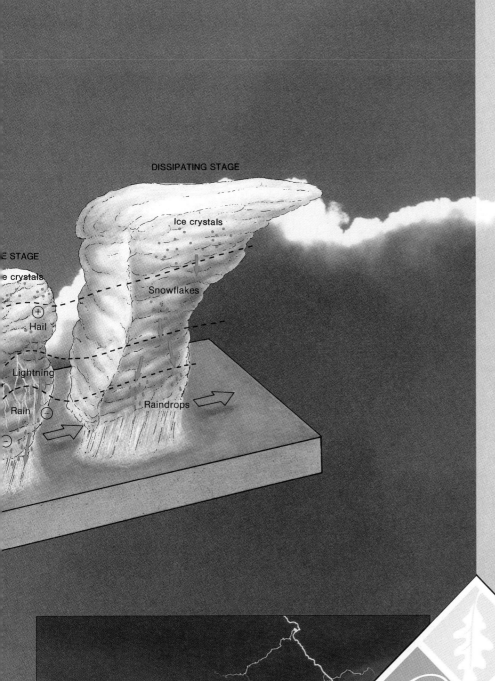

CHAPTER

5

WEATHER SYSTEMS

THIS CHAPTER IS ABOUT:

◆ Air masses
◆ Thunderstorms
◆ Midlatitude weather systems
◆ Tropical weather systems
◆ Polar atmosphere
◆ Forecasting weather
◆ Environmental Issue: Hurricanes—Can They Be Tamed?

Combinations of heated and cooled air, containing varied quantities of water, determine our weather. By studying weather events, scientists piece together distinctive, repeated patterns. A thunderstorm is one type of weather event that occurs all around the world and often has lightning issuing from the cloud base. The features of midlatitude and tropical weather patterns make it possible to predict future weather.

The weather elements of temperature, wind, clouds, fog, and rain or snow combine in distinct, repeated patterns called **weather systems** that range in size from a few square kilometers to thousands of square kilometers. Weather arrives in a series of organized events that make it possible to forecast what is coming during the next few days.

Meteorologists have observed atmospheric phenomena for many years and have built up records of temperature and rainfall in particular. As the amount of data and their global coverage increased in the twentieth century, scientists recognized repetitions of typical weather patterns (Figure 5-1). They devised models to summarize both the similarities and the variability occurring in each distinctive pattern. Meteorologists use these models to predict the weather from the patterns of weather data plotted on maps. The fact that weather forecasts, though increasing in accuracy, are not always "right on," demonstrates that there is still much to be learned about the behavior of Earth's atmosphere.

This chapter begins with a study of air masses, which with their distinctive characteristics of temperature and humidity are the building blocks of the atmosphere and an important key to the understanding of weather patterns. This study leads to an investigation of the main weather systems beginning with small thunderstorm units. The main part of the chapter focuses on weather systems that are characteristic of midlatitude and tropical atmospheres. The chapter ends with a short account of the ways in which these descriptions of weather systems are used in weather forecasting.

AIR MASSES

Large bodies of air covering tens of thousands of square kilometers in the troposphere are almost uniform in conditions of temperature and humidity. These **air masses** are the basic building blocks of the atmosphere. An air mass acquires its characteristics in a source region where air is fairly stable and remains in contact with the ground surface for several days. Typical source regions include polar land surfaces, the ice-covered Arctic Ocean, hot deserts, and tropical oceans. The midlatitudes do not produce their own air masses but occupy the zone where tropical and polar influences confront each other. Four major types of air mass dominate the troposphere. They all affect parts of North America (Figure 5-2), which is used as an example of air mass origins and influences.

Continental polar (*cP*) air is cold, dense, and dry. It forms where insolation is low. Prolonged cooling of the air occurs when heat is lost to space by radiation and is not replenished by the sun. The source territory for *cP* air affecting North America is the Arctic Ocean, where the ice cover is a "continental" type of surface that keeps humid influences away from the overlying air. In winter the source territory extends southward to the frozen north of Canada. On several occasions during a typical winter, *cP* air masses move

Figure 5-1 Weather systems. Each weather system has a distinctive size and geographic distribution. Some systems are confined to the tropics, midlatitudes, or polar atmospheres; but others occur more widely.

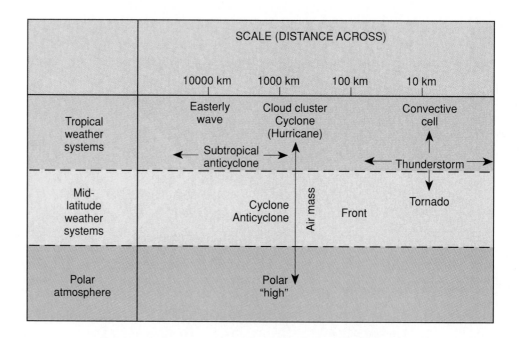

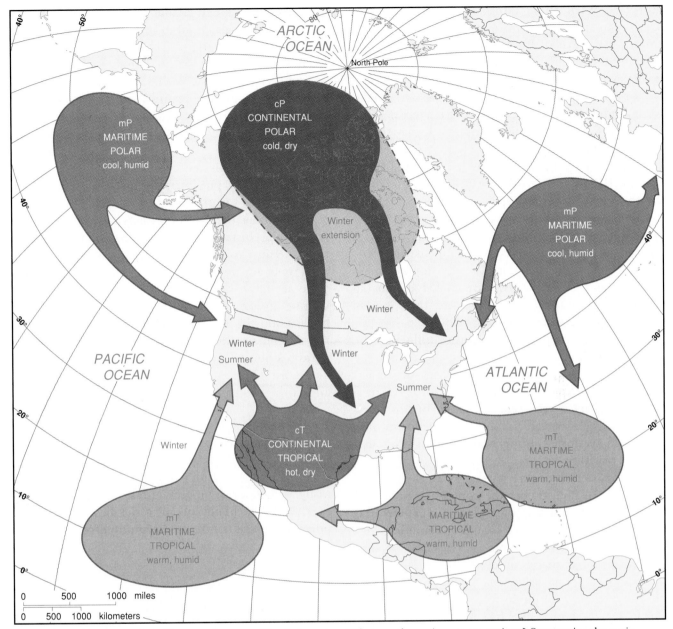

Figure 5-2 Air masses that affect weather over North America. What are the major source regions? Summarize the main weather systems affecting northern Canada, southern Canada, northern U.S.A., southern U.S.A., and the Caribbean.

south into central and eastern Canada and the United States bringing freezing temperatures. They even extend as far as southern Florida and Texas for short periods. Other world sources of *cP* air are Siberia in winter and Antarctica.

Continental tropical (*cT*) air is hot and dry. In North America it forms over hot, arid areas such as northern Mexico and the southwestern United States. Descending air in a high-pressure system warms adiabatically causing the air to dry out as the relative humidity decreases. Air flowing from the

desert in the Southwest brings bursts of hot, dry weather to the central United States in some summers and threatens crop yields when the effect is prolonged. Other world sources of *cT* air include central Australia and the Sahara of northern Africa.

Maritime tropical (*mT*) air is warm and humid. Its source area is over the tropical oceans where evaporation is intense. The *mT* that affects North America acquires its characteristics in the subtropical zones of the North Atlantic and North Pacific oceans or over the Caribbean Sea. Since *mT* air masses migrate westward with the trade winds, Atlantic and Caribbean *mT* air is more important to North America than that of Pacific origin. Worldwide, *mT* is the dominant type of air mass, affecting 53 percent of Earth's surface in January and 47 percent in July.

Maritime polar (mP) air is cool and has a high relative humidity. Such air acquires its characteristics as it passes over the polar margins of oceans. It may start as *cP* air that takes up moisture when it reaches the ocean or as *mT* air that cools as it moves poleward. Most of the *mP* air masses forming east of North America move toward Europe.

The four types of air mass move away from their source areas and are modified as they enter areas with different surface conditions. Tropical air masses transfer large bodies of heated air from the tropics toward the poles, and polar air masses are cooled air moving in the reverse direction. Air masses are, thus, important in the global processes of heat transfer.

On a more local scale, air masses influence the weather in ways that cannot be explained solely by seasonal shifts in the global distribution of solar radiation. Variations in winter weather in North America are often caused by cold snaps of *cP* air or unseasonably warm spells of *mT* air. Cool summers and mild winters in western Europe result from a predominance of maritime air.

Air masses move along regular paths, as shown in Figure 5-2. In midlatitudes the contrasting polar and tropical air masses come into contact along **fronts,** which are narrow zones of changing conditions between two air masses. The mixing of differing air masses on either side of the front is slow and the contrasts may continue for several days. Meteorologists first used the term *front* in 1917 when people were conscious of the almost static battlefronts in eastern France. Atmospheric fronts are zones where converging air masses conflict with each other.

The interactions between contrasting air masses along fronts produce much of the weather commonly experienced in midlatitudes. When two air masses of different temperatures, humidities, or both converge, the warmer air mass rises over the colder, denser air. If warm air moves faster than cold air ahead of it, the warm air is forced up a low-angle sloping surface at its leading edge; this is known as a **warm front** (Figure 5-3). If the cold air is moving faster, it pushes underneath the warm air forcing it off the ground and forming a **cold front.** A cold front is often much steeper than a warm front (Figure 5-4). Under both circumstances, warm air rises and clouds and precipitation are produced. In stable air the clouds are shallow and produce little, if any, precipitation. If the warm air is unstable, deep nimbostratus (warm front) or cumulonimbus (cold front) clouds form and produce heavy rain. Instability is more common along cold fronts because of their greater steepness.

Fronts are not so obvious in the tropics because there are fewer contrasts between air masses in converging airflows. It is common for *mT* air masses to meet along the intertropical convergence zone, but the differences do not show up on surface weather maps. When *cT* and *mT* air masses meet, the less dense, dry *cT* air may rise over the denser humid *mT* air but this produces little cloud.

THUNDERSTORMS

Thunderstorms are the most frequent and widespread of all weather systems. Some 16 million thunderstorms occur around the world each year and as many as 2000 are in progress at one moment. Such frequency makes thunderstorms a significant part of the processes of heat transfer in the atmosphere.

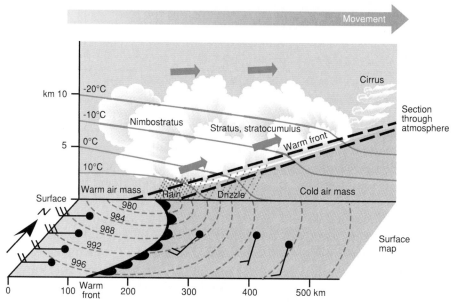

Figure 5-3 Warm front. A warm air mass is forced to rise over the slow-moving cold air in front of it. How does this affect the formation of clouds and rain?

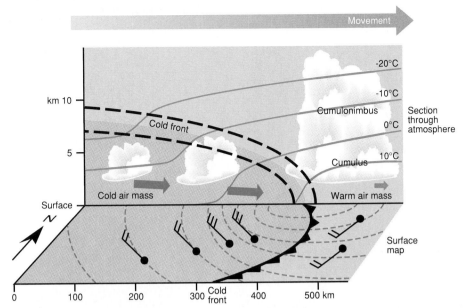

Figure 5-4 Cold front. A fast-moving cold air mass pushes beneath warm air, producing a steep boundary that forces warm air to rise sharply.

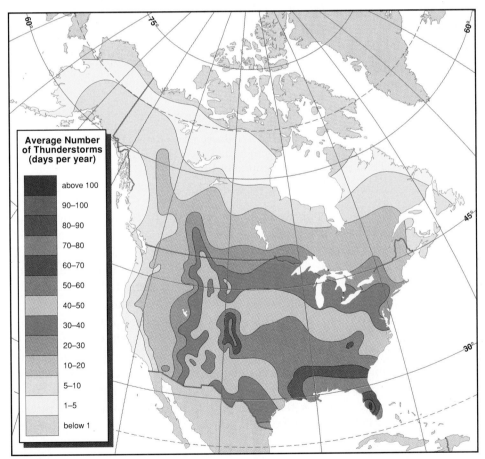

Figure 5-5 Thunderstorm frequency in the United States and Canada. Where are thunderstorms most frequent? Why?
Source: NOAA.

Thunderstorms are particularly frequent in equatorial regions. Kampala, Uganda, averages 242 thunderstorm days per year, and some parts of Indonesia over 300. Florida has the most in North America, followed by the southern Rocky Mountains (Figure 5-5). Such distribution shows that thunderstorms are characteristic of areas where either surface heating or mountain slopes encourage the rapid lift of humid air.

There are three stages in thunderstorm cell development, as shown in Figure 5-6. In the *initial stage,* cumulus clouds form in updrafts of humid air and continue to grow upward in an unstable atmospheric environment. Strong uplift along a steep cold front or mountain range may combine with the surface heating of air bodies to enhance the uplift. After the rising air becomes saturated, large volumes of water vapor condense to cloud droplets, producing towering cumulonimbus clouds with icy tops. Conversion of water vapor to cloud droplets releases latent heat, reinforcing instability and the convectional ascent of air. Air is drawn in at the surface and fuels the upward currents. Deep cumulonimbus clouds develop and precipitation begins.

In the *mature stage* of a thunderstorm, the cloud top rises to a level where cloud droplets and water vapor turn to ice particles. Strong winds spread out the top forming a diffuse, anvil-like feature just below the tropopause—the upper limit of ascending air.

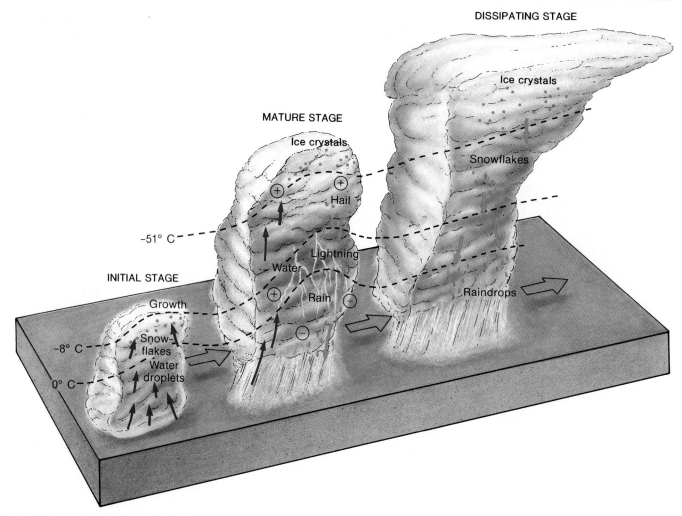

Figure 5-6   Thunderstorm: the development of a cell. Cutaway clouds show how updrafts of air produce tall clouds and icing conditions at the top. Downdrafts of ice and water produce intense rain and lightning and eventually destroy the cumulonimbus cloud. It is common for thunderstorms to include several cells that grow and dissipate by turns.

Precipitation increases as the cloud grows, reaching a maximum after the ice phase forms. A strong downdraft of cold air begins, caused by masses of falling ice crystals and raindrops dragging air down with them. Lightning and thunder occur when conflicting updrafts and downdrafts of air create an electric field. Lightning flashes start soon after the main precipitation begins.

Eventually, the downdraft becomes so strong that it shuts out the updraft of air and destroys the thunderstorm cell. During this *dissipating stage,* the cell disintegrates as it loses its supply of moisture and energy. Thunderstorms, however, often include several cells. As one dissipates, another develops next to it and the activity carries on.

At the height of a thunderstorm cell's development, several dramatic types of weather may occur. The downdraft during the period of heaviest rain may become a powerful wind known as a **downburst.**

Large downbursts can last up to thirty minutes and have surface winds blowing at up to 210 kph (130 mph). They may cause damage over a swath 4 km (2.5 mi) long. Some downbursts have been blamed for aircraft crashes during thunderstorm weather.

Hailstones often fall from thunderstorms in which the ice crystals move up and down in the drafts of air. Ice particles alternately fall to levels where they partially melt and then rise again to the freezing section of the cloud. Hailstones grow layer by layer to as large as 5 cm (2 in) across when they reach the ground (Figure 5-7). Continental interiors where cold and warm air masses meet favor the formation of hail, and the upper Mississippi valley has up to ten devastating hailstorms per year. Fewer hailstones fall in the southeastern United States despite its greater number of thunderstorms.

Lightning is a visible flash resulting from electrification within the cloud (Figure 5-8). Concentrations

Figure 5-7 Hailstones. Large hailstones damage crops, cars, and other surface features. The baseball provides a scale.

of positive charges in the upper parts of the cloud and negative charges at its base develop as the updrafts and downdrafts increase in strength. These charges build up by friction between the air drafts, like a static charge that results from feet rubbing on a rug. A positive charge develops on the ground beneath the cloud and surges up tall buildings, trees, and other high points. Since the atmosphere is a poor conductor of electricity, charges build up to as much as 100 million volts before lightning occurs, momentarily connecting the two areas of opposed charges by an electric arc. The lightning flash discharges the electrical field, but charges may build again. **Thunder** is the sound generated by the discharge of electricity. It is a sonic boom created by the heating and rapid expansion of air along the lightning path. The sound carries up to 10 km (7 mi) from the source. Since light waves travel more rapidly than sound waves, the gap between a lightning flash and a thunderclap increases as the thunderstorm moves away from an observer.

Lightning causes forest fires and is underestimated as a killer. Figures given by public agencies for lightning deaths are underestimates because not all incidents are recorded. Most lightning deaths occur outdoors in rural areas. Those who take shelter under trees are involved in a quarter of all lightning deaths; those who play golf, farm, swim, or boat are also particularly at risk. Four times more males are killed than females.

Lightning occurrence should be an important factor in building design in areas prone to thunderstorms. In 1991 the Florida Department of Transportation had to replace a weigh station built in the mid 1980s on Interstate 75 because it had suffered repeated lightning strikes. Although located in Marion County, which records the highest number of lightning strikes in the United States, lightning was not considered in the design of the building, causing the waste of several million dollars.

TORNADOES

Tornadoes are small but extremely violent rotating storms in which a distinctive funnel-shaped column, or vortex, descends to the ground from the base of a thundercloud (Figure 5-9). The vortex is commonly around 100 meters (300 ft) in diameter at its base, whirls in a counterclockwise direction in the northern hemisphere, and has winds that may exceed 400 kph (250 mph). Each funnel exists for only a few minutes and moves in an erratic path at 50 to 60 kph (35 mph) for up to 15 km (10 mi); a few travel as far as 200 km (125 mi). Waterspouts form in a similar way over the sea or a large lake but are usually less energetic and last for a shorter time.

Tornadoes form in severe thunderstorms. Extreme temperature and pressure contrasts between the surface and upper troposphere generate very strong updrafts of air. Pressure at the center of a tornado funnel falls to around 900 mb. Wind speeds increase sharply with altitude. The difference in wind speed between the ground and higher levels begins a horizontal rolling motion in the air. The

and spirals downward to the ground as a funnel. The narrowing increases the wind speeds to violent levels, so tornado winds commonly churn up soil and destroy buildings. The conditions causing a tornado last only a few minutes; then the funnel dissipates and drops any debris being carried. Other tornadoes may form from the base of the same cloud, however, producing a stop-and-go progress across the land.

Tornadoes are more common in the central United States than anywhere else in the world (Figure 5-11). The central United States are flat and provide a likely zone for the confrontation of *cP* and *mT* air known as "tornado alley." In February most tornadoes occur near the Gulf Coast, but the zone shifts northward during spring and summer, and in June the main area of occurrence is west of the Great Lakes.

Tornadoes cause intense local damage because of their high winds and upward suction as the rotating funnel passes overhead. The forward movement of the tornado enhances the high speeds of the rotating winds. The greatest force impacts a building at the point where the tornado strikes. As the roof lifts, the walls collapse or are blown in.

The National Weather Service in the United States forecasts probable tornado conditions and locally tracks the formation and progress of the storms. A *tornado watch* is announced when conditions make severe thunderstorms and tornadoes possible. A *tornado warning* is issued when a tornado is detected and people need to take shelter. The United States averages just under 800 tornadoes each year, but the most violent, with winds over 333 kph (200 mph), make up only 2 percent of this total.

Figure 5-8 Lightning flashes from the base of a cumulonimbus cloud over mountain foothills near Tucson, Arizona.

strong updraft of air shifts the rotating air into a vertical position high above the ground. A vertical, rotating cylinder of air 10 to 20 km (7 to 12 mi) in diameter builds upward and downward and strengthens the updraft (Figure 5-10). At a particular stage—not fully understood because it occurs only about half the time—the rotating cylinder then narrows

Figure 5-9 Tornado. A well-developed funnel descending from a cloud and moving across country in the Midwest of the United States.

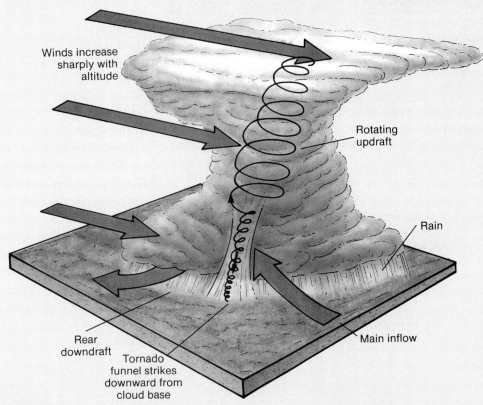

Winds increase sharply with altitude

Rotating updraft

Rain

Rear downdraft

Tornado funnel strikes downward from cloud base

Main inflow

Figure 5-10 Basic features of a tornado. The sharp increase in wind speed with height produces a rolling motion in air near the ground that is turned into a vertical cylinder of rotating air within a thunder cloud. Once developed, the vertical rotation may extend down to the ground in a narrower tornado funnel.

Adapted from "The Tornado" by John T. Snow. Copyright © 1984 by Scientific American, Inc. All rights reserved.

Figure 5-11 Geographic distribution of tornadoes in the United States, 1950-85. Numbers on the map are the average annual state death tolls from tornadoes.

Source: NOAA.

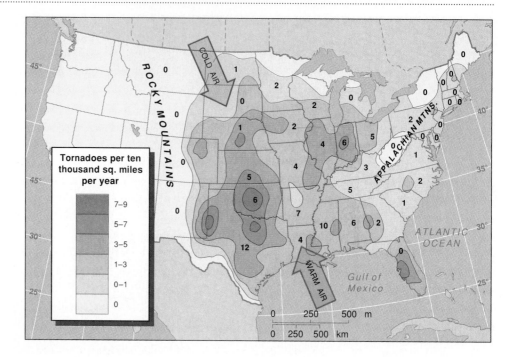

Weather systems are distinct and repeated patterns of weather. Models identified by integrating information from daily weather records help in their recognition. Weather forecasters base their predictions on the models of weather systems.

An air mass is a large body of air that has similar conditions of temperature and humidity throughout. Typical source regions include polar and arid land areas and subtropical oceans. Moving air masses meet along fronts designated "warm" or "cold" in midlatitudes according to whether they introduce a warm or cold air mass into the region over which they pass.

Thunderstorms are localized weather systems and occur in all parts of the world. They result from convectional uplift of moist air and produce heavy rain, hail, lightning and thunder. Tornadoes occur in the most active thunderstorms.

MIDLATITUDE WEATHER SYSTEMS

In the midlatitude atmospheric environment, contrasting air masses from tropics and poles meet along fronts, and the polar-front jetstream (see Chapter 3) plays a major part in controlling the surface weather. The effect of Earth's rotation is strong and midlatitude weather systems are marked by strong circulations of air in rotating patterns.

The main weather systems of midlatitudes are low-pressure cyclones and high-pressure anticyclones. **Midlatitude cyclones** are weather systems in which rotating surface winds converge at a center of low pressure—often called a "low"—where air rises. Surface winds spiralling into a cyclone flow counterclockwise in the northern hemisphere and clockwise in the southern hemisphere (see Figure 3-29). **Midlatitude anticyclones** are weather systems in which air descends in centers of high pressure—often called "highs"—and diverging winds spiral outward. The diverging winds flow clockwise in the northern hemisphere and counterclockwise in the southern hemisphere.

Midlatitude cyclones and anticyclones move from west to east between latitudes 35° and 60°, showing up on surface weather maps as patterns of roughly circular isobars (see Figure 3-26). The recognition of these two weather systems and their distinctive patterns of weather laid the basis for modern meteorology and weather forecasting.

The low surface pressure in midlatitude cyclones often reaches 970 mb but is rarely below 950 mb. The high surface pressure in midlatitude anticyclones commonly reaches 1040 mb but seldom exceeds 1060 mb. Cyclones have close isobars and moderate to strong winds; anticyclones have isobars that are farther apart and weak winds. Cyclones are associated with the meeting of contrasting air masses along fronts. Uplift of air along fronts causes clouds to form and precipitation to fall from them. Descending and diverging air in anticyclones is more likely to result in clear skies and sunny weather.

MIDLATITUDE CYCLONES

Most midlatitude cyclones form along fronts. When this is the case, they are called *frontal lows.* The model of the development of a frontal low (Figure 5-12) summarizes the observed features of these weather systems. The total sequence may take from one to seven days, and the average life of such a system is three days. In practice no particular system follows all these stages without some variation. Each system is a partial replica of the model (Figure 5-13).

In the northern hemisphere, the first stage begins with contrasting air masses—for instance, mP and mT air—blowing in parallel but opposite directions (Figure 5-12a). The front between the two air masses is almost a straight line, and the contrasting air types are in contact along it.

In the second stage, interactions between the opposing winds lead to eddies forming so that the front curves into a *frontal wave* a few hundred kilometers across (Figure 5-12b). These waves look like an ocean wave in profile on weather maps. A low-pressure cell is centered at the wave's crest, which is surrounded by almost circular isobars. Winds blow in a counterclockwise spiral toward the low-pressure center. A warm front and a cold front now bound a *warm sector* that widens to the south. The warm front moves eastward. The cold air at the back of the frontal low blows from the west behind the cold front, pushing beneath the warm sector.

The wave is at its largest in the third stage, the mature cyclone (Figure 5-12c). As the frontal low moves eastward, the faster-moving cold front catches up with the warm front, reducing the size of the warm sector.

The fourth and final stage occurs after the cold front catches up with the warm front. The entire warm sector is lifted off the ground, where a single front remains (Figure 5-12d), called an **occluded front.** The occluded, or hidden, front is shown on weather maps as a combination of the warm and cold frontal lines and symbols. During this stage, the low-pressure cell "fills in" as surface pressure rises, the pressure gradient reduces, and wind strength declines.

The weather sequence produced by a mature frontal low crossing an area includes changes every few hours. As the cyclone moves from west to east in the northern hemisphere, the changes of weather are caused by the passing of, first, cold sector air, second a warm front, third, warm-sector air, fourth, a cold front, and fifth, another sector of cold air.

The *cold sector* ahead of the low often has winds blowing from the south. It forms a sluggish mass of dense air that the rest of the weather system pushes against.

A *warm front* (see Figure 5-3) slopes gently upward at 0.5 to 1 degree over the cold sector. The faster movement of the warm sector air causes its gradual ascent up the warm front. High-level cirrus clouds often signal the arrival of a warm front some twelve hours before the front crosses at ground level. As the front approaches, the cirrus clouds give way to cirrostratus and altostratus clouds. If the warm air rising along the front is unstable, vertical movement of air above the frontal surface increases and produces thick, dark nimbostratus clouds giving steady precipitation. If the warm air is stable, light-gray sheets of stratus clouds form from which drizzle may fall.

The *warm sector* often contains mT air and its arrival behind the warm front brings a rise in temperature and a clearing of the skies. Local showers may occur from cumulonimbus clouds if the air in this sector is unstable.

The *cold front* (see Figure 5-4) generally has a steeper profile than the warm front. The active undercutting of the warm sector by a northwesterly stream of cold air forces strong upward movements of warmer air. In unstable conditions cumulonimbus clouds and thunderstorms are common. Tornadoes occur along a cold front under the special conditions described earlier in this chapter.

The *cold sector* at the rear of the frontal low provides another break in the cloudy and rainy weather after the cold front passes. It is often marked, however, by lines of cumulus or showery cumulonimbus clouds within the cold northwesterly airstream.

Small *secondary lows* form along the cold front behind a frontal low. This situation occurs commonly over the ocean in winter, where warming of the moist air generates local instability. Rising air lowers the surface pressure, and the secondary low grows and becomes part of the main circulation. Secondary lows may develop strong winds and rain so rapidly that weather forecasters have difficulty in spotting them.

MIDLATITUDE ANTICYCLONES

Two major forms of midlatitude anticyclones can be distinguished by their size and their relationship to other weather systems. The smaller form occurs between the cyclones that dominate the westerly flow of weather systems in midlatitudes. These anticyclones provide a respite from cloudiness, rain, and high winds but pass through rapidly, often within a

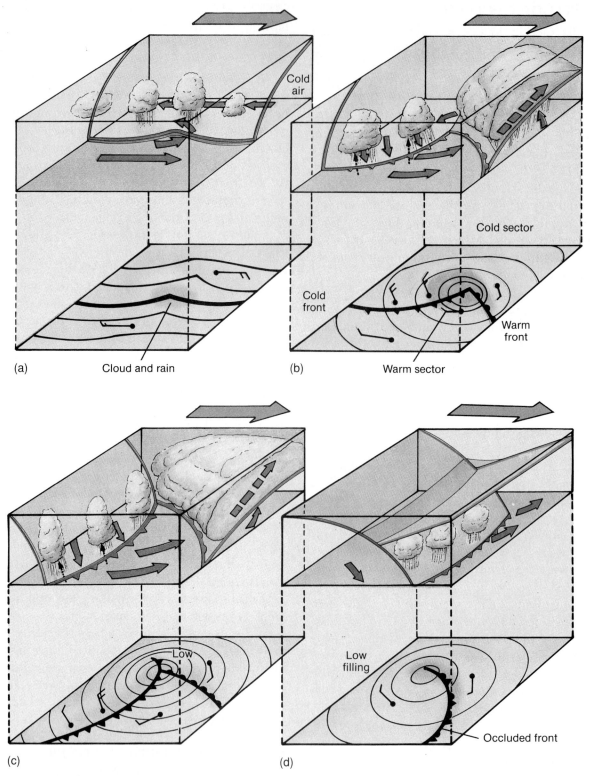

Cold
air

Cloud and rain

(a)

Cold sector

Cold
front

Warm
front

Warm sector

(b)

Low

(c)

Low
filling

Occluded front

(d)

Figure 5-12 Midlatitude cyclone. The evolution of a northern hemisphere example. (a) Initial stage: cold and warm air masses flow parallel to the front between them and in opposite directions. A small wave begins to form. (b) The wave enlarges with the development of a low-pressure center, together with warm and cold fronts. (c) Mature stage. (d) Occluded stage as cold air behind the cold front pushes beneath the warm-sector air, lifting it off the ground.

Figure 5-13   A midlatitude cyclone centered just west of the British Isles on February 3, 1994. The circular pattern of clouds is typical of the occluding stage. To the west there are bands of shower clouds in a northwesterly airstream.

MIDLATITUDE WEATHER SYSTEMS AND THE POLAR-FRONT JETSTREAM

The occurrence of midlatitude cyclones and anticyclones is governed by the polar-front jetstream and its Rossby-wave cycle (see Figure 3-28) in the upper troposphere. It forms a "conveyor-belt" type of mechanism in which the strong upper wind drags along the surface systems of cyclones and smaller anticyclones.

The jetstream waves have alternate sections of converging and diverging air (see Figure 3-29), which force air down toward the surface or draw it upward. When the jetstream is in its west-east pattern with small waves and high windspeeds, a strong surface flow of cyclones and small anticyclones moves almost directly eastward. The jetstream drags the surface systems beneath it at about 30 kph (20 mph) in summer and about 50 kph (30 mph) in winter. When the Rossby waves become more extreme, the cyclone "conveyor belt" meanders in north-south and south-north patterns. Large blocking anticyclones dominate the cores of the ridge waves (see Figure 3-28c).

The jetstream also controls the development of cyclones. In the developing stages, the cyclone center is to the south of the jetstream, and air is drawn upward to a section of diverging flow. The center of the cyclone moves to a position poleward of the jetstream at the occluded stage. This cuts off the surface cyclone from its link to the jetstream and the cyclone ceases to exist.

The main midlatitude weather systems are cyclones (low-pressure centers) and anticyclones (high-pressure centers). Midlatitude cyclones have converging air that rises in the central zone of low pressure. They often include fronts between contrasting air masses. They have strong winds, cloudy conditions, and precipitation.

Midlatitude anticyclones are centers of descending air that flows outward at the surface. They generally have calmer, clearer, and drier weather. The polar-front jet and its Rossby-wave cycle control the midlatitude sequences of cyclones and anticyclones.

day. The descending and outflowing air produces clear skies, gentler air movements, and an absence of frontal conflicts between air masses.

The second type of midlatitude anticyclone is much larger and may dominate the weather of an area for a week or so bringing clear skies and light winds. Such weather systems appear to deflect the passage of midlatitude cyclones around them to north and south and are termed "blocking anticyclones." In summer such conditions bring prolonged periods of sunshine intensifying the warmth of the *cT* air that is commonly associated with these weather systems at that season. High temperatures and drought may result. In winter, the loss of heat through clear skies during long nights lowers the temperatures in *cP* air to extreme levels. If the air is almost still, fog may result and polluting dust and gases may accumulate. With slightly stronger winds, a combination of low clouds and reduced visibility at ground level causes the bright skies to give way to *anticyclonic gloom*.

TROPICAL WEATHER SYSTEMS

The tropical atmosphere has three major features that distinguish it from the midlatitude atmospheric environment.

1. There are fewer contrasts of temperature or pressure at the ground. The Inter-tropical Convergence Zone (ITCZ) marks a major break between airflows, but is often difficult to locate on surface weather maps because the *mT* air on either side is so similar in characteristics.

2. A vertical Hadley cell dominates the circulation.

3. The effect of Earth's rotation is nil at the equator and minimal within 10 degrees of latitude from the equator. The results are that weather systems close to the equator do not have circulating winds like those in the midlatitude cyclones and anticyclones, and that equatorial weather is dominated by great masses of cumulonimbus clouds. Tropical cyclones occur between 10° and 20° North and South of the equator.

CLOUD CLUSTERS

Cloud clusters are the main weather systems in the equatorial regions of the tropics. They consist of groups of cumulonimbus thunderclouds with each cloud approximately 5 km (3 mi) across. The whole system covers a belt up to several hundred kilometers wide. Up to 30 cloud clusters encircle the globe close to the ITCZ at any one moment, as Figure 4-15 shows. Each cluster contains fifty to eighty cumulonimbus clouds towering to 17 km above the ground. Cloud clusters are several times larger than midlatitude thunderstorms and deliver many times the amount of rain.

A cloud cluster system develops in rising unstable air along the ITCZ. Cumulus clouds with rounded tops form over rising air currents and grow vertically, icing at the top (Figure 5-14). The heaviest rain from a cloud cluster occurs in the growing convective cells as they reach their maximum development. The iced upper sections of clouds develop horizontal anvil-shaped heads as strong upper troposphere winds move the ice away from the upward current sources. These anvils show up in the space shuttle photo, Figure 5-15. Once the anvil section forms, the rainfall rate lessens, but rainfall becomes more uniform over a large area. At the far end of the cluster, the cloud disperses as it becomes "rained out."

Cloud clusters are responsible for most of the rain in tropical areas, much occurring in heavy downpours. Two or three clusters make the difference between an adequate and a poor rainy season. When the ITCZ coincides with mountainous islands, as happens in the East Indies for much of the year, daytime heating over the island intensifies the

Figure 5-14 Cloud cluster. Cumulus clouds over tropical oceans and rainforests develop to deep cumulonimbus towers with an extensive downwind anvil.

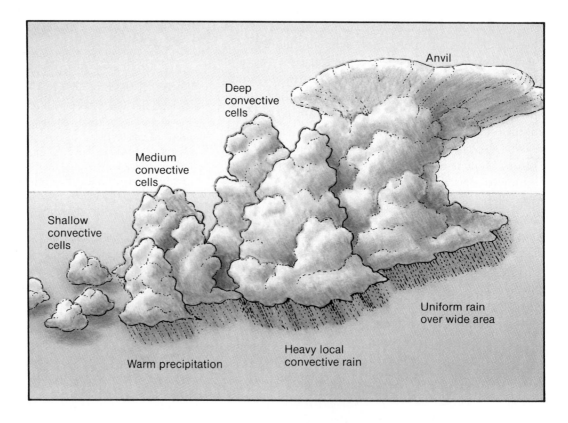

Figure 5-15 Space shuttle view of cloud clusters over Zaire, Africa. How do the clusters show the features of Figure 5-14?

uplift resulting from the convergence of moist air along the ITCZ. Strong sea breezes result in a regular daily pattern of cloud formation and cluster development over the islands. Late afternoon downpours of rain follow clear mornings during which the clouds begin to build.

SUBTROPICAL HIGH-PRESSURE ZONES

Semipermanent belts of high pressure occur over the oceans in the subtropics and are centered at approximately 30° North and South. **Subtropical high-pressure zones** are huge, covering 3500 to 5000 km (2000–3000 mi) from west to east and over 1500 km (1000 mi) from north to south. They have a major role in the weather of the large areas they dominate and in the wider global circulation.

Narrower areas of low pressure occur between the subtropical high-pressure zones. The low-pressure areas provide passages for air movement—equatorward on the eastern margins of the "highs" and toward midlatitudes on the western margins (see Figure 3-31). The equatorward flows form the trade winds that blow around the low-latitude margins of the highs. The flows on the western margins of the highs link into the midlatitude westerly circulation on the poleward side of the highs.

The subtropical high-pressure zones form beneath the westerly subtropical jetstream on the poleward side of the tropical Hadley cell (see Figure 3-32). The descending air under the jetstream is strongest on the eastern sides of the high-pressure

zones in winter. The air warms by compression as it descends but does not always reach the ground. A surface layer of air, cooled over the cold ocean surface, often produces a marked temperature inversion. Such conditions make it difficult for tall clouds and rainfall to develop on the eastern margins of the high-pressure zones. Convective heating affects a deeper layer on the western margins of the high-pressure zones, where clouds grow taller, bringing thunderstorm weather to islands and coasts.

Over land the subtropical high-pressure zones are marked by aridity. Insolation from a vertical sun through cloudless skies produces daytime temperatures that can reach up to 45° C (115° F) in the Sahara. Such surface heating results in air rising, lowering surface pressure, but the dominant downward movement of air in the high-pressure zones restricts the rising air to a shallow layer.

EASTERLY WAVES

Easterly waves are disturbances of the pressure patterns that move westward in the trade wind zone around the equatorward margins of the subtropical high-pressure zones. Easterly waves occur between 5 and 20 degrees of latitude. They measure 2000 to 4000 km (1200–2500 mi) from east to west, last for one to two weeks, and travel 6 to 7 degrees of longitude per day. Such waves are common from June to September in the tropical western Atlantic and Caribbean, where some 50 such waves are recorded each year. Some continue across Central America into the eastern Pacific Ocean, where easterly waves may provide a transitional state to a tropical storm.

An easterly wave forms when converging air becomes subject to weak turning by the effect of Earth's rotation, which is not very strong in these latitudes. The result is a low-pressure trough in which the isobars bend away from the equator, giving the wave form. Rising air in this trough produces clouds and rain mostly on the eastern side of the trough. The movement of an easterly wave within the trade-wind flow helps to intensify rainfall on the east-facing slopes of islands and coasts in their path. For instance, the east-facing slopes of Pacific Ocean islands have some of the world's highest rainfall totals, while the west-facing lee coasts remain relatively dry. Kawai, Hawaii, receives 12,000 mm (480 in) of rain a year.

TROPICAL CYCLONES AND HURRICANES

The hurricanes with which Americans are familiar are one of a group of tropical weather systems known collectively as **tropical cyclones.** The official definition of such systems is that wind speeds average over

115 kph (75 mph) for at least one minute. They are called **hurricanes** in the western Atlantic, **typhoons** in the western North Pacific, and simply cyclones in the Bay of Bengal and northern Australia (Figure 5-16).

Most tropical cyclone weather systems form between 10 and 20 degrees of latitude, and 70 percent occur in the northern hemisphere. They resemble midlatitude cyclones in having low atmospheric pressure at the center and a counterclockwise wind circulation in the northern hemisphere, but hurricanes also have distinctive features and origins. Although about one-third the size of a large midlatitude cyclone, tropical cyclones are much more intense, involving the considerable quantities of energy available in the tropical atmosphere. Pressures in the centers of tropical cyclones fall below 950 mb and are often close to 900 mb.

Tropical cyclones are the most dramatic of a series of tropical weather systems that may develop from less to more intense storms. *Tropical storms* have average wind speeds between 60 and 115 kph (40-75 mph); *tropical depressions* have wind speeds up to 60 kph (40 mph). The destructive nature of tropical cyclones and storms led to attempts to monitor and modify them. Weather satellites and views from space (Figure 5-17) make it possible to follow progress from tropical depression to cyclone and predict the future path for some hours ahead, but attempts to reduce their impacts by dissipating their energy have met with little success (see Environmental Issue: Hurricanes—Can They Be Tamed?, p. 118).

Tropical cyclones are not very common. They occur several times a year in summer or early fall

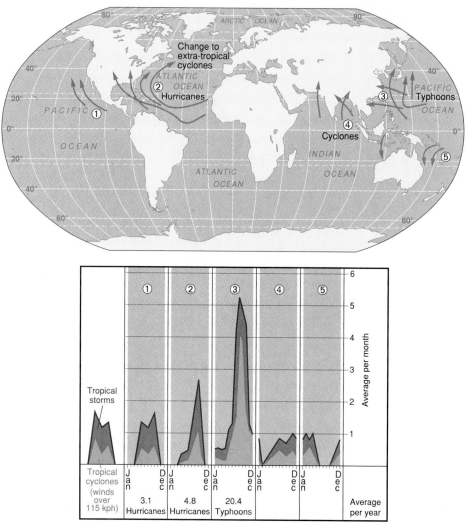

Figure 5-16   Geographic distribution and seasonal occurrence of tropical storms and tropical cyclones. Where are they most frequent? In which seasons of the year do they occur?

and vary in number from year to year. The Florida-Caribbean area has between four and fourteen per year, but greater numbers occur in the western Pacific Ocean. Weather systems of tropical storm intensity are also not very frequent and have sufficient impact to be given individual names, like the full-blown tropical cyclones.

Hurricanes form in the tropical North Atlantic Ocean. They develop from initial disturbances such as a tropical depression, cloud cluster, or easterly wave into tropical storms and hurricanes and then die away. The hurricane stage lasts for a few days. The initial disturbances begin when the ITCZ moves away from the equator in summer. Converging air becomes subject to the effect of Earth's rotation, instigating a weather system with circulating winds. Such disturbances move westward in the trade wind flow but will only develop into a hurricane if all the atmosphere and ocean conditions are just right. There must be a coincidence of high ocean-surface temperature and high-speed winds aloft. The ocean temperature must be at least 26° C (80° F) in a layer of water 60 to 70 m deep so that overturning by the strong winds does not bring up colder water to chill the air.

A crucial transformation converts the cool-cored tropical storm into a warm-cored hurricane. In the cool-cored system, the central low-pressure area has a relatively steep fall in temperature above it. As the hurricane develops, warm, moist air from the surface rises higher, assisted by an outflow of air above the core that drags the warm air upward. Tall cumulonimbus clouds form. The release of latent heat in 100 to 200 of these towering clouds at the heart of the weather system is sufficient to warm the core and change the environment.

The warming of the core intensifies the airflow aloft and leads to a convective circulation in which air descending in the center of the hurricane is surrounded by a circular wall of uprising and condensing air (Figure 5-18). Further heating results from the descent and compression of air in the center and forms the calm and clear central core, or eye, which is the main feature of a hurricane. The eye shrinks in diameter as the hurricane intensifies. The steep pressure gradient between the high pressure of the eye and the low pressure of the wall results in horizontal winds of up to 300 kph (200 mph) at the outer edge of the eye—the most dangerous part of the hurricane. The heaviest falls of rain occur from the inner ring of cumulonimbus clouds, but rain also falls from the spiralling arms of clouds that extend out from the center.

Tropical storms that develop into hurricanes follow paths westward across the Atlantic, but their paths diverge as they reach the western margin of the ocean. Some carry on through the Caribbean, some turn north and remain offshore, and some turn northwest, hitting the Gulf Coast, Florida, or the southern Atlantic coast of the United States.

Hurricanes exist as long as the conditions of surface inflow of hot moist air and divergence aloft continue. They dissipate when these conditions cease as they move over land or cooler ocean water. When a hurricane reaches about 30° N, it meets colder air and fronts develop as it changes to a midlatitude cyclone.

POLAR ATMOSPHERE

Cooling and sinking of air are the main features of the polar atmosphere resulting in static atmospheric conditions, rather than the dynamic air movements that produce distinctive weather systems. The lack of weather observations in these regions prevents a fuller understanding of polar weather systems.

In winter, the continuing darkness results in losses of heat, descent of cold air, and the formation of intense high-pressure systems with outflowing easterly winds. In the northern hemisphere, this extremely cold zone extends southward in winter over the Asian and North American continents, effectively expanding the polar atmospheric environment. In summer, the cold polar core is weaker and midlatitude weather systems invade polar regions more frequently. This is particularly common when the waves in the polar-front jetstream are at their most extreme.

FORECASTING WEATHER

Weather systems act as models for the process of **weather forecasting**—the attempt to predict the weather of a place for the next few hours or days. The U.S. federal government funds billions of dollars to the National Weather Service, which has four functions: to provide severe weather warnings, weather observations and forecasting, education, and aviation briefings. Along with the general forecasts broadcast on television (Figure 5-19) and radio and published in the newspapers, the Weather Service provides specialized reports to such people as farmers and pilots.

Figure 5-17

Figure 5-17 Hurricane Elena, western Atlantic. The space shuttle photo shows the typical cloud bands and the central eye.

Figure 5-18 Mature hurricane. The weather system has spiral bands of cumulonimbus clouds producing intense rain. The upward flow of air is concentrated in the clouds surrounding the central eye. Descent and warming in the eye keeps it clear. Source: NOAA.

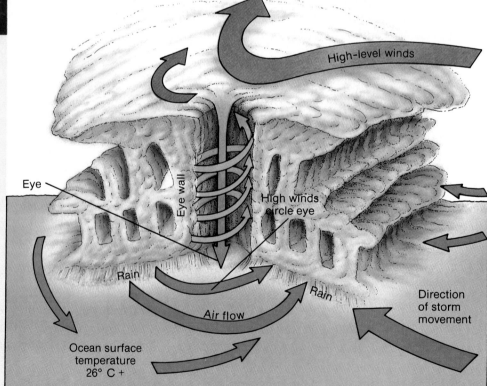

Figure 5-18

The business of weather forecasting begins with the collection of weather data such as temperature, pressure, wind speed, wind direction, cloud forms, and rain. The data are plotted on maps and make it possible to analyze the general atmospheric conditions. The visual models of the weather systems described in this chapter are converted to numerical computer models. In midlatitudes weather systems such as cyclones with their fronts and anticyclones are the main features of weather maps. They are the basic models that forecasters use to predict the weather after computing the speed, direction, and internal features of each system. Since the continuous flow of data indicates that they are changing, sometimes in unexpected ways, forecasters must continuously update their predictions.

At present, forecasts in midlatitudes cover only a few days ahead since the systems themselves only last that long—exceptions being larger blocking anticyclones. Forecasting weather for longer periods of time will require a better understanding of how the polar-front jetstream functions and links to surface weather systems. The forecasts are not perfect, but improving technical facilities and understanding have led to great improvements in the past thirty years.

There is less circular horizontal motion and more vertical movement in the tropical atmosphere because of the reduced effect of Earth's rotation; tropical weather systems are dominated by vertical movements. Air masses in the tropics vary little in temperature. The main tropical weather systems are cloud clusters formed of a series of thunderstorms, subtropical high-pressure zones, easterly waves, tropical storms, and tropical cyclones (hurricanes).

Air masses take their characteristics from contact with the ocean (**atmosphere-ocean environment**), cold or dry land surfaces (**surface-relief environment**), and vegetation cover (**living-organism environment**). Hurricanes illustrate the impact of weather systems on other Earth environments. Besides their impacts on human activities, they may flatten forests, change the features of shorelines, and enhance river erosion.

Weather forecasting involves the collection and plotting of weather information together with the examination of weather-satellite images. Forecasters use this information but still rely on their own experience in making the daily forecast.

SUNDAY, OCTOBER 9, 1994

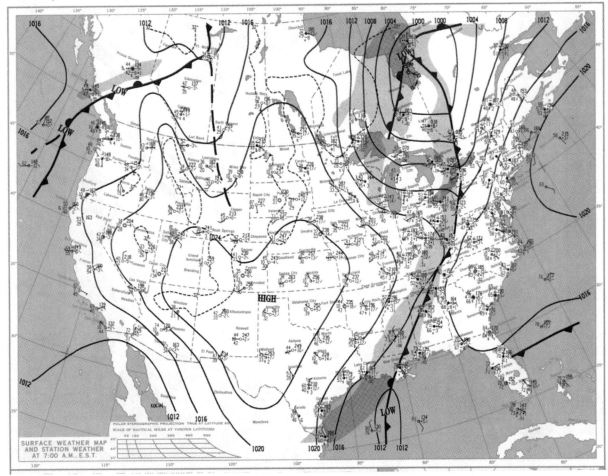

Figure 5-19   A weather map that shows conditions across the United States on October 9, 1994. Much of the country was dominated by a large blocking anticyclone (high). Cyclones (lows) with fronts and rain belts moved around the northern margins of the anticyclone.

ENVIRONMENTAL ISSUE:

Hurricanes—Can They Be Tamed?

Some of the issues in this controversy include:

Do we fully understand what causes hurricanes and how they function?

Is human ingenuity able to control the power of a hurricane?

How can we reduce destruction of property and loss of life from hurricanes?

Tropical cyclones, including hurricanes, bring death and destruction to coastal areas. In May 1991 a severe tropical cyclone hit Bangladesh similar to the one that killed over a quarter of a million people in 1970. The combination of high winds, heavy rainfall, and especially the surge of water several meters above the normal level, which drowned the huge, low-lying Ganges delta, devastated the whole country. In a developing country such as Bangladesh, where the local geography combines with poverty, there is little that the people can do to avert major loss of life in the face of such a massive natural event.

Hurricanes also batter the United States. In 1991, Hurricane Andrew cut a swath of destruction across southern Florida, but most people had evacuated the area and there were few deaths. The costs of repairing the damage, however, are immense, and recovery of the affected areas will take years.

Hurricane Hugo (September 10–22, 1989) provides an example of the progress of a hurricane and how people and agencies reacted to it. Weather

satellites monitored its approach across the Atlantic from September 10, as the map shows. When it was east of Guadeloupe, winds in the core were over 250 kph (150 mph) and wall pressures fell to 918 mb, placing it in the highest category of hurricane strength. On September 14 a hurricane watch was posted for Puerto Rico and the Virgin Islands. It hit St. Croix, Virgin Islands, on September 18, destroying over 90 percent of the island's buildings and stripping the vegetation. It then turned northward and slowed for a while, but the winds picked up over the warm Gulf Stream. People evacuated the barrier islands and beaches from Georgia to southern North Carolina. When Hurricane Hugo crossed the coast near Charleston early on September 22, it brought strong winds, heavy rain, and a 6-meter storm surge that inundated large areas. Although Hurricane Hugo caused destruction of property estimated at $9 billion, including the homes of over 200,000 families on Puerto Rico, the Virgin Islands, and the mainland, just 49 people died—a relatively small number compared with Bangladesh. In the United States improvements in monitoring and path prediction techniques combine with timely warning and public preparedness for evacuation procedures. Although more and more people in the United States live in hurricane hazard zones and property damage continues to rise, fewer people die than was the case in the early part of the twentieth century.

Disasters caused by hurricanes led the United States to institute a program known as Project Stormfury to try and tame the force of hurricanes. The project failed in the sense that attempts to cause the clouds to rain earlier, and so prevent a tropical storm turning into a hurricane, had little effect. The associated research, however,

coupled with the development of weather satellites, led to a fuller understanding of hurricane dynamics and paths across the ocean. It is now possible to monitor hurricanes closely and give a few hours warning of where they will cross the coast, enabling people to evacuate.

Although similar monitoring systems are used in developing countries such as Bangladesh, internal communications and transportation systems are not as good. There are few places for the poverty-stricken people living in the vast low-lying and flood-prone area of the Ganges delta to move to at short notice.

As we begin to understand more fully how tropical cyclones work, it is clear that humans cannot modify these powerful natural systems. Although it would be best to live in areas where they do not occur, many humans choose to live in hurricane-affected areas. Many others have little choice as to where they live. Improved monitoring, communications, and transportation are the keys to reducing loss of life and property.

Linkages

Hurricanes form as a result of interactions between the tropical atmosphere and ocean (**atmosphere-ocean environment**) and die out after reaching land (**surface-relief environment**) or moving over colder water.

Questions to Debate

1. Are the millions of dollars spent on hurricane research worthwhile if hurricanes cannot be tamed?

2. Why do hurricanes cause great loss of lives in some countries, but mainly loss of property in others?

(a) Damage caused by Hurricane Hugo on the South Carolina coast in 1989 (left), and by Hurricane Andrew in Florida in 1992 (right).

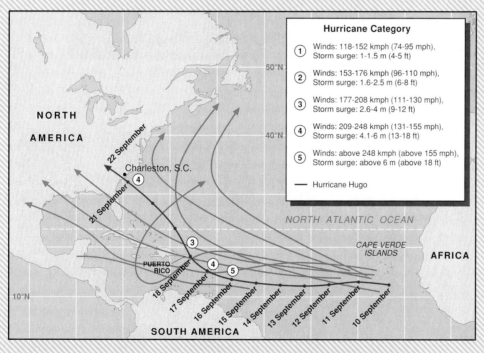

(b) Hurricane paths, North Atlantic Ocean. Some typical paths and the one taken by Hurricane Hugo in 1989. Satellite views of clouds and rain in Hurricane Hugo.

Source: (b) Data from NOAA.

ENVIRONMENTAL
ISSUE

(Continued)

(c) Hurricane Hugo (1989).

(d) Hurricane Andrew (1992).

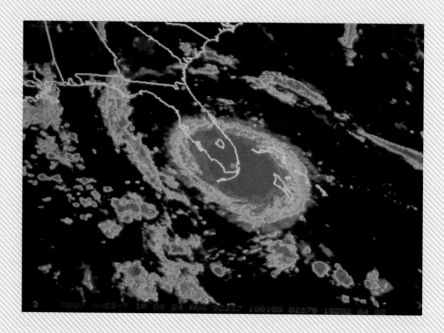

CHAPTER SUMMARY

1. Weather systems are distinct and repeated patterns of weather. Models identified by integrating information from daily weather records help in their recognition. Weather forecasters base their predictions on the models of weather systems.

2. An air mass is a large body of air that has similar conditions of temperature and humidity throughout. Typical source regions include polar and arid land areas and subtropical oceans. Moving air masses meet along fronts designated "warm" or "cold" in midlatitudes according to whether they introduce a warm or cold air mass into the region over which they pass.

3. Thunderstorms are localized weather systems and occur in all parts of the world. They result from convectional uplift of moist air and produce heavy rain, hail, lightning, and thunder. Tornadoes occur in the most active thunderstorms.

4. The main midlatitude weather systems are cyclones (low-pressure centers) and anticyclones (high-pressure centers). Midlatitude cyclones have converging air that rises in the central zone of low pressure. They often include fronts between contrasting air masses. They have strong winds, cloudy conditions, and precipitation.

5. Midlatitude anticyclones are centers of descending air that flows outward at the surface. They generally have calmer, clearer, and drier weather. The polar-front jet and its Rossby-wave cycle control the midlatitude sequences of cyclones and anticyclones.

6. There is less circular horizontal motion in the tropical atmosphere because of the reduced effect of Earth's rotation; tropical weather systems are dominated by vertical movement. Air masses in the tropics vary little in temperature. The main tropical weather systems are cloud clusters formed of a series of thunderstorms, subtropical high-pressure zones, easterly waves, tropical storms, and tropical cyclones (hurricanes).

7. Air masses take their characteristics from contact with the ocean (**atmosphere-ocean environment**), cold or dry land surfaces (**surface-relief environment**), and vegetation cover (**living-organism environment**). Hurricanes illustrate the impact of weather systems on other Earth environments. Besides their impacts on human activities, they may flatten forests, change the features of shorelines, and enhance river erosion.

8. Weather forecasting involves the collection and plotting of weather information, together with the examination of weather-satellite images.

Forecasters use this information but still rely on their own experience in making the daily forecast.

KEY TERMS

weather system   *100*

air mass   *100*

front   *102*

warm front   *102*

cold front   *102*

thunderstorm   *102*

downburst   *104*

lightning   *104*

thunder   *105*

tornado   *105*

midlatitude cyclone   *108*

midlatitude anticyclone   *108*

occluded front   *109*

cloud cluster   *112*

subtropical high-pressure zone   *113*

easterly wave   *113*

tropical cyclone   *113*

hurricane   *114*

typhoon   *114*

weather forecasting   *115*

CHAPTER REVIEW QUESTIONS

1. What are the distinctive features of the four main types of air mass? How do they relate to their source regions?

2. Describe the development and dissipation of a thunderstorm.

3. Compare and contrast the weather received from midlatitude cyclones and anticyclones.

4. To what extent is a tropical cloud cluster merely a large thunderstorm?

5. What determines whether a tropical storm develops into a hurricane?

SUGGESTED ACTIVITIES

1. Follow the daily sequence of weather in your locality for a few weeks. Note the occurrence of midlatitude cyclones and anticyclones. How long do they last? How fast do they move? Where do they go? What types of weather occur as warm and cold fronts pass overhead? Count the number of thunderstorms and note any unusually dramatic weather.

2. Assess the prospects for human attempts to modify destructive weather systems such as tornadoes, hail storms, and hurricanes.

CHAPTER
6

CLIMATE: PRESENT, PAST, AND FUTURE

THIS CHAPTER IS ABOUT:

◆ The importance of climate
◆ Climatic environments
◆ Tropical climatic environments
◆ Midlatitude climatic environments
◆ Polar climatic environments
◆ Highland climatic environments
◆ Climates of the past and future
◆ Environmental Issue: The Future—Global Warming or Cooling?

Deserts form in very dry parts of the world where rain and snow seldom occur and plants exist with difficulty. Such dry areas are one of many types of climatic region that give distinctive characters to each part of Earth's surface, from California's Death Valley to the dunes of eastern Algeria (inset).

THE IMPORTANCE OF CLIMATE

Climate is the long-term behavior of the atmosphere-ocean environment. Weather varies over hours or days; climate concerns periods of years. The climate at a location is not merely "average weather"; it is also the diversity of temperature, moisture, and atmospheric pressure conditions experienced through the seasons and over a number of years. For instance, Chicago has almost the same average annual temperature as Valentia Island (on the southwest coast of Ireland) but a much greater range of monthly temperatures (Figure 6-1). Chicago has humid summer heat with thunderstorms and freezing, snowy winters; Valentia Island experiences cool summers and few winter frosts although it is 10 degrees of latitude farther north than Chicago.

The differences between Chicago and Valentia Island are just one example of crucial climatic differences around the world that affect other aspects of physical geography and the lives of people. People are increasingly conscious of the changing nature of climatic conditions where they live and are concerned about what future changes may mean for their way of life.

Physical geographers apply their understanding of climate and climate change to the planning and management of human activities. Climate is particularly important to farmers and agricultural development. Crops require specific conditions of temperature, growing season, and water availability. A variety of corporations and agencies, including the Defense Department, employ physical geographers to assess the significance of climatic conditions. During World War II the planning of large scale amphibious invasions and jungle combat required an improved knowledge of climate. Advances begun then continue in the design and construction of buildings for use in varied climates from tropics to tundra and in determining the best sites and orientation for airport runways. The space program requires knowledge of climate for rocket launchings and space shuttle landings. The potential of warfare in arctic conditions and the desert location of the Gulf War in 1991 require modifications of military equipment (Figure 6-2).

For the physical geographer, climatic considerations are important in an understanding of the origins and distributions of landforms, soils, and living organisms. Differences in climate strongly influence the ways in which rocks break down, the flow of rivers and glaciers, and the formation of soils. Physical geographers are particularly concerned with trying to work out the relative roles of natural climatic change and human intervention in such phenomena as desertification—the degrading of land quality in dry regions.

This chapter first examines today's climates at the global scale. It then shows how scientists chart past climate change and how they use this knowledge in their attempts to predict future changes.

CLIMATIC ENVIRONMENTS

A **climatic environment** is the distinctive group of factors that produce the climate of an area (Figure 6-3). These factors include the heat balance (Chapter 3), the water budget (Chapter 4), the surface type (land, ocean, ice), and the weather systems (Chapter 5) that affect the area. Climatic environments occur at different geographic scales, from the major global divisions that will be the main focus of this chapter to areas as small as a field or town street.

LOCAL-SCALE CLIMATIC ENVIRONMENTS

The smallest units of climatic study include the immediate growing environment of a plant, a farm field, a stand of trees, or a built-up area (suburb or city center). Each has a heat budget, water balance, ground surface, and a habitual series of weather systems. At this scale, however, the environment is smaller than most weather systems.

Urban areas are growing throughout the world and constitute significant local climatic environments. Seventy-five percent of the U. S. population lives in large cities where the ground surface includes houses, paved roads, factories, shopping centers, and tall office or apartment blocks. These structures provide a surface that contrasts with the ground cover of rural areas (Figure 6-4). Major

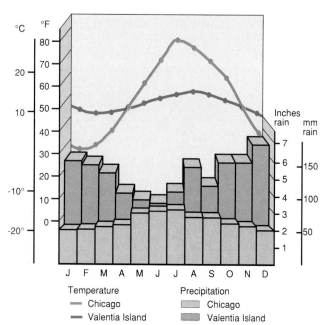

Figure 6-1 Climate averages and ranges. How do Chicago (41°56′N) and Valentia Island (51°50′N) compare in terms of ranges of temperature and precipitation? Both have temperature averages of 10° C (50° F).

Figure 6-2 Climate and warfare. This tank had to be repainted and have engine modifications for use in the 1991 Gulf War.

Figure 6-3 Climatic environment. Weather data are measured, recorded, and averaged over 30 years. The four sets of interacting conditions—heat balance, water budget, surface type, and weather systems—cause the measured weather conditions.

Figure 6-4 Local climates. What are the contrasting atmospheric conditions in urban and rural areas?

Figure 6-2

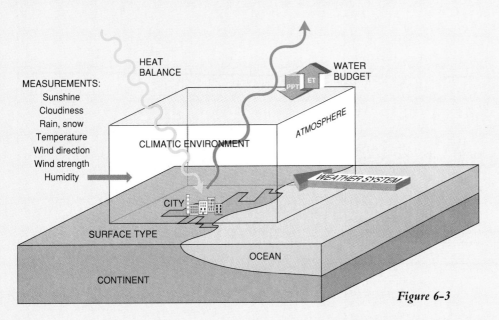

Figure 6-3

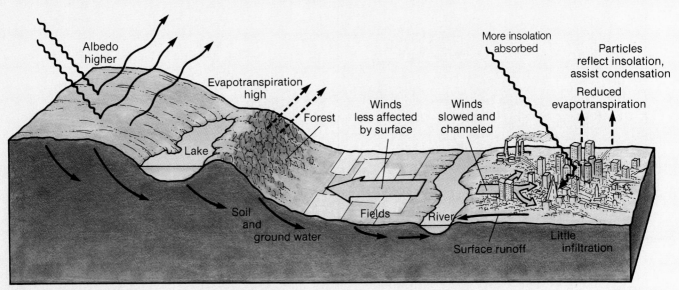

Figure 6-4

urban areas absorb more insolation than rural areas because of the predominance of concrete, brick, asphalt, and stone surfaces. Furthermore, the air above cities has more dust particles. The heat absorbed by these surfaces is radiated into the lower atmosphere and combines with the heat released by power stations, factories, vehicles, and uninsulated buildings. Given these factors and the reduced contribution of the cooling effects of evaporation and plant transpiration, it is clear that the greenhouse effect is greater in the city atmosphere. The temperatures recorded in major urban areas are higher than those in the surrounding countryside (Figure 6-5). Cities produce a "heat island" in a cooler rural "sea." Winter temperatures commonly show the greatest contrasts between urban and rural areas.

Other distinctive features of the climates of large urban areas are higher levels of pollution, a greater incidence of thunderstorms, and generally lower windiness than in rural areas. Winds, however, may be channelled along narrow streets between high buildings, and they may have higher local velocities. Smog is endemic to cities as the combined result of higher levels of dust, car exhaust fumes, and smoke pollution.

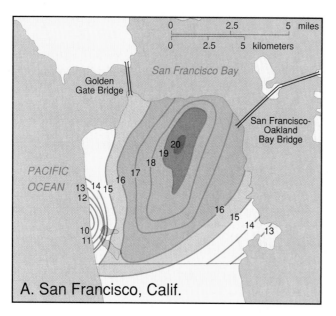

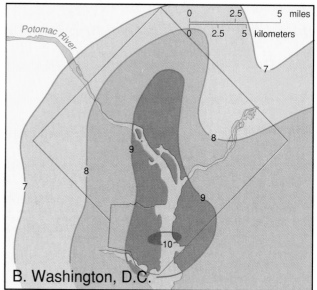

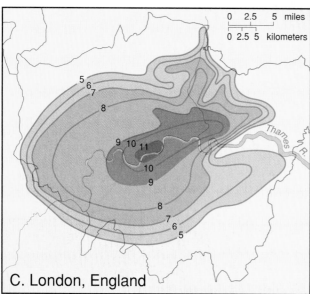

Figure 6-5 Urban heat islands. Examples are drawn from three cities. (a) San Francisco on a spring evening. (b) Washington, D.C.: mean annual temperatures. (c) London, England: mean temperatures, May 1959.

Adapted from "The Climate of Cities" by William P. Lowry. Copyright © 1967 by Scientific American, Inc. All rights reserved.

KÖPPEN CLASSIFICATION OF WORLD CLIMATES

Climatic conditions such as temperature and precipitation distinguish climatic environments. Measuring climatic conditions makes it possible to draw world maps of climate regions. One example of such a world climate map was produced by Austrian botanist Vladimir Köppen in 1918 (Figure 6-6 and Table 6-1), who based the divisions between climate regions on the isotherms and precipitation characteristics that marked the boundaries of the major types of natural vegetation. This approach makes it possible to relate climate type to other aspects of the natural environment, including vegetation and soils, and to present an integrated whole. The classification has continued to be used in physical geography courses for this reason.

There are, however, several difficulties that arise from using the **Köppen system.** First, it limits climate classification to land areas when the oceans not only cover 71 percent of Earth's surface but also have a considerable effect on climate. Second, it is based on effects rather than causes. This makes it difficult to link Köppen climate regions to weather system occurrence and frequency. Third, it provides a detailed breakdown of climate regions but uses a complex method of calculating the positions of boundary lines that involves some apparently arbitrary choices of data and calculations to fit the vegetation divisions. These complexities tend to obscure the significant features arising from an understanding of climate differences around the world.

Köppen's five main climate groups are designated by letters: *A* for tropical rainy; *B* for dry; *C* for "mesothermal" (midlatitude rainy with mild winters); *D* for "microthermal" (midlatitude rainy with cold winters); and *E* for polar climates. He added an *H* group for highland climates but did not show this on the map. Each main class divides into smaller groups with additional letters. These are drawn on the world map, explained in the table, and referred to in the following regional analyses of climatic environments.

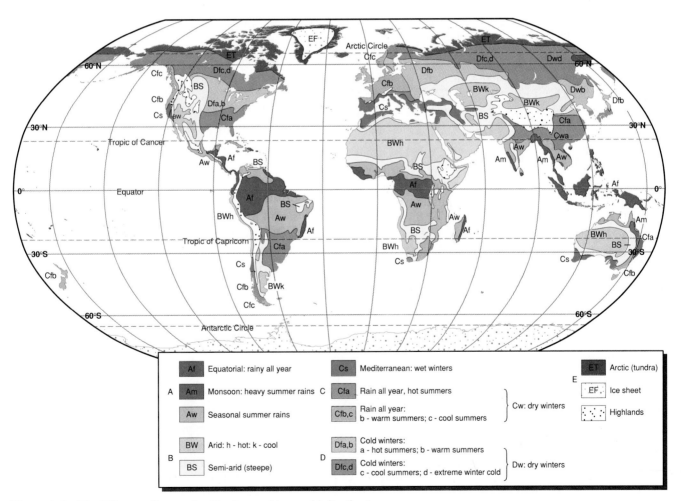

Figure 6-6   The Köppen climate classification and its world distribution.

TABLE 6-1   THE KÖPPEN CLASSIFICATION OF CLIMATE

Criteria	Major Category	Subdivisions
Coolest month 18° C (64.4° F) and above	**A** **Tropical Rainy**	f – rain all year (6 cm or over in driest month) m – monsoon (major contrast between very high rains in summer season and marked winter dry season) w – dry winter, summer rain (less than 6 cm in driest month)
Annual evaporation greater than precipitation	**B** **Dry**	Boundaries between humid and dry climates, and between BS and BW climates determined by formula (see below) W – arid; subdivided further by termperature (BWh = average annual temperature of 18° C or over; BWk = average annual temperature of under 18° C) S – steppe (semi-arid)
Coolest month less than 18° C but –3° C (26.6° F) or more; warmest month at least 10° C (50° F)	**C** **Mesothermal Climates** (Midlatitude rainy with mild winter)	f –wet all year: Cfa – hot summers; Cfb – warm summers; Cfc – cool summers w – dry winter (driest month with less than 10% of wettest summer month precipitation): Cwa – hot summers; Cwb – warm summers; Cwc – cool summers s – dry summer (driest month with less than 33% of the wettest month and less than 4 cm precipitation)
Coldest month less than –3° C; warmest month at least 10° C	**D** **Microthermal Climates** (Midlatitude rainy with cold winters)	f – wet all year: Dfa – hot summers; Dfb – warm summers; Dfc – cool summers; Dfd – extreme winter cold w – dry winter (driest month with less than 10% of wettest summer month precipitation): Dwa – hot summers; Dwb – warm summers; Dwc – cool summers; Dwd – extreme winter cold s – dry summer (driest month with less than 33% of the wettest month and less than 4 cm precipitation)
Warmest month less than 10° C	**E** **Polar Climates**	T – warmest month average 0–10° C F – warmest month average less than 0° C
	Highland Climates	Subdivisions too small for world map

Formula for B climates (P = mean annual precipitation in centimeters, T = mean annual temperature in °C):

BW, arid: $P < T + 7$ e.g., for $T = 20°$, $P < 27$ cm; for $T = 10°$, $P < 17$ cm

BS, steppe: $P < 2(T + 7)$, $> T + 7$ e.g., for $T = 20°$, $P = 27$–54 cm; for $T = 10°$, $P = 17$–34 cm

[Humid A, C, D climates: $P > 2(T + 7)$ e.g., for $T = 20°$, $P > 54$ cm; for $T = 10°$, $P > 34$ cm]

MODELS OF GLOBAL CLIMATE

Just as records of temperature, pressure, humidity, wind force, wind direction, and precipitation provide a basis for weather-system models, so long-term records of weather data provide a basis for models of climate and the prediction of future changes. The extension of atmospheric measurements to all parts of the global surface and several kilometers into the atmosphere, the use of weather satellites, and the invention of powerful computers that can process huge quantities of data has made it possible to generate models that come closer to coping with the extreme complexity of the atmosphere.

The most sophisticated models of climate are known as **general circulation models.** They collect data on a grid covering the globe in which each cell measures several hundred kilometers square (several degrees of latitude by several degrees of longitude). Such a model may also include as many as nine vertical layers up to 35 km (22 mi) above the ground. Data on temperature, humidity, winds, and soil moisture are entered for each cell on the three-dimensional grid. At the scale adopted, some important features, such as individual clouds, are too small for inclusion except in a generalized fashion. Despite these potential drawbacks, tests comparing the results of these models with past changes in climate have been reasonably successful and provide a basis for predictions of the future state of the atmosphere and its climatic environments.

GLOBAL SCALE CLIMATIC ENVIRONMENTS

On the global scale, the consideration of heat balance, water budget, surface type, and weather systems leads to a two-level division of climatic environments. First, the tropical, midlatitude, and polar zones provide major differences in climatic environment.

Second, within these major divisions differences of heating, water availability, surface type, and weather systems produce variations. The distribution of global climatic environments is shown on Figure 6-7.

Tropical climatic environments extend between approximately 30° North and South of the equator. Convergence of trade winds, uplift of air along the intertropical convergence zone, and descent of air in the subtropical high-pressure zones on their poleward limits characterize these regions. The air masses within this circulation are mainly *mT*, with *cT* forming over the land masses beneath the subtropical high-pressure zones. Cloud clusters develop along the ITCZ, which migrates seasonally north and south with the vertical sun. Easterly waves, tropical storms, and tropical cyclones also form, mainly over the ocean, between 10 and 20 degrees of latitude.

Tropical climatic environments divide in a mainly north-south pattern. Close to the equator, surface convergence and an absence of the effect of Earth's rotation produce a climatic environment characterized by cloud clusters and heavy rains with few breaks throughout the year. North and south of the equatorial climatic environments are those produced by seasonal shifts of the ITCZ with the vertical sun (Figure 6-8), so land areas have wet summers and dry winters. In some places the ITCZ moves over land poleward of its more usual path and brings extreme monsoon rains with it. Offshore in the tropics the ocean realm of the trade winds produces another distinctive climatic environment that experiences easterly wave and tropical cyclone weather systems at some distance from the equator. At the northern margin of the tropics over land are arid zones where subtropical high-pressure zones dominate surface weather as a result of descending air beneath the subtropical jet.

Midlatitude climatic environments occur between approximately 30° and 65° North and South. Conflicts between tropical and polar air masses produce the polar front and the high-level, polar-front jetstream. At ground level in both hemispheres, there is a west-to-east sequence of midlatitude cyclones and anticyclones girdling the globe. This sequence and the interaction of continental and oceanic

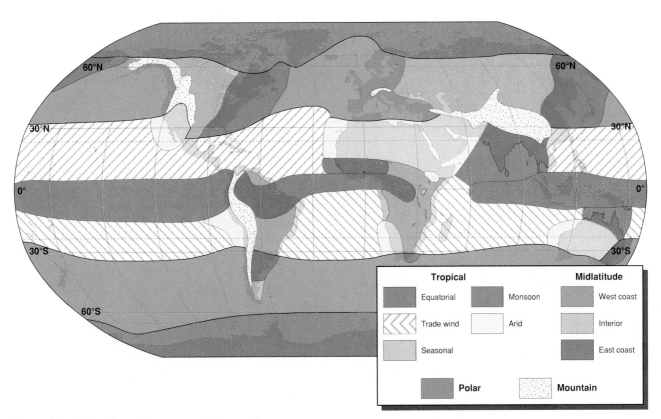

Figure 6-7   Climatic environments of the world.

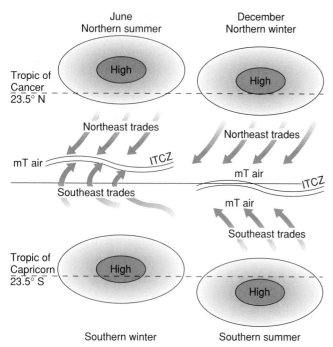

June
Northern summer

December
Northern winter

Tropic of
Cancer
23.5° N

Northeast trades

Northeast trades

mT air   ITCZ

mT air
ITCZ

Southeast trades

mT air

Southeast trades

Tropic of
Capricorn
23.5° S

Southern winter

Southern summer

Figure 6-8   The tropical climatic environment. The changing seasonal patterns of pressure and the position of the Inter-Tropical Convergence Zone.

surfaces set the context for a distinctive west-to-east pattern of climatic environments in these latitudes.

Midlatitude west coast areas receive most of the moisture carried from the oceans by maritime air masses. The same air masses cool the air in summer and keep it mild in winter. Continental interiors are drier because of their distance from such oceanic influences, and many of the weather systems passing through the interiors have been "rained out." The continental areas are more affected than the coastal areas by rapid heating and cooling of the air just above the ground by solar insolation and terrestrial radiation. They exhibit some of the greatest summer-winter temperature contrasts on Earth. The east coasts of midlatitude continents have cold winters dominated by *cP* air masses from the continental interiors, but in summer *mT* air flows in from the ocean bringing warm and rainy weather.

The *polar climatic environments* in latitudes above 65° North and South have air that is cooling and sinking and then flowing outward at the surface as *cP* air masses. The surface of most polar climatic environments is permanently covered by snow. Temperatures

remain very low. In summer midlatitude weather systems occasionally penetrate these regions bringing milder temperatures and some precipitation.

The analyses of climatic environments that follow contain graphic summaries of average measurements for selected stations and some water budget diagrams. Each environment is illustrated by several **climographs,** diagrams that bring together climatic data for a station. Figure 6-9 describes the symbols used.

The study of climate is important to other aspects of physical geography and to the planning and management of a wide range of human activities. A climatic environment is defined by local conditions of heat balance, water budget, surface conditions, and the weather systems that affect an area. Climatic environments occur at a variety of geographic scales from local to global.

TROPICAL CLIMATIC ENVIRONMENTS

EQUATORIAL CLIMATIC ENVIRONMENTS

. . . no matter what month, daily temperature variations are greater than monthly averages. This daily cycle of temperature plays a role in the scheduling of agricultural work. During my 1973–74 fieldwork in Vila Roxa, most colonists native to the Amazon began work around 06.30 hours and stopped by 11.00 hours. Between 11.00 and 15.00 to 16.00 hours they would go home, rest in the shade, and carry out moderate types of activities such as sharpening tools, feeding livestock, and visiting neighbours. This is the hottest time of day, when humidity levels go above the already high mean of 85 per cent. Cooling is difficult during strenuous work since humidity hovers between 86 and 90 per cent from March through October, the months of most intense work effort. In the course of strenuous work such as forest clearing and cutting, for instance, four men each consumed an average of a gallon of water in a half day . . .

The volume and relative constancy of precipitation in the humid tropics can blind one to the significant patterns of variation. Altamira has a marked dry season of four months, during which the rainfall is below 60 millimetres per month. The mean annual precipitation is 1705 mm, although variability from year to year can be great. Annual potential evaporation is 1595 millimetres. Variation is not

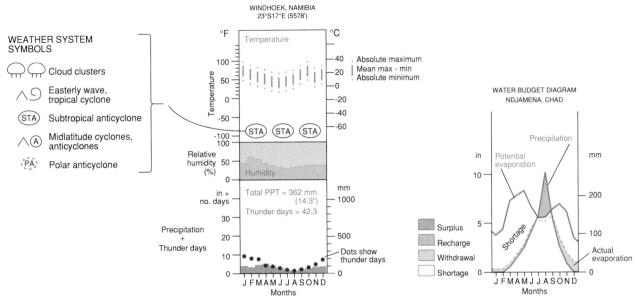

Figure 6-9   Climographs and water budget diagrams: The data used were collected for aviation purposes at airports around the world.

only seasonal, but also daily. One third of the annual rainfall at Maraba in 1974 came down in a 24-hour period. In Altamira I noted that in March 1974 there were 24 consecutive rainy days, in which 522 mm of rain fell—an amount greater than any monthly total and 30 per cent of the annual mean.

Moran, E. 1981. *Developing the Amazon*. Bloomington: Indiana University Press.

Equatorial climatic environments are hot and humid, with rains possible throughout the year. The opening extract describes one person's experience of living in the Amazon rainforest region. Figure 6-10 shows the combination of heat and humidity that cause the fatigue described in the excerpt. Areas with similar climates occur in areas up to 10 degrees of latitude from the equator (Figure 6-11). This band includes an important expanse of ocean.

The Köppen classification designates these areas as *Af*, signifying a coolest month with temperature above 18° C (64° F) (*A*) and rain all year (*f*). The four weather stations in Figure 6-11 demonstrate the small annual temperature ranges of this climatic environment. They are less than 10° C (15 to 20° F) with virtually no recognizable seasons. Humidities remain high all year (over 75 percent) and rain occurs in every month. Annual rainfall totals exceed 1770 mm

(70 in) for inland stations, and over 3000 mm (120 in) is common at coastal stations. The water budget diagram for Singapore shows that precipitation exceeds evaporation throughout the year.

At the equator Earth's rotational effect is nil and so there is little tendency for air movements to develop into circular systems. The convergence of trade winds at the ITCZ brings together warm, moist air masses and results in uplift, condensation in cloud clusters, and heavy rain with downdraft winds. Convergence on its own does not produce strong winds everywhere or always, and the light surface winds alternate with sudden squalls.

Most of the rain in equatorial climates falls from the thunderstorms that compose cloud clusters. One place on Java has recorded an average of 322 days per year with thunderstorms. This high number is caused by the coincidence of the ITCZ and strong sea breezes on the mountainous island, but many places with equatorial climates have over 100 thunderstorm days per year. Intensive deluges are common. *Cloudbursts* (over 1 mm of rain per minute for at least 5 minutes) make up nearly one-fourth of rainstorms, compared to less than 1 percent in midlatitude humid regions. The high annual rainfall produces massive runoff in the rivers of

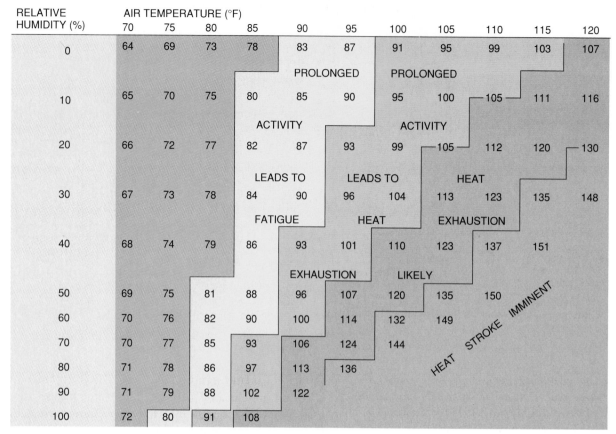

RELATIVE HUMIDITY (%)	AIR TEMPERATURE (°F)										
	70	75	80	85	90	95	100	105	110	115	120
0	64	69	73	78	83	87	91	95	99	103	107
10	65	70	75	80	85	90	95	100	105	111	116
20	66	72	77	82	87	93	99	105	112	120	130
30	67	73	78	84	90	96	104	113	123	135	148
40	68	74	79	86	93	101	110	123	137	151	
50	69	75	81	88	96	107	120	135	150		
60	70	76	82	90	100	114	132	149			
70	70	77	85	93	106	124	144				
80	71	78	86	97	113	136					
90	71	79	88	102	122						
100	72	80	91	108							

PROLONGED ACTIVITY LEADS TO FATIGUE

PROLONGED ACTIVITY LEADS TO HEAT EXHAUSTION

HEAT EXHAUSTION LIKELY

HEAT STROKE IMMINENT

Figure 6-10 The Apparent Temperature Index. It combines humidity and temperature data to provide an indication of personal comfort.

these regions; the Amazon and Zaire rivers have the two largest flows in the world.

Even though total rainfall is high, there is a precarious balance between inputs and outputs of moisture. Short drier spells occur, during which the rainfall does not exceed the continuously high potential evapotranspiration rate. Although it is rare to experience more than thirty days without rain, the distribution of rain is uneven from one local area to another.

Mean daily temperatures vary little throughout the year since the sun is always high in the sky, daylight varies little from periods of twelve hours, and winds from the oceans bring a constant supply of warm, moist air. Diurnal temperature variations are greater than seasonal differences, and frosts may occur on clear nights in upland areas. The highest surface temperatures do not reach the levels experienced in tropical arid regions because the cloud cover of equatorial areas reduces insolation while evaporation takes up

much energy at the surface. This energy is released high above the ground in the cloud clusters and fuels the Hadley cell circulation.

The high levels of temperature and rainfall in equatorial climatic environments provide virtually continuous growing conditions for plants throughout the year. The natural vegetation is rainforest, and it might be thought that such an environment with plenty of heat and water would provide an almost ideal basis for the development of human activities. These regions have generally proved difficult to develop, however, because of poor soil conditions, and they mostly support small populations except for some of the East Indian Islands and Malaysia.

TROPICAL SEASONAL CLIMATIC ENVIRONMENTS

Although British South Africa extends almost 19° of latitude it has so many features of topography and climate in common that it may well be treated as a whole.

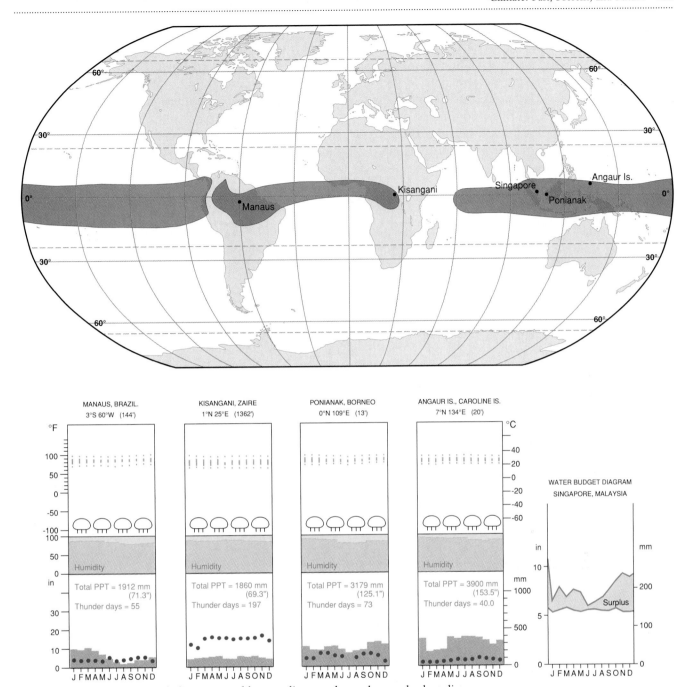

Figure 6-11 Equatorial climates: world map, climographs, and water budget diagram.

During the summer of the southern hemisphere atmospheric pressure is relatively low, the low pressures having migrated south with the sun from the equator. There is a fairly steep gradient from the subtropical anticyclones of the South Atlantic and South Indian Oceans, and the wind blows in to the continent, from the north-east, east, south-east, and (probably owing to topographical influences) south-west in Natal, from south and south-east in the east of the Cape Province, bringing rain and cloud over the whole country except the neighbourhood of Cape Town.

In winter anticyclonic conditions establish themselves over the land. The wind tends to the light, the weather fine and rainless, and the sky clear.

Kendrew, W. G. 1937. *The Climates of the Continents*. 3d ed. Oxford: Oxford University Press.

Land areas just outside the equatorial climatic environments experience an alternation of dry, low-sun seasons and wet, high-sun seasons. These regions form a transition from the rainy equatorial regime to the dry regime of the tropical arid areas. Such climatic environments occur in the southern continents of South America—Venezuela, interior southeastern Brazil, and Paraguay—and Africa—the belt of countries along the southern margin of the Sahara, eastern Africa, and most of southern Africa—as shown on Figure 6-12.

The Köppen system labels these regions *Aw* (*A* for a mean coolest month over 18° C (64° F), and *w* for a dry winter and rainy summer) or *BSh* (*B* for a potential evaporation excess over precipitation, *S* for a precipitation total over half the evaporation rate, and *h* for an annual temperature above 18° C).

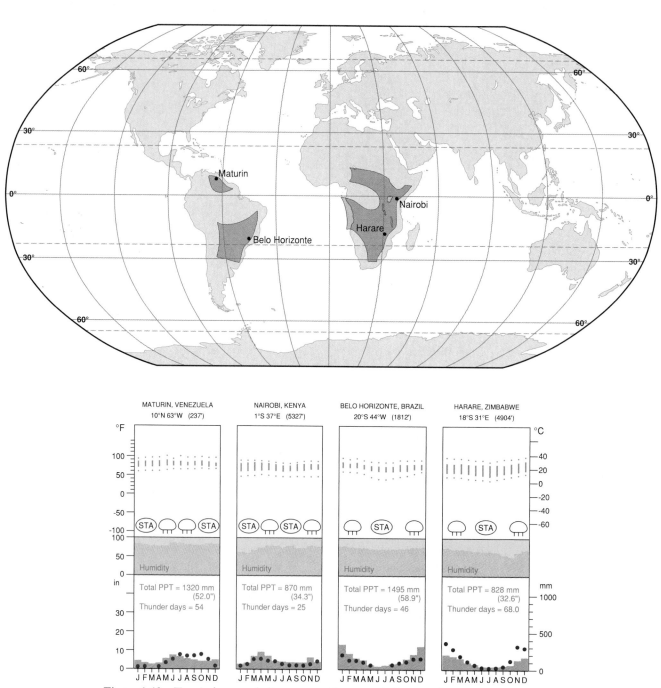

Figure 6-12   Tropical seasonal climates: world map and climographs.

The four weather station climographs show small seasonal fluctuations in temperature. In most cases the range of temperature is greater than that in the equatorial regions. Humidity levels at 60 to 75 percent are lower than those in the equatorial regions and fluctuate between distinct wet and dry seasons, each lasting several months. The temperature ranges of the dry seasons tend to be greater than those in the wet season. Nairobi is an exception to the summer rainy season, since it has two rainy seasons, in April and November. This city is close to the equator and the rain belt associated with the ITCZ passes overhead twice, just after each equinox. Nairobi is included in the tropical seasonal climate because its altitude on the East African plateau causes the total precipitation to be much less than that of equatorial climatic environments. This gives Nairobi an environment close to that of the tropical seasonal regions, having long dry periods between rainy ones.

The seasonal movement of the ITCZ rainbelt (Figure 6-13) brings precipitation when the sun is at its highest to extensive areas in the tropical sections of the southern continents. The same convergence responsible for downpours in equatorial climates causes the summer rains of seasonal tropical climates. The precipitation totals on the poleward margins of this belt are extremely variable from year to year, particularly in areas distant from the oceanic moisture sources. Long dry periods may also occur within the rainy season. The uncertainties of the rainy season causes the marginal areas to have a tendency toward aridity.

The dry winter season is an extension of desert climatic conditions toward the equator as the ITCZ shifts into the opposite hemisphere and high-pressure systems cover the belt. Figure 6-14 shows the vegetation typical of this seasonal climate, wooded savanna, which is an adaptation to the long dry season.

The Sahel Zone   The drier margins of the seasonal tropics immediately south of the Sahara in Africa are known as the **Sahel** zone (Figure 6-15). They have been the subject of intense concern owing to the effect of prolonged drought on the lives of the inhabitants since the early 1970s (Environmental Issue: Desertification—Human or Climatic Disaster?, p. 408).

Recent climate research demonstrates strong links between the dry conditions of the Sahel and higher temperatures in the ocean waters off West Africa. Further understanding of these links will provide a better basis for future land use and water management decisions.

TROPICAL MONSOON CLIMATIC ENVIRONMENTS

(At Calcutta) At length in the early part of June, the clouds gather more thickly, while the barometer falls to a lower point than it has reached since the beginning of the year; and in the first or second week heavy and continuous rain ushers in the monsoon. This first burst of the rains usually accompanies a cyclonic storm, formed either at the head of the Bay (of Bengal) or over the delta (of the Ganges) itself . . . Its immediate effect is a great fall of the day temperature; and the comparative coolness, supervening on many weeks of close oppressive weather, brings a sense of relief . . . When, however, in September the rainless intervals become longer, and the day temperature begins to rise, while the air, still highly charged with moisture, is almost motionless, the relaxed energy of the human system fairly rebels against this further trial of its endurance, and all who are not by their vocations compelled to remain at their post hasten to escape to the temporary refuge of a hill station. September and October are thus the most trying and unhealthy season of the year.

Blanford, H. F. 1889. *The Climates and Weather of India, Ceylon and Burma*. London.

"Monsoon" is derived from the Arabic *mausim,* meaning "season." The wind shifts from northeast to southwest every six months in the Arabian Sea and

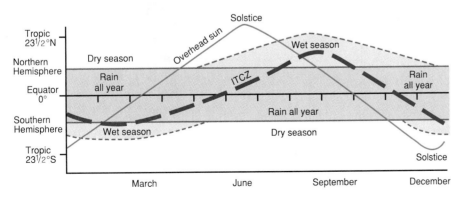

Figure 6-13   The Inter-Tropical Convergence Zone, overhead sun, and seasonal rains. Equatorial climates have rain throughout the year with slightly more or less in some seasons. Seasonal climates have wet summers and dry winters.

Figure 6-14 Seasonal tropical climate: Zimbabwe, Africa. A farmed area near the end of the dry season. The hills in the background are covered with vegetation that tolerates a long dry season.

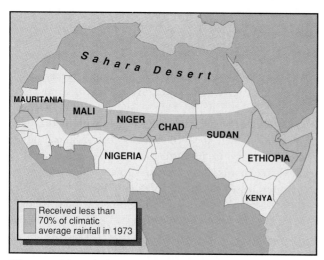

Figure 6-15 Sahel, North Africa. This zone is along the southern margin of the Sahara. Note the countries of the Sahel zone (yellow) and the area where the 1973 annual rainfall was under 70 percent of the long-term average.

causes contrasting rainy and dry seasons. The term is commonly extended to refer to the extreme summer concentration of precipitation in the Indian subcontinent (India, Pakistan, Bangladesh, Nepal, and Bhutan), southeast Asia, northern Australia, and West Africa (Figure 6-16). The opening extract highlights the seasonal contrasts experienced. The monsoon climatic environment affects some of the most densely populated parts of the world where the timing and total of the monsoon rains are crucial for the livelihoods of millions of people.

The Köppen classification distinguishes monsoon climate from tropical seasonal climates only by the intensity of contrast between wet and dry seasons. The *Am* designation signifies heavy summer rains by the letter *m*. The tropical monsoon climatic environment, however, is fundamentally different from the seasonal type. Each of the monsoon regions occurs on a land mass opposite a large expanse of ocean stretching into the opposite hemisphere. This arrangement affects the circulation of heat and moisture and creates the distinctive monsoon conditions.

The four weather stations selected show strong similarities. The Asian and Australian stations have a high summer peak of precipitation with many thunderstorms—except at Bombay on the coast. In West Africa the high summer rainfall peak is not so marked, but thunder days are. At all stations relative humidity peaks at the time of the summer rains but falls below 70 percent in the drier half of the year. The water budget diagram emphasizes the seasonal contrast in water availability caused by the variations in precipitation and potential evapotranspiration.

The Indian Subcontinent Monsoon From December through June in India, Pakistan, Bangladesh, Nepal, and Bhutan, northeast winds produce sunny

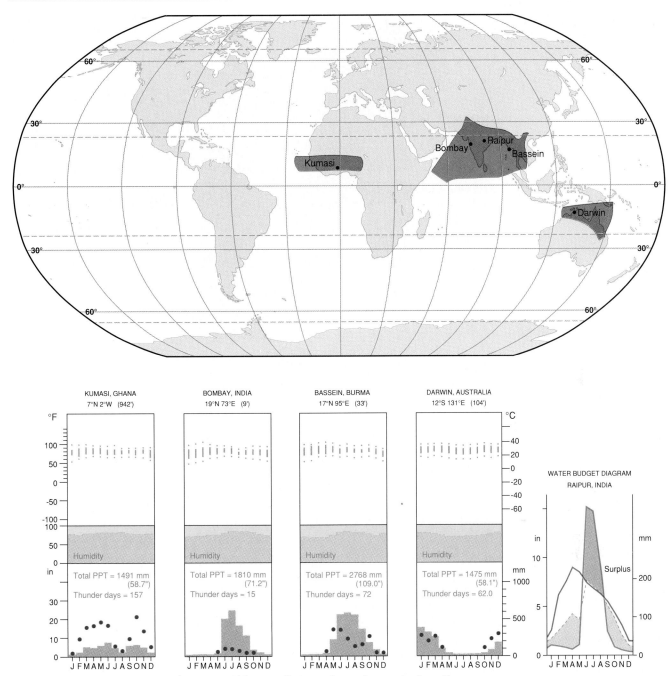

Figure 6-16   Monsoon climate: world map, climographs, and water budget diagram.

and dry weather with increasing heat. The reversed airflow of the southwest monsoon usually begins in June with a sudden burst of rainy weather, and continues with heavy rains throughout the summer. The arrival of the rainy weather, however, may come earlier or later than June, and this uncertainty affects farming productivity. The summer abundance of water favors growing rice, which needs large amounts of water during its early growth (Figure 6-17).

A sequence of explanations of this wind reversal illustrates how the understanding of atmospheric processes is developing. At first the monsoons were regarded as a giant seasonal form of land-sea breeze (see Figure 3-21). In summer the land heats up causing air above it to rise. Moist oceanic air flows in from the southwest to replace the rising air. In winter the continent cools more quickly than the ocean, and the high atmospheric pressure established over the land causes winds to blow from land to sea.

Figure 6-17 Rice paddy in India. The intense monsoon rains make it possible to grow rice.

A second explanation offered in the early twentieth century related the monsoon reversals to shifts of the ITCZ. As the ITCZ moves into southern Asia in summer, it draws in the trade winds from the ocean in the southern hemisphere. The huge supplies of humidity imported in this way provide the basis for the exceptional monsoon rains.

A third and most recent interpretation links the reversals to changes in the upper troposphere winds. In winter (Figure 6-18a) the westerly polar-front jetstream moves south and splits in two around the Himalayan mountains. The southern arm intensifies the surface high pressure in northern Pakistan by forcing air to descend. As this situation persists, descending and outflowing air dominates most of the Indian subcontinent keeping it dry. The Himalayas prevent the cold air of central Asia from reaching India during winter, and Calcutta has January temperatures that are 7° C (15° F) higher than those at Hong Kong on the same latitude.

In the change to summer conditions, the entire westerly polar-front jetstream shifts north of the Himalayas, and an easterly equatorial jetstream establishes itself across the northern Indian Ocean (Figure 6-18b). The southwest monsoon hits India as a surface convergence forms beneath this easterly jetstream, drawing in air from south of the equator. The moisture picked up over the Indian Ocean, western Arabian Sea, and the Bay of Bengal makes the air very humid. The mountainous west-facing coasts of India and Burma force the humid air to rise forming huge thunderclouds and producing high rainfall totals. While the western coasts of India and northern hills receive rains during the southwest monsoons, tropical cyclones in the Bay of Bengal occasionally bring death and destruction to the low-lying Ganges delta region. Much of the plateau land in between lies in a rain shadow and receives less rain.

The rest of the summer in the Indian subcontinent has periods of greater and lesser weather activity. Days of heavy rain give way to quieter conditions with clear weather when the easterly jetstream weakens. The distribution of rain is patchy and uncertain over the subcontinent since heavy rains in one place result in fewer disturbances and less rain elsewhere. While rainfall is unevenly distributed, evaporation remains high throughout the subcontinent in summer. Even areas that receive plentiful rain dry out rapidly and few places in India or Pakistan have a large overall surplus in their water budget despite the intense monsoon rains.

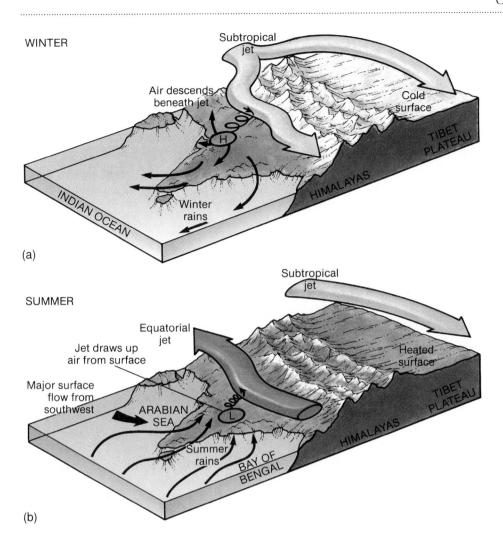

WINTER

Subtropical
jet

Air descends
beneath jet

Cold
surface

TIBET
PLATEAU

HIMALAYAS

H

Winter
rains

INDIAN OCEAN

(a)

SUMMER

Subtropical
jet

Equatorial
jet

Jet draws up
air from surface

Heated
surface

Major surface
flow from
southwest

ARABIAN
SEA

TIBET
PLATEAU

L

Summer
rains

HIMALAYAS

BAY OF
BENGAL

(b)

Figure 6-18 Monsoon in the Indian subcontinent. The surface and upper troposphere winds.

West Africa and Australia   In West Africa the summer-winter wind reversal occurs with open desert to the north instead of a high mountain chain. This lack of a mountain barrier enables cT air from the Sahara to blow southward in winter so northeasterly flows of desert air, including the Harmattan wind, dominate most of West Africa north of the coastal zone. In summer, however, the ITCZ moves deeply into the continent and southwesterly flows from the equatorial Atlantic Ocean bring clouds and rain.

The Australian situation is more like that of West Africa than of the Indian subcontinent since the interior desert is not shielded by any mountain barrier. In summer winds bring moist air from the Pacific Ocean north of the equator, and the heaviest rain occurs along the northern coastal area. In winter dry winds blow from the desert toward the ocean.

TRADE-WIND CLIMATIC ENVIRONMENTS

When it grew light next morning, a thick mist lay over the coast of Peru, while we had a brilliant blue sky ahead of us to westward. The sea was running in a long quiet swell covered with little white crests, and clothes and logs and everything we took hold of were soaking wet with dew. It was chilly, and the green water around us was astonishingly cold for 12 degrees south. There was a rather light breeze, which had veered from south to south-east.

And the wind came. It blew up from the south-east quietly and steadily. Soon the sail frilled and bent forward like a swelling breast, with Kon-Tiki's head bursting with pugnacity. By the late afternoon the trade wind was blowing at full strength. It quickly stirred up the ocean into roaring seas which swept against us from the stern. We knew that from now onwards we should never get another onshore wind or a chance of turning back. We had got into the real trade wind, and every day would carry us farther and farther out to sea. The only thing to do was to go ahead under full sail; After a week or so the sea grew calmer, and we noticed that it became blue instead of green. We began to go west-north-west instead of due north-west.

Heyerdahl, T. 1948. *The Kon-Tiki Expedition*. London: Allen & Unwin.

Trade-wind climatic environments dominate the tropical oceans, where the ocean surface affects the atmospheric environment in ways that set this climatic environment apart. Just as the trade winds are the most reliable and constant in the world, the climates they produce contain less variation from year to year than other climatic environments.

Although this is the most extensive climatic environment of all, many climatic classifications including Köppen's ignore it because it is not land-based. It occurs almost exclusively over the oceans but also affects strips of land on east-facing tropical coasts (Figure 6-19). Köppen classifies such coastal areas under the same head as the equatorial climatic environments.

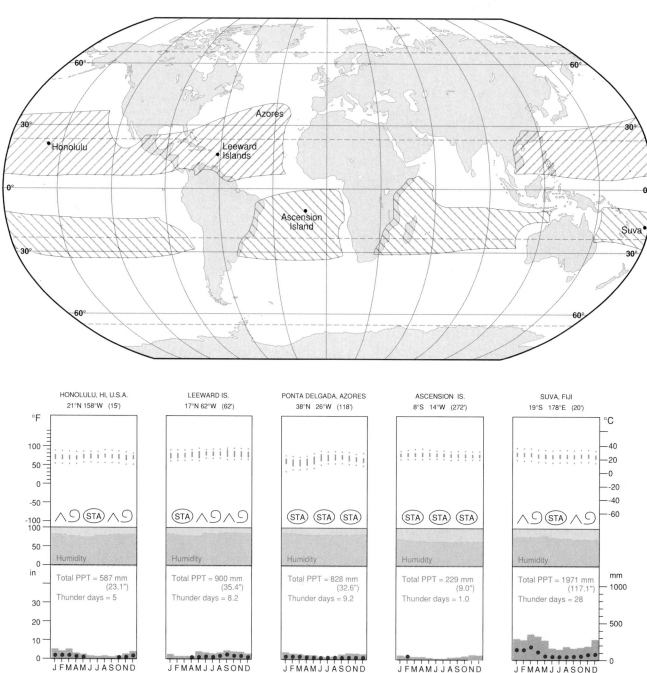

Figure 6-19 Trade-wind climates: world map and climographs.

The selected weather stations demonstrate some typical features of the trade-wind climatic environment. First, they have the small monthly and annual ranges of temperature characteristic of coastal and island locations. Most trade-wind stations have extreme temperature ranges of less than 20° C (36° F). The Azores at the poleward limit of the trade-wind region, have the greatest annual range. Second, the relative humidity is generally high, from 70 to 80 percent. Third, each station, apart from Ascension Island, has considerable precipitation. Precipitation occurs at most stations throughout the year, but there is a tendency to a summer maximum close to the western margins and a winter maximum close to the eastern margins of the oceans. It is noticeable that amounts of precipitation and thunderstorm activity increase toward the western margins of oceans.

The trade winds that dominate this climatic environment blow toward the equator from the eastern margins of the subtropical high-pressure zones. They are northeasterly in the northern hemisphere and southeasterly in the southern hemisphere. These winds are among the most constant features of global atmospheric circulation, and the area they affect shifts little with the seasons. The opening extract gives some idea of the strength of the southeast trade winds in the South Pacific Ocean. Weather systems occurring within this environment include easterly waves and tropical cyclones.

East-West Contrasts   Within the trade-wind climatic environment, there are major contrasts in conditions between the eastern and western areas of oceans. These differences occur, however, as part of a common dynamic atmospheric system lying between the ITCZ and the subtropical high-pressure zones (Figure 6-20).

The atmosphere on the eastern margins of the subtropical high-pressure zones is dry because it either descends in the high-pressure zones or blows from the continental deserts. Contact with the ocean causes high rates of evaporation that soon raise the relative humidity of the air. The cold ocean currents on the eastern sides of oceans cool the air above, resulting in condensation as fog. Clouds and rain seldom occur, however, because the cold water reduces the environmental lapse rate. In addition, the descending air above is warmer than the surface layer of air, producing an inversion that traps the humid air and fog near the ocean surface.

In the western parts of tropical oceans, the trade winds are less constant, the air has a higher relative humidity, and the surface waters are warm. The inversion layer rises, and the greater instability in the atmosphere increases convectional lift. Weather systems, such as easterly waves, tropical storms, and tropical cyclones also develop to their full potential bringing heavy rains, especially in summer and early fall.

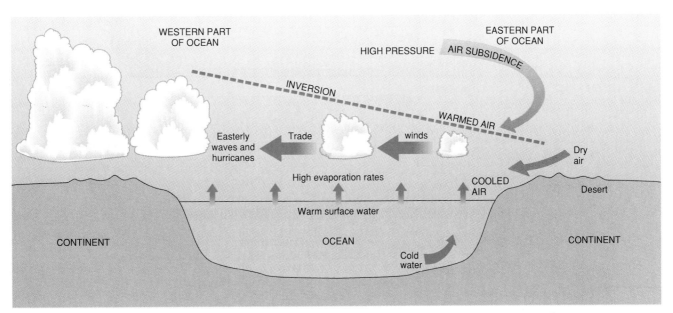

Figure 6-20   Trade-wind climatic environment: the differences between the eastern and western parts of oceans.

The relatively low totals of precipitation for most of the stations representing trade-wind climatic environments do not tell the full story. The heaviest rains fall on windward slopes of islands and continents, and there are often strong rain shadow effects on lee slopes. Many weather stations are sited near towns in the more protected parts of these islands. The island of Hawaii, for instance, receives 3000 mm (120 in) of rain per year on the northeastern side from the trade winds, but less than a tenth of this falls on the far side of the same island.

The trade-wind climatic environment also affects coastal areas such as northeast Brazil, eastern Africa, and Vietnam. Northeast Brazil has heavy rains on the coastal hills, but the plateau behind the hills lies in a rain shadow and has great swings between arid and humid years. Life is uncertain for the inhabitants, who have few good years. The poverty of the people led to massive emigrations to cities such as São Paulo and to other parts of the country. In response to political pressures, some were resettled in the empty Amazon basin. That, in turn, led to environmental pressures on the Amazon rainforest.

El Niño   Studies of the El Niño phenomenon since the 1970s emphasize the central significance of the tropical ocean climate environment in the global climatic system. The **El Niño current** is a southward flow of warm water along the coast of northwestern South America. Around Christmas—El Niño means Christ child—the warm current pushes the cold Peru Current southward along the Peruvian coast (see Figure 3-33). In some years (e.g., 1953, 1957–58, 1965, 1972–73, 1976–77, 1982–83, 1986–87, 1991–92) the Peru Current retreats farther south and warmer water replaces it for much longer. This may last for over a year and brings disaster to the Peruvian fishing industry, whose catch depends on the nutrients brought to the surface in the cold current.

The ultimate causes of this "enhanced El Niño" are not known, but the sequence of associated events is clearly complex and embraces much of the tropical atmosphere. Under normal circumstances during the latter half of the year, the subtropical high-pressure zone causes trade winds blowing from the eastern Pacific toward low pressure over Indonesia (Figure 6-21a). The consistent trade winds blow warm surface

Figure 6-21   The El Niño effect. How do the changes in wind direction affect the surface waters off the coast of western South America?

Source of information courtesy of National Geographic Society.

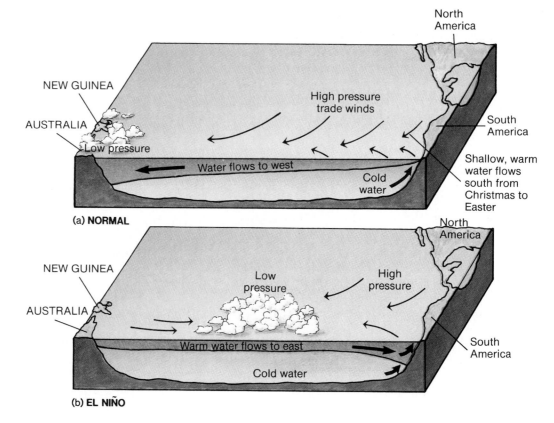

water westward away from the Americas, reducing the depth of warm water off the South American coast. Cold water and nutrients rise from deeper water to the surface along that coast and maintain an ecosystem rich in plankton, fish, and sea birds. The shallow, warm El Niño current remains farther north off Colombia and Ecuador except from December to April when the winds weaken and the current extends southward.

Every few years the normal situation breaks down and the low pressure over Indonesia moves to the middle of the Pacific Ocean. The subtropical high-pressure zone weakens, and westerly winds replace the trade winds over a large part of the tropical Pacific. These winds blow the surface waters eastward, so that the warm water "piles up" against South America (Figure 6-21b). The cold water off the Peruvian coasts is covered by a deeper mass of warm water, and the long-term stoppage of upwelling is the *enhanced* El Niño. The atmospheric low pressure also brings clouds and rain to the coasts of Ecuador and Peru, while the replacement of the Indonesian low by high pressure results in drought in Australia.

El Niño is part of an atmosphere-ocean environment phenomenon that affects the whole of the tropical Pacific Ocean. As might be expected, events that affect a fourth of the tropics have wider repercussions. Some suggest linkages with the Asian monsoons and equatorial convection centers over the Amazon and Zaire basins, but the linkages are not clearly worked out. Further research into these possible connections may help in making predictions of large-scale world weather changes.

TROPICAL ARID CLIMATIC ENVIRONMENTS

Suddenly, after hardly any twilight, the sun rises into the clear sky. In this dry atmosphere its rays are already scorching in the early morning, and under the influence of the reflection from stone and sand the layer of air next to the ground is warmed rapidly. There is no active evaporation to moderate the rising temperature. After 9 o'clock the heat is great and goes on increasing till 3 or 4 in the afternoon, when the quivering mirage is sometimes seen, produced by the vibration of the air, heated as in an oven. It gets slowly cooler towards evening, and the sun, just before it sets, suffuses the cloudless sky with a glow of colour. In the transparent night the rocks and sand lose their heat almost as rapidly as they acquired it, and the calmness of the atmosphere, which is so still that a flame burns without a tremor, also favours the cooling of the air. We shiver with cold, and it is no uncommon thing in winter to find water on the surface of the ground frozen in the morning.

Schirmer. 1937. "Le Sahara." In *The Climates of the Continents*. 3d ed., W. G. Kendrew. Oxford: Oxford University Press.

Arid climatic environments are marked by a shortage of surface water. Arid conditions occur in the southwestern United States, northern Mexico, the Sahara-Arabia-Pakistan region of northern Africa, western Asia, the Atacama of southern Peru, northern Chile, the Kalahari of southwestern Africa, and the Australian desert (Figure 6-22). Although arid zones are sometimes subdivided into semiarid, arid, and extremely arid, they share the conditions of low precipitation and high rates of evaporation that create long-term arid climatic environments.

The Köppen system defines tropical arid climates as *BWh* and *BSh* areas. These two classifications affect the degree of aridity. They both have an excess of evaporation over precipitation (*B*) and an average annual temperature above 18° C (*h*), but precipitation is less than half the potential evaporation in *BWh* and 50 to 75 percent of it in *BSh*.

The station climographs in Figure 6-22 all exhibit high summer temperatures and a general shortage of water. The three northern hemisphere and two southern hemisphere stations all have low precipitation totals, generally low humidity, and ranges of monthly and annual temperatures that are greater than other tropical climates. Local factors cause variations from station to station. For instance, the maximum temperatures all rise to over 40° C (104° F), except at Iquique and Windhoek. Iquique is a coastal station with a cold offshore current that keeps the air temperatures low, aided by a high incidence of fog that prevents direct insolation reaching the ground on many days. Windhoek is in uplands at 1704 m (5578 ft) above sea level, where the temperatures are reduced by altitude. The relative humidities of the stations are less than 50 percent, except for Iquique and Bahrain, both of which are in coastal locations. Precipitation is less than

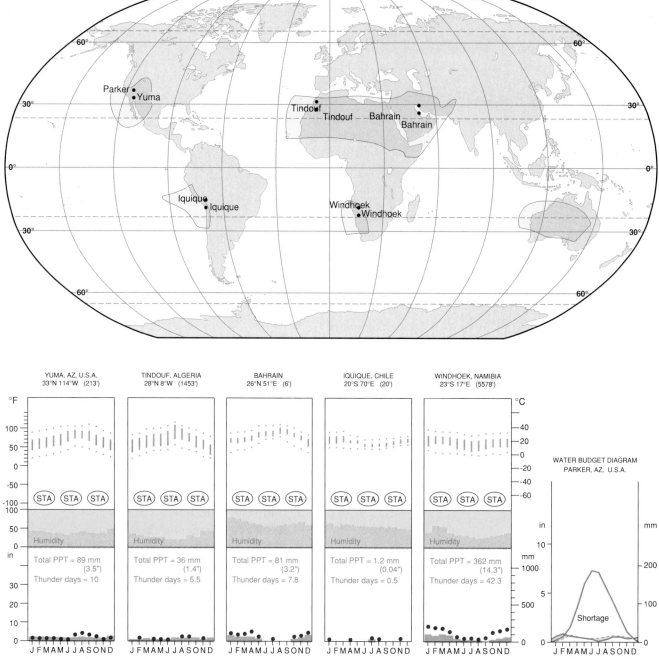

Figure 6-22 Tropical arid climates: world map, climographs, and water budget diagram.

250 mm (10 in) at all stations except Windhoek, where it is 362 mm (14.3 in) because the upland environment encourages the uplift of air that produces forty-two days with thunder.

Aridity is not so much a matter of low rainfall as of the balance between precipitation inputs and evapotranspiration outputs in the water budget. The water budget diagram for Parker, Arizona, demonstrates the massive deficit of water in such locations. Tropical arid areas have rainfall totals that come nowhere near the amount needed to compensate for the high potential evaporation rates caused by intense insolation and descending dry air. Although some places in these regions may have rainfalls of over 750 mm (30 in) per year and a few up to 1270 mm (50 in), the evaporation rates are much higher and the rainfall is so irregular that there is insufficient moisture for crop growth without irrigation. This is effective aridity.

The tropical arid climatic environments constitute one of Earth's "difficult" environments and support few plants or animals and hardly any people unless water can be brought to them. Once water is available, however, the high insolation produces good crop yields, and large cities such as Los Angeles, Baghdad (Iraq), and Karachi (Pakistan) can develop.

Low Precipitation Conditions   In arid climatic environments, atmospheric conditions work against the occurrence of rainfall. The widespread descent of air beneath the subtropical jetstream causes adiabatic warming and drying of the air. Atmospheric stability is dominant. Over land, winds blow outward from the surface high-pressure cells, preventing the invasion of moisture-bearing air from the oceans. Even when clouds form and rain falls as the result of local heating or the invasion of an atmospheric disturbance, much or all of the rain evaporates before reaching the ground.

Few parts of tropical arid climatic environments are absolutely without rain, however. When rain falls in arid regions, it is often in short-lived showers, seldom as heavy as those in humid regions. Such showers are most common in upland areas, such as around Windhoek, Namibia. There may be a winter maximum of such falls on the higher latitude margins of arid areas and a summer maximum on the equatorward margins, but there is often little pattern in the distribution throughout the year or from year to year. Rain is essentially unpredictable and there are long dry periods between the showers.

The driest parts of the tropical arid areas occur on west coasts, such as southern Peru, northern Chile, northwestern Mexico, and southwestern Africa. Iquique in northern Chile has one of the lowest average annual rainfalls on Earth—only 1.3 mm (0.05 in)—and the Namibian coast in southwestern Africa has places with only 18 mm (0.7 in). Such conditions extend several hundred kilometers offshore as strong, dry winds blow out from the land, particularly in winter, and drag the surface waters with them causing cold water to well up along the coast. The cold water surface cools the overlying air, reducing the environmental lapse rate and resulting in greater atmospheric stability. In summer moister mT air moves across the cold waters producing advection fog or fine drizzle as sea breezes bring it onshore. Such fog is familiar to coastal Californians as far north as San Francisco.

Temperature Fluctuations   The combination of intense insolation through cloudless skies and adiabatic warming of the descending air leads to high daytime temperatures. Shade temperatures commonly exceed 30° C (86° F). The world's highest recorded shade temperature of 58° C (136° F) occurred at Azizia in the Libyan Sahara, and 57° C (134° F) has been measured in Death Valley, California. Temperatures of 82° C (180° F) occur in desert air above exposed rock or sand surfaces out of the shade. These temperatures would be even higher, however, were it not for the high albedos of sand and bare rock, which reflect away a high proportion of insolation. Further, the low water content of the atmosphere diminishes the amount of heat that the air absorbs. At night, temperatures fall rapidly as terrestrial radiation continues through clear skies and air in contact with the ground cools rapidly. The differences in temperature between day and night cause diurnal ranges of as much as 40° C (72° F). The cooling at night can result in dew or winter frost despite the low daytime relative humidity.

MIDLATITUDE CLIMATIC ENVIRONMENTS

MIDLATITUDE WEST COAST CLIMATIC ENVIRONMENTS

This is our present situation, truly disagreeable. . . . O, how tremendious is the day. This dredful wind and rain continued with intervales of fair weather, the greater part of the evening and night. . . . The emence Seas and waves which break on the rocks and Coasts to the southwest and north-west roars like an emence fall at a distance, and this roaring has continued ever Since our arrival in the neighbourhood of the Sea Coast which has been 24 days Since we arrived in Sight of the Great Western; (for I cannot Say Pacific) Ocian as I have not seen one pacific day since my arrival in its vicinity, and its waters are forming perpetually breake with emence waves on the Sands and rocky coasts, tempestuous and horiable.

W. Clark. 1804. In *Passage Through the Garden,* by J. L. Allen. 1975. Urbana: University of Illinois Press.

Storms, high winds, and changeable weather are features of midlatitude west coast climatic environments, as Clark recorded about the northwest coast of the United States. But, such areas also have the smallest annual average temperature ranges of midlatitude climates. Climates with small differences between summer and winter temperatures are called **equable.** The tropical climatic environments are all fairly equable, but the west coasts stand out in comparison with other midlatitude climatic environments which are not generally equable.

Midlatitude west coast climatic environments affect land mainly in the northern hemisphere and particularly in western Europe (Figure 6-23). Only small sectors of the southern hemisphere continents extend into these latitudes, and mountain ranges restrict the

extent of this climate type in North America. Large parts of the midlatitude oceans also have this climatic environment.

Köppen designated the continental parts of such regions as *Cfb*, denoting a coolest month between 18° C (64° F) and −3° C (27° F) (*C*), rain all year (*f*), and warm summers (*b*). The climographs of the five selected weather stations demonstrate the small monthly and annual temperature ranges (especially when compared to other midlatitude climatic

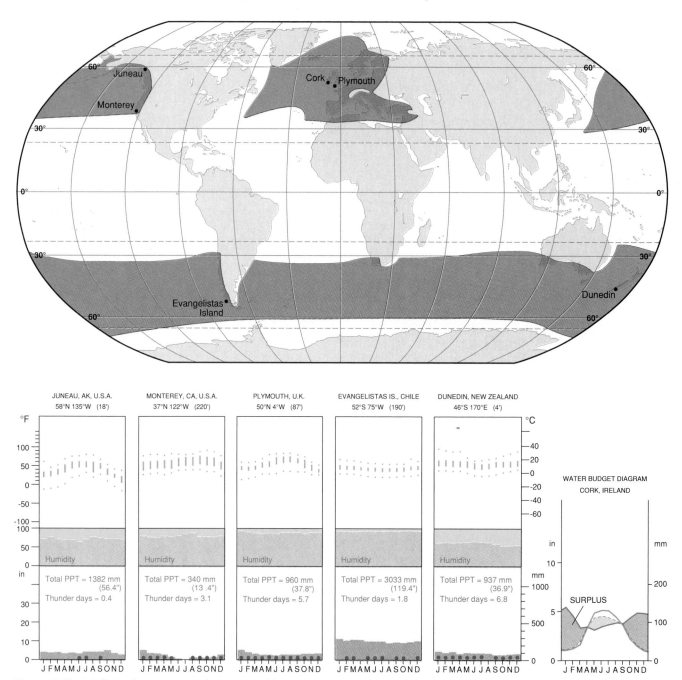

Figure 6-23 Midlatitude west coast climates: world map, climographs, and water budget.

environments). Extreme temperatures seldom exceed 40° C (104° F) or fall below −12° C (10° F), and mean monthly temperature ranges are generally between lows of −5° C (23° F) and highs of 16° C (60° F). Periods of extreme cold or heat lasting for more than two to three weeks are rare. Precipitation occurs at all seasons, with totals of 760 mm (30 in) and more. The average figures for precipitation and temperature, however, do not provide a full picture of the characteristics of these climates, which receive the wide range of weather types associated with a succession of midlatitude cyclones and anticyclones.

The movement of frontal cyclones and anticyclones beneath the polar-front jetstream brings together mainly *mP* and *mT* air masses as they reach the western margins of the oceans. At any season of the year when the westerly jet is in the low-amplitude west-to-east phase of its cycle, the sequence of cyclones and associated fronts bring periods of rain every few days. The establishment of larger and more persistent "blocking" anticyclones halts this flow. These occur in the later stages of the jetstream cycle when large high-pressure systems form in the ridges south of the jetstream. Such anticyclones may contain a *cP* air mass in winter or a *cT* air mass in summer (see Chapter 5).

Seasonal changes consist mainly of temperature differences rather than precipitation. In summer the combination of fourteen to eighteen hours of daylight and a higher sun angle bring direct heating and warm weather. Even then, the hottest days occur with the invasion of *cT* or *mT* air masses. The arrival of *mP* air can lower the maximum temperatures on many summer days below 20° C (68° F), especially near the coasts. Winter temperatures may be very low (below 0° C, 32° F) for several weeks if *cP* air flows in from the east or north. Milder winter temperatures up to 10° C (50° F) prevail when *mP* air comes from the ocean. Most of the warmth in winter comes from this maritime air, rather than directly from the sun.

Precipitation is mostly from thick layer clouds above warm fronts and cumulonimbus clouds along cold fronts. Deep snowfalls accumulate on higher land in winter. The highest precipitation totals occur on west-facing mountainous coasts where the moist air lifting over a front rises farther on encountering high relief. The heavy showers from cumulonimbus clouds along cold fronts increase in frequency during summer months and are especially numerous inland where surface heating is more intense.

On the equatorial side of the midlatitude west coast climatic environment, there is a transitional zone between the all-year humid conditions and the tropical arid climates. This zone is most extensive around the Mediterranean Sea, which has given its name to the type of climate. Elsewhere in the world the *mediterranean* climate type affects central California, central Chile, the southwest tip of South Africa, and southwest Australia.

Köppen labelled such areas *Cs*, signifying that the *C* temperature conditions typical of midlatitude west coasts combine with dry summers (*s*) and winter rain. Extremely dry summers are caused by the subtropical high-pressure zone extending into higher latitudes, whereas wet winters result from the southward movement of the belt of cyclones beneath the polar-front jetstream.

MIDLATITUDE CONTINENTAL INTERIOR CLIMATIC ENVIRONMENTS

As temperatures in this remote Siberian city (Yakutsk) dropped this week to minus 50 degrees Centigrade [−60° F]—about three times below that in the average domestic freezer—it became easy to understand why the new Kremlin leadership is fighting a losing battle to remedy Siberia's acute labour shortage.

Visibility was reduced to a few eerie yards by the swirling tuman, or freezing fog that never lifts at such extremes and is thickened by the fumes from thousands of vehicle engines kept running round the clock. Eyelashes often freeze together and outsiders are told to rub themselves with snow at the first tell-tale signs of frostbite.

All buildings are erected on stilts above the permanently frozen subsoil, and triple-glazed in an effort to keep out the winter that lasts for eight months of every year. During the rest, temperatures soar to a sweltering 32 degrees Centigrade [90° F] and attract swarms of vicious mosquitoes.

Yakutsk, one of the coldest inhabited spots on the globe, is also one of the main administrative centres involved in the costly Soviet drive designed to persuade an estimated one million workers to join the search for the forbidding region's massive deposits of minerals.

"For us, this is a relatively mild day, the schools are still functioning above the fourth grade and men are still out working on the construction sites", explained the mayor. "It is only the weak who cannot face it and leave."

The Times. 31 January 1986. London.

The North American and Eurasian continents are large enough to isolate extensive interior regions from oceanic influences (Figure 6-24). In North America the western mountain ranges form a barrier to oceanic influence from the west. In Eurasia, the interior of the continent is even more distant from the ocean. Similar conditions occur over a smaller area in southern Argentina, east of the Andes Mountains. The southern parts of the midlatitude continental interiors are arid because of the lack of a moisture supply and high levels of evapo-

transpiration in very hot summers. Farther north, however, an excess of precipitation over potential evapotranspiration results when humid air gets in or because evapotranspiration rates are not so high.

Köppen recognized several climate types in the midlatitude continental interiors. The arid and semiarid regions are designated *BWk* and *BSk* respectively, with the *k* referring to the cool temperatures that distinguish midlatitude arid regions from tropical arid

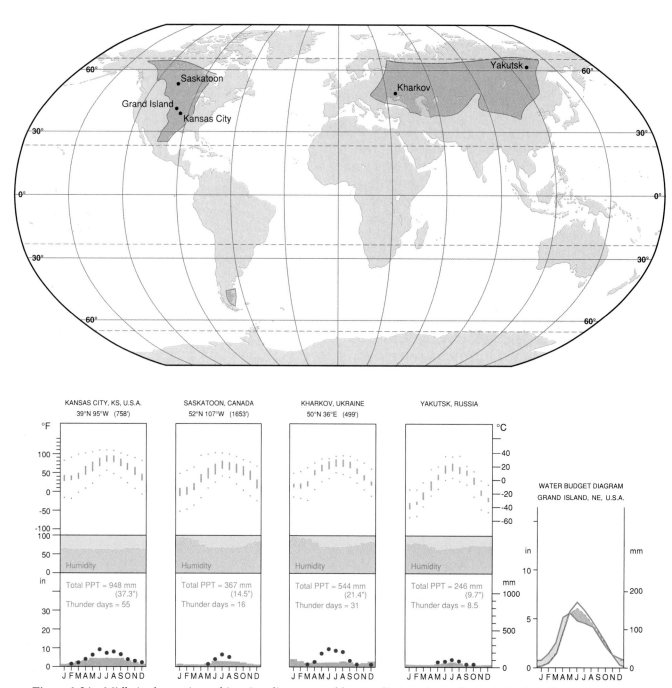

Figure 6-24   Midlatitude continental interior climates: world map, climographs, and a water budget diagram.

regions. Other climate types in continental areas are named *Df* or *Dw*. They all have cold winters with a coldest month less than −3° C (27° F) (*D*), but some have precipitation all year (*f*) and others have dry winters (*w*). Further letters (*a, b, c, d*) are added to indicate the summer temperatures (*a* is hot, *c* is cool) or extreme winter cold (*d*). Köppen's climate types reflect the range of natural vegetation from desert in the south and northward through grassland to forest. The dynamic atmospheric conditions, however, are similar and so this account includes them all in one climatic environment.

An extreme summer–winter temperature contrast is the major feature of midlatitude continental interior climatic environments. All four selected weather stations show this feature having extreme summer temperatures that rise to 40° C (104° F) and extreme winter temperatures that fall well below freezing for several months. Parts of Siberia have the coldest winter temperatures recorded on Earth outside Antarctica. Yakutsk has extreme temperatures ranging from 38° C (100° F) to −56° C (−70° F), and its annual range of mean monthly temperatures is over 56° C (100° F). The opening extract gives some idea of the everyday impacts of such low winter temperatures.

Temperature extremes in North America are not quite so great as those in Siberia. This is partly because of the difference in the size of the two continents and partly because the Mississippi valley makes it possible for tropical air to move into the heart of North America, while the Himalayas prevent such air from reaching central Asia. The four stations also show that a summer maximum of precipitation is general, apart from Kharkov, which is close to the Black Sea moisture source.

The summer–winter temperature contrasts in the interiors of Eurasia and North America are due partly to seasonal differences in insolation and partly to dynamic changes in the atmosphere. In summer, the high-angle sun and longer days result in heating of the lower atmosphere. Evaporation and transpiration are highest during this season, and large water bodies—such as the Great Lakes in North America or the Black Sea, Aral Sea, and Lake Baikal in Eurasia—supply significant quantities of water vapor to the atmosphere. Maximum precipitation occurs in summer, often from thunderstorms caused by local convection enhanced by uplift along cold fronts in the midlatitude cyclones crossing the region. Tornadoes often form in the central United States when there is a large contrast between *mT* and *cP* air masses.

Winter is a season of little effective insolation and excessive heat loss from the ground in midlatitude continental interiors. Average temperatures remain well below freezing for three to six months depending on latitude. Northern Canada and central Siberia become extensions of the polar climatic environment and source areas for *cP* air masses. At this season, Siberia is isolated from warming influences because the outflow of cold air pushes cyclones southward. Central North America experiences cyclones passing through from west to east, but the air in them dries out as it descends from the Rockies. Precipitation is low during this season and falls mainly as snow, which often stays on the ground for several weeks or months before melting. Relative humidity levels, however, may be high since it takes little water vapor for the air to reach saturation point at such low temperatures. The most extreme conditions, as described in the opening excerpt, make life extremely difficult.

The *windchill factor* lowers effective temperatures further (Figure 6-25). As wind speeds increase, the insulating layer of air around a person gets thinner and the rate of heat loss rises. At an air temperature of 6° C (43° F) a light breeze reduces the effective atmospheric temperature by a degree or two, but a strong wind reduces it to −10° C (14° F). The effect becomes greater as air temperatures fall further and is very noticeable in midlatitude continental interiors.

MIDLATITUDE EAST COAST CLIMATIC ENVIRONMENTS

> While yet it is cold January, and snow and ice are thick and solid, the prudent landlord comes from the village to get ice to cool his summer drink; impressively, even pathetically wise, to foresee the heat and thirst of July now in January,—wearing a thick coat and mittens! when so many things are not provided for. It may be that he lays up treasures in this world which will cool his summer drink in the next. He cuts and saws the solid pond, unroofs the house of fishes, and carts off their very element and air, held fast by chains and stakes like corded wood, through the favoring winter air, to wintry cellars, to underlie the summer there.
>
> Thoreau, H. 1854. *Walden, or Life in the Woods*. Boston: Tickner and Fields.

Midlatitude east coast climatic environments have warm-to-hot summers and cold winters as a result of a seasonal wind reversal akin to the tropical monsoon climates of southern Asia, but caused by different processes. Winds blowing outward from the icy continental interiors cause the cold winters, and oceanic

Wind (km/hour)	Thermometer reading (°C)						
	5.0	2.5	0	−2.5	−5.0	−7.5	−10.0
Calm	5.0	2.5	0	−2.5	−5.0	−7.5	−10.0
Light (15) breeze	4.0	−5.5	−7.0	−9.5	−12.0	−15.0	−18.0
VERY COLD							
30	−7.0	−10.0	−14.5	−18.0	−21.5	−25.0	−28.5
			DANGER			EXTREME	
45	−10.0	−14.0	−18.5	−21.5	−24.5	−27.5	−30.5
Strong winds						DANGER	
60	−13.0	−16.5	−20.0	−23.5	−27.0	−30.0	−33.5
Little further effect above 60 kmph							

Figure 6-25   Windchill equivalent temperatures, measured as the effect of wind on an unclothed body.

influences produce warm and rainy summers. This climatic environment affects the densely populated areas of eastern Asia (eastern China, Korea, Japan) eastern North America, and smaller areas in the southern hemisphere (Figure 6-26). Thoreau's ice cutters in the opening excerpt capitalized on the winter–summer contrast in New England.

Köppen designated such areas *Cf* and *Df*, signifying mild (*C*) or cold (*D*) winters and rain all year (*f*). This, however, places them in the same category as much of the west coast and continental interior midlatitude climatic environments and creates problems in using the Köppen system because it does not discriminate between very different climatic environments.

The selected weather stations in Figure 6-26 have similar annual temperature regimes. Summer extremes rise to over 40° C (104° F), and winter extremes go below freezing. These extremes are not as great as those in continental interiors but are greater than those on the midlatitude west coasts. In the eastern United States (Boston and Atlanta), precipitation occurs throughout the year without a noticeable seasonal maximum. Stations in other parts of the world show a summer rainfall peak. Atlanta is farther south and has a longer warm season than Boston. Its mean minimum temperature of over 16° C (61° F) lasts for six months, compared to three at Boston; Boston's winter temperatures are lower for longer. Tokyo has the most pronounced summer rainfall maximum. Brisbane in Australia also has a summer rainfall maximum, but the southern hemisphere stations do not experience the cold winters that those in the northern hemisphere do.

In winter the westerly midlatitude circulation brings a series of cold winds and *cP* air masses to eastern Asia. Temperatures remain low. Precipitation is also low except on the coast and offshore islands (such as Japan), where the *cP* air meets maritime air masses causing clouds and drizzle. The heaviest belt of winter precipitation occurs beneath the polar-front jetstream.

In spring more midlatitude cyclones cross Siberia bringing atmospheric disturbances and rain to much of China between March and June. By summer this westerly flow moves northward, and a southwesterly monsoon flow that begins in the Indian Ocean dominates southern and central parts of China. Thundery rain is common and typhoons may affect the coastal areas of southern China in late summer and fall.

Eastern North America is noted for its wide range of weather during the year. Although this climatic environment is no more changeable than that on midlatitude west coasts, the swings of temperature between summer and winter are much greater. A pattern of weather systems and seasonal changes similar to that in eastern Asia brings cold *cP* air masses from the northern interior in winter. Snowfalls are heavy on upland areas of New England and the Middle Atlantic states, especially just east of the Great Lakes.

Heat, humidity, and thundery rain characterize the eastern summers from north of the St. Lawrence River to Florida. The duration of such summer weather increases from three months in the north to six months in the south. Hurricanes affect the southern coasts in late summer and may divert northward producing heavy rainfalls over the Appalachian Mountains as they dissipate.

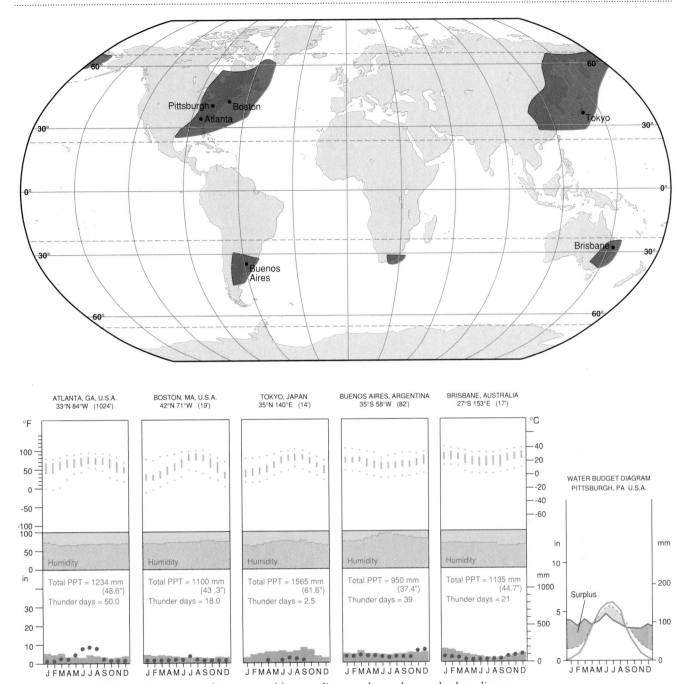

Figure 6-26   Midlatitude east coast climates: world map, climographs, and water budget diagram.

Eastern Australia lies in a belt of eastward-moving, high-pressure cells that bring dry *cT* air masses from the continent's interior. The dry air meets *mT* air on the east coast and the disturbances that result cause heavy rains, especially in the summer. Eastern Australia has lower seasonal temperature differences than the northern continents because the center of the continent is warmed beneath a subtropical high-pressure zone rather than receiving cold winds and a winter *cP* air mass.

POLAR CLIMATIC ENVIRONMENTS

Eight men watched the Theron sail away, and as they watched, with the ship less than a mile distant, the pack ice closed solidly. Already the temperature was falling below zero [-18° C] and their first need was to establish themselves securely and organise their daily existence.

As February advanced the temperatures continued to fall and work was usually started in the morning with from thirty to fifty degrees of frost. Occasionally the blizzards blew for a day, causing much waste of time while everything was dug out. To handle nuts and bolts it was necessary to work with bare hands, and frostbitten fingers became a problem.

Each night they retired to their tents, but comfortable sleep was now a rare thing, for the temperature ranged down to −45° F [−43° C] and the sleeping bags were becoming heavy with frozen condensation, cracking and creaking as the men crept into them.

On 20th April the sun set for the last time for four months, but work continued by the light of Tilley lamps. The rays of the sun, hidden by the horizon, would colour the clouds with red, orange and green that faded into the pale moonlit sky. . . .

May was the coldest month, when the temperature really began to fall. The average was −35° F [−37° C], which meant that on many days it was in the minus fifties and sixties. At the beginning of the month there was another blizzard lasting ten days, during which nothing but essential work could be done outside.

Sir V. Fuchs, Sir E. Hillary. 1958. *The Crossing of Antarctica*. London: Cassell.

The polar climatic environments include the areas that lie approximately within the Arctic and Antarctic circles (Figure 6-27). The climate is dominated by winter cold and darkness, as the opening extract emphasizes. Such environments cover the Antarctic continent and its surrounding waters, Greenland, the Arctic Ocean, and the northern coasts of North America and Eurasia.

Köppen places these areas in the *E* category of cold climates. They are divided into places with a warmest month averaging 0 to 10° C (32–50° F) (*T*—mainly tundra areas) and those where it falls below 0° C (32° F) (*F*—mainly ice sheet areas).

The selected weather records show that the lowest temperatures occur at the South Pole. All the stations, however, have very long winters of six months or more with extremely low temperatures. Relative humidity remains high at such low temperatures, but precipitation and absolute humidity are low everywhere. Much of this climatic environment is arid because of the low precipitation and low availability of water.

Dense, cold air remains in contact with the ground. The sun has little direct influence on temperatures, being at a low angle even in summer, when there are twenty-four hours of daylight at latitudes

greater than 66.5°. Water, snow, and ice surfaces reflect most summer insolation directly out to space, and radiation losses are high from the ground and atmosphere. In winter there is no insolation during the twenty-four hours of darkness in these latitudes, but heat losses by outward radiation continue. Whereas summer temperatures over melting snow may rise a little above 0° C (32° F), the coldest winter temperatures descend to −50° C (−58° F) over ice in the Arctic Ocean, and to −70° C −94° F) in the coldest parts of the high Antarctic plateau.

Polar regions make their own climatic environment. The cold surface air masses are so dense that they resist the intrusion of warmer and moister air. Precipitation is minimal because of the small amount of humidity in the atmosphere and the general sinking of air. The small quantities of precipitation come mainly from the occasional frontal lows that intrude during the north-south phases of the polar-front jet wave cycle. Snow that falls in the low-temperature environment seldom melts, but winds may blow or drift the dry grains that do not stick together. The snow in blizzards may be newly precipitated or merely older snow whipped up by winds. Polar areas with very low precipitation may have only a thin covering of snow rather than ice in winter, and summer melting exposes bare rock and soil.

HIGHLAND CLIMATIC ENVIRONMENTS

Highland climatic environments vary in scale with the size of the upland area under consideration. At one end of the scale are the huge masses of the Himalayas and adjacent Tibetan Plateau, the Andes Mountains of South America, the western ranges of North America, the Alps of Europe; at the other end are smaller areas such as Ethiopia or individual peaks. Mountains at all scales have effects related to altitude. Within mountain systems rapid changes in height and orientation to the sun produce a multitude of local climates, but the largest mountain masses affect the distribution of climate at the global scale.

Köppen identified highland climates but did not show them on his world map. The largest areas of highland climate, however, are large enough to show on such a map (Figure 6-28). Although the four

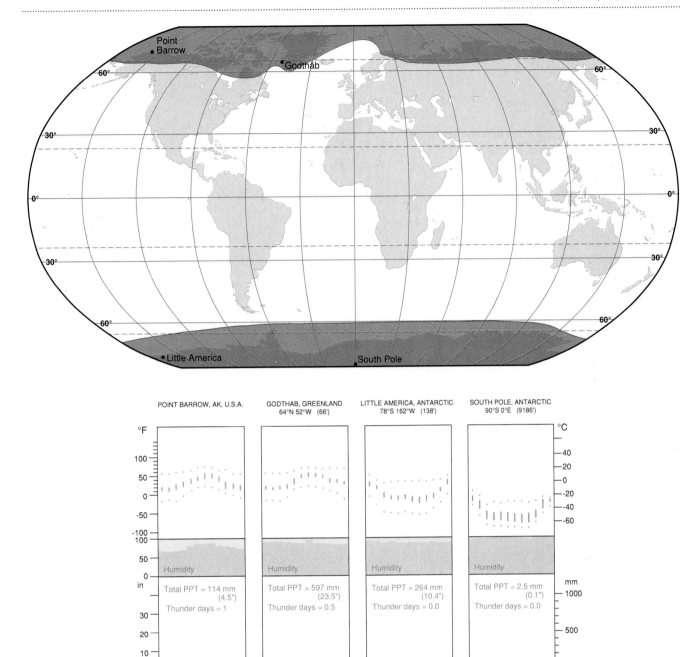

Figure 6-27 Polar climates: world map and climographs.

selected weather stations show that highland climatic environments relate to conditions in adjacent climatic environments, they also exhibit a number of common features. Thus, Dillon in the Rocky Mountains has a similar seasonal regime to the midlatitude continental interior, Zermatt in the Alps to the midlatitude west coast, Simla in the Himalayas to the monsoon, and El Alto in the Andes to the seasonal tropical climate. When compared with other stations in these categories, each of the highland climate stations has greater monthly temperature extremes and more thunderstorm activity than the surrounding areas.

Highland climatic environments have a number of distinctive features in common. The atmosphere at

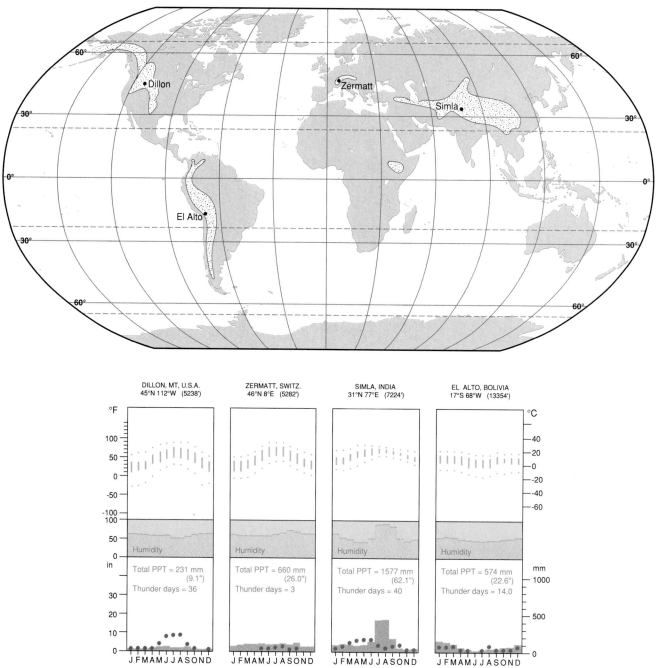

Figure 6-28   Highland climates: world map and climographs.

high altitudes is less dense than that near sea level and contains less moisture and dust. Temperature falls with height, whereas the diurnal temperature range increases (Figure 6-29). The lower density and lower relative humidity of the high-altitude air result in less daytime absorption of terrestrial radiation and rapid heat loss at night. Exposed slopes and valley winds intensify the heat loss. Winds are stronger on exposed mountains and the wind-chill factor is often high.

Seasonal temperature regimes in highland areas vary with latitude. Mountains near the equator have slight changes in monthly temperatures throughout the year. The summer–winter contrast increases toward midlatitudes and then narrows again in polar regions, which are cold all year.

Highland areas affect precipitation falling on them and adjacent climatic environments. When air masses encounter an upland area, they are forced to

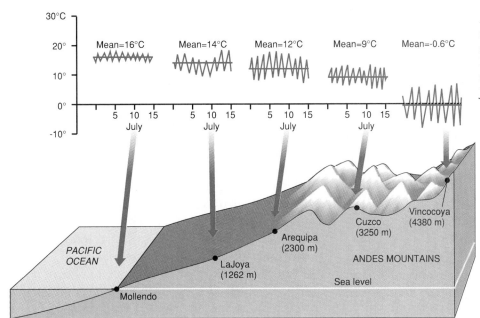

Figure 6-29 Highland climates: changing day-night temperature ranges with increasing altitude in Peru, South America. How do the fluctuations during the first half of July change at different altitudes?

rise. Cloud and precipitation increase with height up to about 3000 m (10,000 ft), as on the windward slopes of the northern Andes. When air rises further, most of its moisture has been precipitated and totals decrease. Tops of mountains, even on the equator in such countries as Ecuador and Kenya, are often so cold that water precipitates as snow and lies on the surface all year. The **snow line** is the altitude of the lower limit of permanent snow cover. When heavy snowfalls accumulate, the snow converts to ice and forms glaciers, as described in Chapter 10. On lee slopes descending air warms adiabatically, and arid conditions often occur in the rainshadow of a mountain range, as in Nevada, USA, and Patagonia, Argentina.

The alternation of high peaks and deep valleys in upland areas causes dramatically different climatic conditions to occur in adjacent areas. There are contrasts between valley floor, midslope, and mountaintop; between slopes facing north and those facing south; and between slopes facing into, or away from, dominant winds. Valley floors, although generally warmer than the upper slopes, are often the sites for the formation of cloud or fog after cold air from upper slopes sinks to the valley bottom (see Figure 3-24). It is common to look down from upper sunny slopes on the top of cloud or fog in the lower part of a valley.

The **aspect** of a slope—the direction it faces—also causes local differences. Slopes facing the sun receive more insolation at higher angles than slopes that face away. South-facing slopes in the Swiss Alps are often terraced for vineyards, pastures, or are village sites. North-facing slopes are cool and damp and left in forest; few people live there.

Local winds also feature in upland climates. There is a daily regime of upslope winds generated by daytime heating and downslope winds following cooling on the upper slopes at night. Major valleys through upland areas may also channel winds (see Chapter 3).

There are three groups of world-scale climatic environments, divided by the features of the tropical, midlatitude, and polar atmospheres. The tropical climatic environments are the equatorial, seasonal rainy, monsoon, trade wind, and arid. The midlatitude climatic environments are the west coast, interior, and east coast. Highland climates are another distinctive group. When climatic environments and their workings are better understood, it will be more feasible to model them mathematically to help predict future changes.

Climate has major linkages between the **atmosphere-ocean environment** and the other surface Earth environments. Different climatic environments activate distinctive processes of rock breakdown and transfer that produce landforms in the **surface-relief environment.** They also affect the formation of different types of soils and the distribution of plants and animals in the **living-organism environment.** When climate changes, it favors different organisms, soil types, and landforms. It is common to find the old and new together as a result of climate change.

CLIMATES OF THE PAST AND FUTURE

So far, this chapter has been concerned with present climates. But the climatic environments experienced in the late twentieth century are merely a "snapshot" and part of a long sequence of climate fluctuations over billions of years. Climate change is normal.

Climate change occurs when there is a shift from one set of climatic environmental conditions to another at a place. Such changes may occur over periods of tens, hundreds, thousands, or millions of years, but the changes of shorter length tend to be changes of smaller magnitude. For instance, the global temperature has risen by just over 0.5° C (0.9° F) since the mid-nineteenth century, but rises and falls of 5–7° C (9–12.6° F) occurred during the Ice Age fluctuations that lasted tens of thousands of years. Shorter periods of change are merely part of a pattern of fluctuations on the larger scale.

Climate change and the frequency of its fluctuations have implications for both the continuing human occupation of Earth and also the operation of natural processes in Earth environments (see Evironmental Issue: The Future—Global Warming or Cooling?, p. 165).

This section on climate change begins by looking at the historic impacts of past changes and the different types of evidence used in reconstructing the nature of the changes. It then considers the causes of climate change and finishes by assessing the potential for predicting future changes.

CLIMATE CHANGE AND HISTORY

Historians often overlook climate change and its impact on human activities. Few refer to it as a causal factor in human events or culture either because they are unaware of the workings of the natural environment or because the relationship between the climate factor and human activities has sometimes been oversimplified. Growing knowledge of environmental changes suggests a number of links that historians could examine. Studies of past impacts of climate change may assist in planning human responses to contemporary climate change.

Links have been established between increasing aridity of the deserts in Mesopotamia—approximately where modern Iraq is—and northern Africa in the Atlantic phase 5000 years ago and the concentration of people in the Nile and Tigris-Euphrates valleys. During this period, Jacob and Joseph migrated, and the Israelites were enslaved in Egypt. Later, Roman expansion into northern Europe (2000–1500 years ago) occurred during another phase of warmer climate, as did the Viking colonization of Iceland and Greenland (800–1100 years ago). Whereas warmer conditions encouraged the expansion of farming and settlement to higher altitudes, cooler conditions led to retreat of farms from upland areas. Some major periods of political upheaval, such as the Dark Ages in Europe (1100–1500 years ago), also coincided with cooler and stormier climatic conditions.

A detailed study of settlements on Dartmoor in southwest England show that warmer conditions allowed early Bronze Age (4000 years ago) farmers to build villages over 300 m (1000 ft) above sea level—some 100 m above the present limits of settlement. During the Iron Age, beginning about 3400 years ago, the climate was cooler and wetter, and people abandoned the Bronze Age settlements for forts at lower altitudes. Some of the Bronze Age villages were resettled in the warmer phase between 800 and 1100 years ago before being finally abandoned. In the United States there is evidence that abandonment of cliff dwellings and other types of Native American settlements in the southwestern United States occurred during the same period of general warmth.

The greatest impact of climate change is at the margins of habitable areas, particularly when it coincides with other changes in human society that impose pressures on the land. Farming of the hilly areas in the Italian peninsula during warmer Roman and Medieval times led to massive soil erosion. The volume of eroded material was so great that it formed river mouth deltas extending out to sea. Such catastrophic environmental change was probably the result of a combination of overpopulation, natural climate change (to heavier rains and cooler conditions), sandy soils, and steep slopes.

RECONSTRUCTING PAST CLIMATES

Reconstructing past distributions of climatic environments is like assembling a jigsaw puzzle. Climate change affects many different sectors of Earth environments that reflect those changes differently. The person who tries to work out what climates were like in the past might use any, or all, of meteorologic records, historic evidence from farm records or diaries, archaeologic evidence, fossil pollens, and fossils of other plants and animals.

Much of the evidence used to reconstruct past climates is specific to a particular time period. Figure 6-30 shows how this works in a set of expanding time scales of climate change. Changes over the last 100 years or so are reconstructed mainly from meteorologic measurements (Figure 6-30a) and over the last 1000 years from historic records and natural phenomena like tree rings. Changes back to the end of the last glacial phase some 15,000 years ago are reconstructed from archaeologic data, fossil pollens, tree rings, and annual layers in ice sheets (Figure 6-30b). Changes over the last million years are reconstructed from studies of surface landforms and sediments and ocean floor deposits (Figure 6-30c). Older changes are reconstructed from geologic evidence in rocks and fossils (Figure 6-30d).

The Period of Meteorologic Records
This period is short when compared with major climate changes. Thermometers measure atmospheric temperature and are the oldest instruments used to record weather phenomena. The record of temperature at enough locations around the world to provide a picture of climatic change, however, only goes back to around 1860, and most weather stations are located in urban areas, which also gives a bias to their use. Figure 6-31 shows that the global temperature has risen by just over 0.5° C (0.9° F) in that period—a general warming trend with fluctuations every ten to fifteen years. This trend is known as **global warming.** Uncertainty exists, however, because some countries, including the United States, have experienced a slight cooling over the same period. Further research is being designed to find reasons for the warming process and for the disparities among records.

Precipitation totals have also been recorded widely for over 100 years. Like temperature, patterns of precipitation also fluctuate over that time. A period of lower precipitation between 1925 and 1940 in the United States (the main phase of the Dust Bowl wind erosion in Oklahoma) succeed higher totals at the start of the twentieth century. Higher precipitation in the 1940s again gave way to drier conditions in the mid-1950s and mid-1960s. Wetter conditions have existed since.

Climate Changes Since the Last Glacial Phase
Ice ages occur on Earth at intervals of around 250 million years. An **ice age** is a period of a few million years when ice sheets form over polar regions and spread to cover extensive areas in the higher latitudes and even parts of the midlatitudes. Between ice ages there is often no polar ice on Earth for approximately a hundred million years. At present, Earth is in an ice age that began some 20 million years ago when ice started to cover Antarctica. The ice reached its maximum coverage during the last 2 million years.

Within each ice age, the climate fluctuates in colder *glacial phases,* lasting up to 100,000 years, when the ice sheets grow and warmer *interglacial phases,* lasting about 20,000 years, when they shrink. The current distribution of climates is typical of the later stages of an interglacial phase. If the present interglacial phase follows the pattern of the last one, the warmest periods are past and climates will now cool gradually, albeit with many fluctuations, toward the next glacial phase. If this is so, the cooling trend may balance or outweigh any global warming effect.

No meteorological instrument record exists for reconstructing the evidence about the beginning of the present interglacial phase 15,000 years ago. The study of fossil plant pollens trapped in lake floor sediments and peat bogs around the world is the main basis for constructing the sequence of climate change since the last glacial phase. Plants are strong indicators of climate since temperature and water availability primarily determine where they thrive. Pollens for each type of tree, shrub, and grass are microscopic but distinctive. Winds often disperse pollens, but precipitation brings them to the ground and rivers carry them into lakes and seas, where they form part of the sediment layers (Figure 6-32). Ancient pollens collected from such sediments are analyzed in the laboratory and placed in age sequence.

The evidence from pollen records indicates a period of global warming that continued from the end

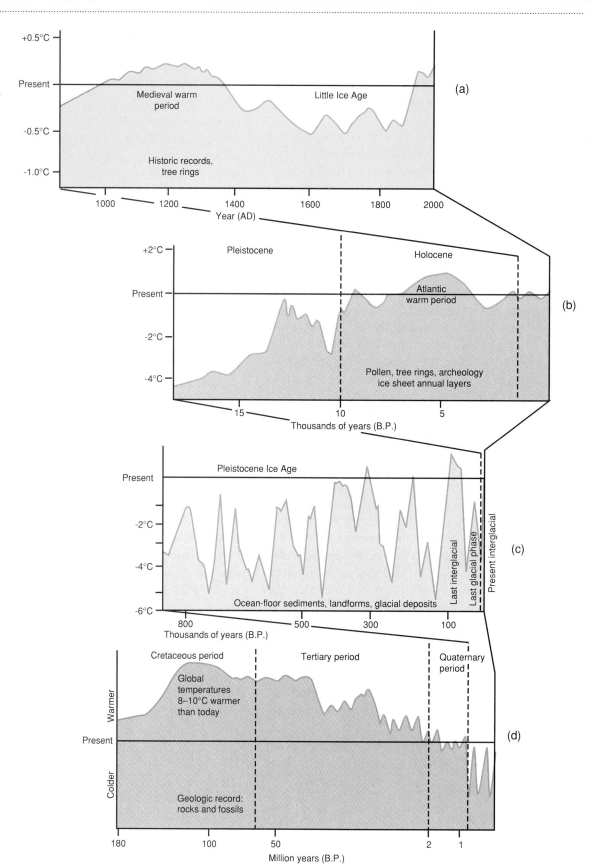

Figure 6-30
Time scale and types of evidence of climate changes. Climatic variability and change work at different scales.

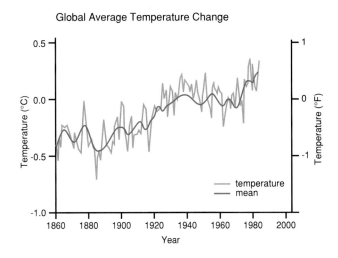

Global Average Temperature Change

Figure 6-31   Global temperature change since 1861. The annual values are shown with a smoothed graph to indicate trends. Note the size of the changes and the fluctuations within the period.
Source: NOAA.

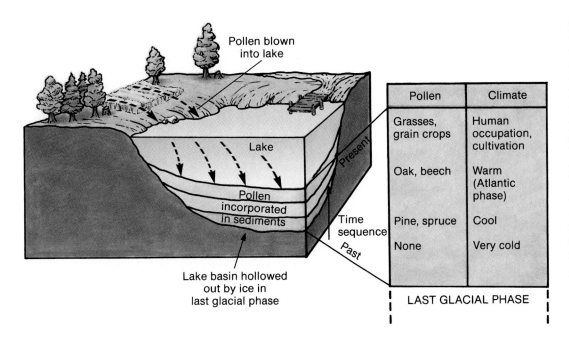

Pollen	Climate
Grasses, grain crops	Human occupation, cultivation
Oak, beech	Warm (Atlantic phase)
Pine, spruce	Cool
None	Very cold

LAST GLACIAL PHASE

Figure 6-32
Reconstructing past climates from pollen deposited in a lake. In this example, a hollow eroded by ice in the last glacial phase provides a starting point. The bottom sediments include pollens that blew into the lake just after the ice melted and vegetation returned. Sediments forming at present include pollens of grasses, trees, and crops.

of the last glacial phase to the warmest point, the *Atlantic stage*. Other evidence makes it possible to date the Atlantic stage as occurring about 5000 years ago. After the Atlantic stage, climate fluctuated. Some periods were warmer, as in medieval times from about 900 to 1300 A.D., but none reached the warmth of the Atlantic stage. Other periods were colder, as in the Little Ice Age from 1400 to 1800 A.D.

Further evidence from a variety of sources supports the record provided by pollen records. Studies of tree rings, glacier fluctuations, archaeologic evidence, and written records from chronicles to farm logs and tax documents, confirm the pattern of change. Such studies also chart the increasing impacts of human activities on changing natural environments.

The global warming immediately at the end of the last glacial phase caused a major rise in sea level as the ice sheets melted and returned huge quantities of water to the oceans. For instance, the rise in sea level along the coast of eastern North America was over 50 m (150 ft). It drowned much of the previously exposed edge of the continent, including the lower parts of river valleys that now form the wide estuaries of Chesapeake and Delaware Bays

(Figure 6-33). This rise ended about 6000 years ago and subsequent fluctuations, as in the Little Ice Age, produced sea-level changes of less than a meter.

Climate Change During the Pleistocene Ice Age

The climates of the Pleistocene Ice Age fluctuated in alternating glacial and interglacial phases over the last 2 million years. Climatologists reconstruct these events from many types of evidence, such as ocean floor deposits, soil studies, and cores drilled through Greenland and Antarctic ice. These sources all utilize records of the environmental impacts of climate change.

The longest sequence of climate changes has been reconstructed from studies of ocean floor sediments and the tiny fossils of surface creatures they contain. The materials used by these creatures in the construction of their shells depend on the temperature of the surface waters at the time when they lived, and the record is virtually continuous through several million years. The results show that the last half million years have had regular fluctuations of glacial and interglacial conditions. During glacial phases ice sheets took up to 100,000 years to build up over North America and northern Europe. At the end of each glacial phase, there was a relatively rapid phase of melting to an interglacial phase lasting about 20,000 years. The Vostok ice core from Antarctica provides records confirming the temperature changes through the last glacial phase and previous interglacial phase (Figure 6-34).

The Pleistocene Ice Age affected the whole globe. Where ice covered the land, its melting deposited rock and soil scraped from other areas. Outside the area covered by ice, frozen soils produced in conditions of extremely low temperature still underlie much of Siberia and northern Canada. Closer to the equator, all the climate belts narrowed, causing changes in vegetation and landform processes. The arid climates extended closer to the equator and the sand dunes formed then are now being uncovered and reactivated during dry periods by the loss of vegetation in the tropical seasonal climatic environment of today. Parts of present arid areas became wetter at various phases within the Pleistocene Ice Age, as is shown by the former extent of the Great Salt Lake of Utah (Figure 6-35).

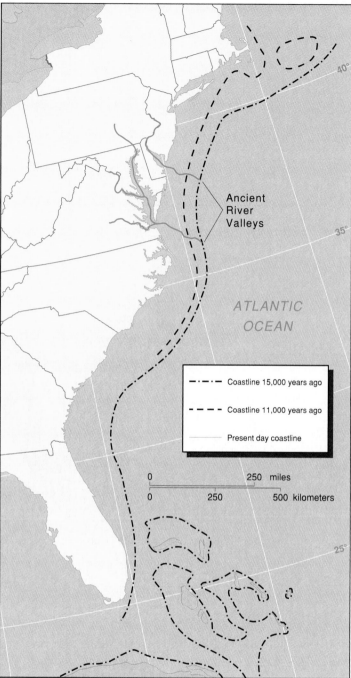

Figure 6-33 Climate change and changing sea level. The shoreline of eastern North America was farther out when sea levels were lower before the end of the last glacial phase some 10,000 years ago. Oceans reached their present level about 6000 years ago.

The alternate growth and melting of the ice sheets resulted in alternate falling and rising sea levels. These events affected the form of present coastlines and allowed plants and animals (including early humans) to migrate between continents. The first humans to settle North America came from Siberia during the last glacial phase when sea level was lower.

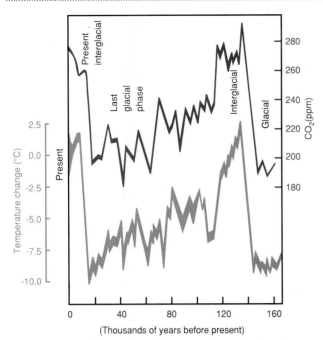

Figure 6-34   Climate change. Temperatures (red) and carbon dioxide concentrations (blue) as measured in a core drilled through the Antarctic ice at Vostok. What were the temperature trends over the last 160,000 years? How did carbon dioxide vary in comparison with the temperatures?

Climate Changes Before the Pleistocene   The period before 2 million years ago experienced even greater climate changes. The evidence for reconstructing these changes comes from rocks and fossils. Earlier ice ages occurred between 250 to 300 million years ago and over 600 million years ago. At other times Earth was free of ice.

The history of living organisms on Earth suggests that conditions were hostile to them at the start and that it took many hundred millions of years to raise the oxygen and ozone levels to the point where organisms could exist safely on the continents (see Figure 2-9). Many climate changes must have occurred during these millions of years, but evidence is too sparse to reconstruct them.

CAUSES OF CLIMATE CHANGE

Climate change occurs because of changes in factors that control the heating of Earth's atmosphere-ocean environment. These factors include the incoming solar radiation, the arrangement of continents and mountain ranges, and interactions of insolation with the surface and atmosphere, and some human activities. All vary in their own way and, therefore, the

Figure 6-35   Bonneville Salt Flats, Utah. The area was covered by the expanded Great Salt Lake during a period of wetter climate several thousand years ago. Evaporation of water left behind the salt.

causes of climate change are complex. Changes in one factor may be balanced by changes in another so that there is little effect on climate, or several factors may work together to enhance the amount and speed of change. Figure 6-36 summarizes the main factors that produce climate change.

Solar Radiation and Its Reception   Solar output of radiation is fairly constant, but variations in the distance from the sun to Earth cause variations in the receipt of insolation. Sunspot cycles lasting eleven to fourteen years produce small changes in solar output, and this has been linked to the short-term fluctuations in temperature or precipitation that are characteristic of many locations. Evidence for such impacts on Earth climates is difficult to interpret, but there are indications of greater numbers of sunspots during the period of medieval warmth and of hardly any sunspots during the Little Ice Age.

Regular fluctuations in the distance that solar radiation travels on its way from sun to Earth are of greater significance than changes in the solar energy output. These fluctuations have three sources.

Earth's orbit around the sun varies from a more circular to a more elliptical shape in a cycle of about 95,000 years. This affects the amount of solar radiation intercepted by locations on Earth.

Earth's axial tilt is presently 23.5 degrees from a line perpendicular to the plane of Earth's orbit. Over time the angle of tilt varies between 21.8 degrees and 24.4 degrees in a cycle of about 42,000 years. The amount of tilt modifies the summer–winter heating contrast.

At present, Earth is closest to the sun (*perihelion*) on January 3, during the southern hemisphere summer. This has little effect on overall Earth temperatures. In about 10,500 years, however, Earth will have its perihelion in June during the northern hemisphere summer. The increased insolation will be magnified because of the greater extent of land in the northern hemisphere.

The combined effect of these three cycles has been linked to the timing of major fluctuations of climate

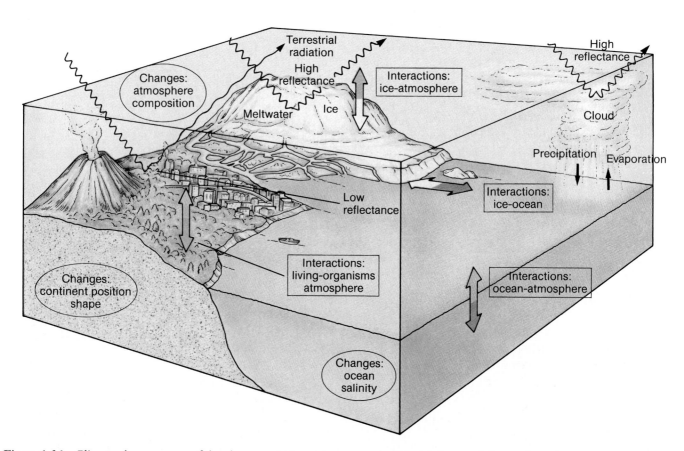

Figure 6-36   Climate change: some of the changes within environments (ovals) and the causal factors (boxes).

during the Pleistocene Ice Age and since. Diminished seasonal heating contrasts caused by a small axial tilt and a perihelion in the southern summer produce glacial conditions. On this basis the next glacial phase may begin in 3000 to 4000 years.

INSOLATION AND EARTH'S ATMOSPHERE

As insolation reaches Earth and passes through the atmosphere, the surface albedo affects the amount of reflected insolation. In particular, cloud cover and volcanic dust following major eruptions increase the albedo. Absorption of radiation in the atmosphere depends on the quantities of greenhouse gases such as water vapor, carbon dioxide, methane, and CFCs. These absorb a little of the incoming insolation and virtually all of the outgoing terrestrial radiation. Increases in the quantities of these gases are associated with rising temperatures in the lower atmosphere; decreases are associated with falling temperatures. The Vostok ice core (see Figure 6-34) demonstrates this relationship. It is not certain, however, whether rising carbon dioxide levels cause temperature increases or vice versa. What is clear is that carbon dioxide is part of the complex processes involved in the cycling of carbon.

Earth Environment Variations   Internal rearrangements within Earth's environments also cause climate change. The arrangements of continents influence the circulation of the atmosphere and ocean waters and the heat they transfer to the midlatitude atmosphere. Changes in the positions of continents and ocean basins and the formation of mountain systems occur over hundreds of millions of years and affect climate on the longest time scale.

Feedback in the Atmosphere-Ocean Environment Although the solar insolation factor and the positions of continents on Earth account for much of the recorded climate change, interactions within the atmosphere-ocean environment affect the nature and extent of such changes. A set of feedback mechanisms operate to amplify or damp down the initial effects. Such feedbacks introduce further complexity to understanding climate change and many are poorly understood.

One example of these feedbacks involves water vapor. The amount in the atmosphere determines cloudiness and precipitation but depends on the atmospheric temperature. A warming atmosphere may hold more water vapor, but water vapor is also a greenhouse gas and more of it in the lower atmosphere helps to trap more heat. This is an example of *positive feedback* in which heating adds more water vapor to the atmosphere and this in turn leads to more heating.

Snow and ice cover provide another example of positive feedback. The increase of snow and ice cover during climate cooling raises the albedo and so reduces the insolation absorbed by Earth's surface. A small decrease in solar radiation that leads to the growth of an ice sheet also results in the lowering of global temperatures. For instance, a 1 percent reduction in insolation at Earth's surface would lower average temperatures around the world by 1.5° C (3° F) if the ice sheets did not grow, but by 5° C (10° F) if the ice sheet area grew by 1 percent. The extension of sea-ice cover, which restricts the exchange of heat and moisture between ocean and atmosphere, is regarded as the main cause of enhanced climate cooling that led to glacial advances in polar regions. The reverse effect, when snow and ice-cover melt, produces a decreasing albedo as rock and soil are exposed. Ocean waters release heat and moisture into the atmosphere as sea ice retreats. Melting of ice thus promotes more melting and results in the rapid end of glacial phases.

The role of clouds in feedback is known to be very important, but the mechanisms are poorly understood. Clouds absorb terrestrial radiation and prevent its escaping by radiation to space. Increasing cloud cover might be expected to increase heating of the lower atmosphere. Clouds also reflect insolation, however, and increasing cloud cover may have a cooling effect.

The oceans also provide feedbacks in the system. They absorb and emit heat and carbon dioxide, exchanging both with the atmosphere. The atmosphere contains only a tiny fraction of the carbon held in the oceans. As carbon dioxide increases in the atmosphere, marine organisms that are particularly active in colder regions absorb it and transfer the carbon to ocean-floor sediments. As the oceans warm, however, these organisms are less active and the atmosphere loses less carbon dioxide. This important greenhouse gas remains and absorbs more heat.

Global warming might also affect the deep ocean circulation that is stimulated by cold currents of water originating off southern Greenland and the coasts of Antarctica. Warming of these source areas would change the pattern of deep current flows. If greater upwelling of cold water occurred off western South America, this might cool the southern hemisphere atmosphere and act as a negative feedback mechanism.

PREDICTING CLIMATE CHANGE

An understanding of possible future climates includes the type of changes, the size of the changes, the timing of the impacts, and the role of human activities in causing the changes. The fundamental causes of climate change are beyond human control, but it may be possible to alter human activities that enhance the bad effects of climate change.

Predicting climate change is regarded as crucial for the future of human existence on an increasingly crowded Earth. The complexity of the causes of climate change render precise predictions over long periods impossible at present despite great efforts. Scientists attempt to evaluate the effect of changing the value of one climatic element, such as temperature, on others, consult past sequences of atmospheric conditions, or use general circulation models on giant computers. Each of these methods has some virtue, but none provides information that is sufficiently clear or accurate to cause governments and large corporations to markedly change their habits.

One attempt to assess the impact of global warming was published by the U. S. Environmental Protection Agency in 1989. The authors combined the results of three GCMs with historic weather data to produce possible scenarios of the climatic and economic consequences of doubling the atmospheric carbon dioxide. The estimated rise in temperature predicted by the three general circulation models varied between 3° C and 5.1° C (5.4 to 9.2° F). The middle value of 4.3° C (7.7° F) formed the basis of conclusions about impacts on natural ecosystems, sea level (to rise between 0.5 and 2 m by 2100 A.D.), farming practices, and water resources. This example illustrates the uncertainties still left by attempts to predict future climates.

Climate changes over short periods of a few years and very long periods of millions of years. Reconstructing past climates uses a variety of evidence. Since meteorologic instrument records cover little more than 150 years, scientists use written historic, archaeologic, botanic, and geologic evidence, as well. These show that Earth is in the grip of an ice age and at present is in one of the warmer interglacial phases. Before the present ice age, there were periods of warmth without any ice on Earth and at least two other ice ages.

Climate changes result from variations in the receipt of solar heating, the composition of the atmosphere-ocean environment, Earth's surface relief, and a set of feedback mechanisms within the atmosphere-ocean environment. The prediction of future climate is still at an early stage of development since the complex processes involved in climate change are not fully understood.

The causes of climate change involve many interactions and feedbacks among Earth environments. There are feedbacks within the **atmosphere-ocean environment,** such as ocean heat transfers and the effect of cloud cover; there are interactions with the **solid-earth environment** as a result of volcanic activity and the movements of continents; and there are linkages with the **living-organism environment** as a result of the exchanges of carbon with plankton and other plants.

ENVIRONMENTAL ISSUE:

The Future—Global Warming or Cooling?

Issues involved include:

What is the evidence for global warming or global cooling?

To what extent are human activities contributing to climate change?

Will Earth's climate change so much that the conditions for supporting rapidly expanding human populations cannot be sustained?

In the 1980s people became aware of the phenomenon of global warming—the impacts of rising temperatures around the world. It was said that the rising temperatures would not only alter climates and affect farming, transport, and other economic activities, but would melt the ice sheets and cause the ocean level to rise. Coastal flooding would bring disaster to the many major port cities.

The human contributions to the rising level of greenhouse gas additions to the atmosphere was assumed to be the major cause of global warming. The mood of panic affected governments, which began to assess the needs for research and possible action. The evidence for the possibility of global warming includes the rising level of world temperatures since the mid-nineteenth century and the parallel rise in carbon dioxide—the main greenhouse gas—in the atmosphere. This is a straightforward link, and extensions of the trend into the future have resulted in bold statements about the inevitability of global warming and its effects.

Within the scientific community, however, there is controversy over how much global temperatures will increase and even about whether they will increase at all. Even if they do, there is also controversy concerning the impacts such as the degree of ice melting that will occur.

The temperature rise around the globe since 1850 is very small, about 1° C (1.8° F), and is not noticed by the ordinary person since the ups and downs of temperature over a few days are greater. Furthermore, the rise has not been the same everywhere. In North America, for instance, there has been a very slight fall in average temperature over the same period. The general rise since 1850 should also be seen in the context of changing world temperatures over longer periods since it followed 400 years of colder-than-average temperatures in the Little Ice Age; temperatures before that were higher than current levels.

The impact of rising temperatures on ice sheets is not straightforward. Most of the ice is on Antarctica, and that land mass dominates its own climate so that major changes in temperature would be necessary to break into the cooling system around the South Pole. Rising temperatures around the world would also cause increased evaporation, condensation, and precipitation. The balance between melting ice and additional snow falling on ice fields would be complex.

Other linkages in the natural environment also affect global warming. The carbon dioxide content of the atmosphere is the outcome of many processes that cycle carbon through land, sea, and air. Carbon combines with oxygen in several chemical reactions, including respiration by animals, to form carbon dioxide in the atmosphere. Volcanic eruptions, industrial processes, cars, and airplanes add carbon gases to the atmosphere. At the same time, plants remove carbon dioxide from the atmosphere and turn it into plant matter, both on land and at sea. The tiny plants at the ocean surface use the carbon to make their shells of calcium carbonate. Those shells sink to the ocean floor when the plants die and stay there for millions of years. Ocean waters contain fifty times as much carbon as the atmosphere and twenty times as much as there is in land plants. The oceans thus control the balance of carbon and may be removing much of the carbon added to the atmosphere by human activities.

In addition to the uncertainties about global warming and its impacts arising from the complexities of and interactions within the natural environment, some scientists even suggest that Earth's trend is toward cooler conditions. The present interglacial phase has lasted over 12,000 years, and the previous one lasted 20,000 years. The warmest period within the present interglacial phase occurred 5000 years ago; since then there have been increasingly cold periods between the warmer ones. It is possible that Earth's climates are going to become cooler during the next few thousand years with warmer centuries punctuating the overall temperature decline.

A Norwegian power plant adds carbon gases to the atmosphere.

Linkages

The controversy over future global temperatures illustrates many linkages within Earth environments. The **solid-earth environment** erupts carbon in volcanoes and is the source of fossil fuels burned in furnaces and cars. The **living-organism environment** exchanges oxygen and carbon dioxide with the atmosphere. The **atmosphere-ocean environment** involves constant exchanges between air and sea.

Questions for Debate

1. Should governments restrict human activities that add carbon gases to the atmosphere?

2. What would be the impacts of warmer or colder conditions on food production or other economic activities?

CHAPTER SUMMARY

1. The study of climate is important to other aspects of physical geography and to the planning and management of a wide range of human activities. A climatic environment is defined by local conditions of heat balance, water budget, surface conditions, and the weather systems that affect an area. Climatic environments occur at a variety of geographic scales from local to global.

2. There are three groups of world-scale climatic environments divided by the features of the tropical, midlatitude, and polar atmospheres. The tropical climatic environments are the equatorial, seasonal rainy, monsoon, trade wind, and arid. The midlatitude climatic environments are the west coast, interior, and east coast. Highland climates are another distinctive group. When climatic environments and their workings are better understood, it will be more feasible to model them mathematically to help predict future changes.

   Climate has major linkages between the *atmosphere-ocean environment* and the other surface Earth environments. Different climatic environments activate distinctive processes of rock breakdown and transfer that produce landforms in the *surface-relief environment.*

   They also affect the formation of different types of soils and the distribution of plants and animals in the *living-organism environment.* When climate changes, it favors different organisms, soil types, and landforms. It is common to find the old and new together as a result of climate change.

3. Climate changes over short periods of a few years and very long periods of millions of years. Reconstructing past climates uses a variety of evidence. Since meteorologic instrument records cover little more than 150 years, scientists use written historic, archaeologic, botanic, and geologic evidence, as well. These show that Earth is in the grip of an ice age and at present is in one of the warmer interglacial phases. Before the present ice age, there were periods of warmth without any ice on Earth and at least two other ice ages.

4. Climate changes result from variations in the receipt of solar heating, the composition of the atmosphere-ocean environment, Earth's surface relief, and a set of feedback mechanisms within the atmosphere-ocean environment. The prediction of future climate is still at an early stage of development since the complex processes involved in climate change are not fully understood.

   The causes of climate change involve many interactions and feedbacks among Earth environments. There are feedbacks within the *atmosphere-ocean environment* such as ocean heat transfers and the effect of cloud cover; there are interactions with the *solid-earth environment* as a result of volcanic activity and the movements of continents; and there are linkages with the *living-organism environment* as a result of the exchanges of carbon with plankton and other plants.

KEY TERMS

climatic environment   *124*

Köppen system of climate classification   *127*

general circulation model   *128*

climograph   *130*

Sahel   *135*

El Niño current   *142*

equable climate   *146*

snow line   *155*

aspect   *155*

climate change   *156*

global warming   *157*

ice age   *157*

CHAPTER REVIEW QUESTIONS

1. List ways in which the use of average climate data often masks important features of a climatic environment.

2. How does an urban area enhance the greenhouse effect?

3. Draw up a table of global climatic environments and the weather systems that are closely associated with each.

4. Referring to the climographs in this chapter, show how coastal or upland locations may have distinctive climatic conditions.

5. Describe the different types of evidence used in reconstructing past climates.

6. What are the differences between weather forecasting and the prediction of future climate?

7. In what ways does a study of climate change emphasize the importance of linkages among Earth environments?

8. Assess the potential human activities have to cause climate change.

SOME ACTIVITIES

1. Prepare a climograph for your home town.

2. Rewrite some of the excerpts at the beginning of climatic environment descriptions giving a technical description of the climate.

3. Locate and analyze temperature, precipitation, or drought-index data for your own area back to 1895 or earlier. Draw graphs of the data and note periods of drier, wetter, hotter, or colder conditions.

Earthquake
Volcano

CHAPTER 7

CONTINENTS, OCEAN BASINS, AND MOUNTAINS

THIS CHAPTER IS ABOUT:

- ◆ Earth's surface features
- ◆ Earth's layers and rocks
- ◆ Dynamic earth
- ◆ Continents, ocean basins, and mountain systems
- ◆ Supercontinents
- ◆ Environmental Issue: Earthquake Hazards—Can Humans Cope?

Earthquakes and volcanoes like Mount St. Helens (inset) show that sources of energy inside Earth influence events at the surface. The world distribution of volcanoes and earthquakes, illustrated on the map, is concentrated along narrow zones where large sections of surface rocks collide or pull apart. The lava welling to the surface on Hawaii is liquid and the bright colors show it is at temperatures of nearly 1000°C.

EARTH'S SURFACE FEATURES

Alongside the study of weather and climate in different parts of the world, the physical geographer's attention focuses on the features of Earth's surface. Distinctive patterns of mountains, plains, continents, and oceans give a unique character to each region of the world. In combination with climatic environments, surface features influence the main areas of human settlement and lines of transportation routes. At the local scale, the angle of slope on a hillside may limit the use of the land as a site for farming, housing, or other human activity.

Earth's surface features are often described by height, or **relief.** Formally, the term relief describes the difference in elevation between two points on Earth's surface, such as a hilltop and valley floor. The height difference is related to a reference level, usually sea level. More generally, the term relief describes the features of Earth's surface—the **landforms**—that create such differences in elevation. The relief of an area comprises landforms of varied height, shape, and size from complex mountain systems to single streams (Table 7-1).

The most significant division of Earth's surface relief is between **continents** and **ocean basins.** The continental areas above sea level cover only 29 percent of Earth's surface. Ocean waters cover 71 percent. Continents have high mountain ranges, broad plateaus, and low plains; ocean basins contain submarine ridges, extensive ocean floor plains, thousands of submerged peaks, and deep trenches. The highest mountain, Everest in the Himalayas, is 8848 m (29,030 ft) above sea level. The deepest point in the oceans is in the Mariana Trench in the western Pacific, 11,022 m (36,163 ft) below sea

level. The difference between the highest and deepest points is 19,870 m (65,193 ft, or 12.3 mi), only 0.15 percent of Earth's diameter. Earth's relief is merely a slight "roughness" on the planet's surface.

Earth's landforms result from interactions between two very different Earth environments—the atmosphere-ocean environment and the solid-earth environment—as shown on Figure 7-1. The *solid-earth environment* consists of the rocky surface and interior of the planet. The nature and world position of most major surface features, such as mountain systems and ocean basins, result from movements powered by heat energy deep in Earth's interior. The mountain systems and the rocks that form them meet the water and gases of the *atmosphere-ocean environment* in a shallow zone at Earth's surface. This shallow zone forms a third major environment, the *surface-relief environment.*

Chapters 7 through 11 provide a study of Earth's surface features in two parts. Chapter 7 focuses on the solid-earth environment, examining the layered structure of Earth's interior, the materials that compose the surface rocks, and the processes to which they are subject. These processes affect surface features of the largest size, moving continents, building mountains, and recycling rock materials in huge quantities. Earthquakes and volcanic eruptions provide short-lived indications of this internal Earth activity. The chapter concludes with an account of how, over millions of years, the processes at work in the solid-earth environment give rise to changing patterns of major relief features on the continents and ocean floors.

Chapters 8 through 11 continue the study of Earth's surface features, where consideration is given to the processes within the atmosphere-ocean

TABLE 7-1   LANDFORMS AND PROCESSES

	Order	Size (sq km)	Example	Dominant Processes
Major Landforms	First	Billions	Continent, ocean basin	Internal processes
	Second	Millions to hundred thousands	Mountain system (Andes, Rockies, Appalachians), major plateau (Colorado), Mississippi lowlands, ocean trench	Internal processes
	Third	Ten thousands	Adirondacks, Blue Ridge Mountains, Mississippi delta	Internal processes plus surface processes
Smaller Landforms	Fourth	Hundreds to tens	Individual peaks: e.g., Mount St. Helens; valley; estuary	Mainly surface processes
	Fifth	Fractions (square meters)	Beach ripple, gully	Surface processes dominant

Figure 7-1   Steep cliffs on the southernmost point of Ireland, Mizen Head. The sea cut into rocks that were part of mountains raised by Earth's internal forces. The surface detail was carved by atmospheric and oceanic processes.

environment that wear down the hills and mountains produced by the solid-earth environment. Changes of temperature, chemical reactions, and the action of plants and animals break solid rocks into small particles, which are moved downslope by gravity (Chapter 8), and are then moved toward the ocean by streams (Chapter 9), glaciers, and wind (Chapter 10). Along coasts winds ruffle the sea surface causing it to pound cliffs and form beaches (Chapter 11).

EARTH'S LAYERS AND ROCKS

To understand the solid-earth environment, it is necessary to know about its structure and composition. Earth is a layered planet formed of a core, the mantle, the crust, and finally the atmosphere (Figure 7-2). A series of "building blocks" at different scales make up Earth's rocky layers. Taking the largest view, there is a single building block, the planet. Within the planet there are groups of building blocks ranging in size from the major surface-relief features and the internal structures beneath them, through the constituent rocks and minerals, to the atoms that compose the minerals. At each scale these materials play important roles in forming and wearing down surface features.

CRUST, MANTLE, AND CORE

Earth's interior is a series of concentric shells of mostly solid rock (Figure 7-3). Neither this arrangement nor the composition of the interior can be viewed from the surface, but they were discovered by studying shock waves produced by earthquakes and explosive charges. Such studies show a threefold division of Earth's interior into core, mantle, and crust.

Earth's **crust** is a relatively thin veneer of rock covering the planet's surface. Its thickness averages 33 km (20 mi) in the continents but may reach 75 km (45 mi). The continental crust has two layers, sima and sial. *Sima* is the lower layer and is composed of dense, dark rocks that contain large proportions of silicon and magnesium. Sima rocks alone underlie the ocean basins. On average, the sima is about 10 km (6 mi) thick. *Sial* occurs only on the continents, where it lies on top of sima and is composed of a variety of less dense rocks containing high proportions of silicon and aluminum. Since the continents are capped by lower-density sial rocks, they stand higher than the sima of the ocean basin floors.

Earth's **mantle** is a mostly solid layer, nearly 2900 km (1800 mi) thick. Temperature increases with depth in the crustal rocks and continues to increase through

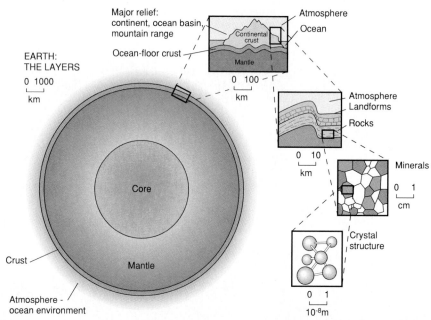

Figure 7-2   The building blocks of planet Earth. The largest building blocks are the concentric layers of the atmosphere-ocean, crust, mantle, and core. The crust's surface is marked by major relief features, which are formed of rocks, which comprise minerals and atoms of elements.

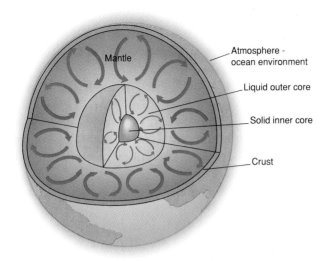

Figure 7-3 Earth's interior. The core is a source of heat energy that is transferred through the mantle by conduction, and possibly by convection (shown by the arrows).

the mantle toward the core. Pressure also increases with depth because of the increasing weight of overlying rocks. As pressure increases, however, so does the melting point of rock. Through most of the mantle, therefore, the pressure is too high for rocks to melt. Only in a thin layer between 100 and 200 km (60 to 120 mi) below the surface does the rate of temperature rise overcome the effect of increasing pressure. In this zone, called the *asthenosphere,* partial melting and flow occur (Figure 7-4).

The two layers within the crust, the sima and sial, together with the uppermost part of the mantle form a single rigid unit above the asthenosphere that is called the *lithosphere.* This composite layer averages 100 km (60 mi) thick, increasing to over 300 km (180 mi) thick beneath some continents and decreasing to only 45 km (30 mi) thick beneath some ocean ridges.

Earth's **core,** formed of very high density rocks, begins 2900 km (1800 mi) below the surface. The outer

section is thought to be liquid because some earthquake shock waves that do not pass through liquids also do not pass through the core. If the outer core is liquid, it is because the very high core temperatures are sufficient to melt the rock, overcoming the high pressures deep in the interior. The temperature of the innermost core is thought to be around 2500° C (4500° F) and solid. A further increase in depth and pressure raises the melting point above the temperature of the innermost rocks.

ROCKS AND MINERALS

Rocks and minerals are the building blocks of Earth's crust. **Minerals** are naturally occurring combinations of chemical elements bonded together in orderly crystalline structures. **Rocks** are coherent masses of minerals. The minerals that compose rocks either form an interlocking mass of individual crystals or are fragments broken from other rocks that become cemented together. Rocks and minerals also comprise Earth's lower mantle and core, but little is known about them.

A small number of elements and minerals dominate the crustal rocks. Despite the great variety of elements present, oxygen and silicon comprise 74 percent of the total volume of Earth's crust. Aluminum, iron, sodium, calcium, potassium, and magnesium make up another 23 percent. This small number of

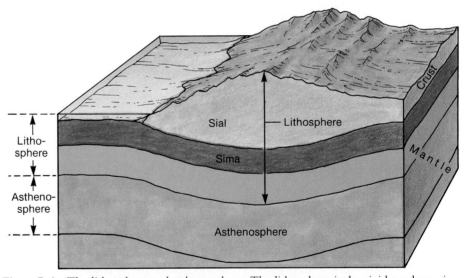

Figure 7-4 The lithosphere and asthenosphere. The lithosphere is the rigid top layer, in which the crust (sial and sima) and uppermost mantle are bonded together. In the asthenosphere the rising temperature overcomes the effect of rising pressure and the rocks begin to melt. The partly melted rocks are mobile and assist movement in the overlying lithosphere.

elements combines in the minerals that form rocks. Most rock-forming minerals are in the *silicate* group, containing high proportions of silicon and oxygen. The darker and denser minerals, such as olivine, that compose the sima, include high proportions of iron and magnesium and are called *mafic*. Lighter and less dense minerals, such as quartz, are more common in the sial and are called *felsic*.

Each mineral has a distinctive internal structure that determines its chemical and physical properties such as color, shape, density, and resistance to being broken down or worn away. Crystals of the mineral olivine, for instance, form at high temperatures (over 1000° C) as molten rock cools, often under the high pressures beneath Earth's surface. Olivine is, therefore, susceptible to change in the lower temperatures and pressures when rocks containing it are exposed at the surface. Quartz, however, does not solidify until the molten rock reaches much lower temperatures. When exposed to the atmosphere, quartz is one of the most resistant minerals to chemical breakdown and physical disintegration. Most sand is quartz that has been exposed to the atmosphere for thousands or millions of years without change.

The solid-earth environment consists of the internal layers of the planet and its rocky surface. The internal layers are the core and mantle. The crust and upper mantle include the rigid lithosphere and the underlying, weaker asthenosphere. The layers of the solid-earth environment are composed of rocks and minerals containing a small range of chemical elements.

DYNAMIC EARTH

Changes in the solid-earth environment occur slowly, often in durations of a few million to a few billion years. But changes do occur. On a larger time scale, the solid-earth environment is as dynamic as the atmosphere-ocean environment. Rocks appear static to humans, but they are part of an ever-changing system powered by internal Earth energy. This energy is sufficient to move huge sections of the lithosphere and to create massive mountain systems.

There are two major sets of ideas that help an understanding of the changes in the solid-earth environment—*plate tectonics* and the *rock cycle*. The former explains the origins of ocean basins, continents, and mountains and the short-term action of earthquakes and volcanic eruptions, and the latter summarizes how rocks form in Earth's crust by a variety of processes linked to the results of plate tectonics.

PLATE TECTONICS

The theory of **plate tectonics** explains the origins of Earth's major relief by mobile and interacting plates. **Plates** are rigid sections of Earth's lithosphere. Although suggested earlier in this century, the idea of plate tectonics developed fully in the 1960s from studies of the ocean floors, magnetic evidence in rocks, and radiometric dating. The theory accounts for shifting continents, opening and closing ocean basins, and rising mountain systems.

According to the theory, new plate material is added along the boundary between plates that move apart, and old plate material is destroyed where two plates collide (Figure 7-5). The plates include the upper mantle and crustal sima rocks that form ocean

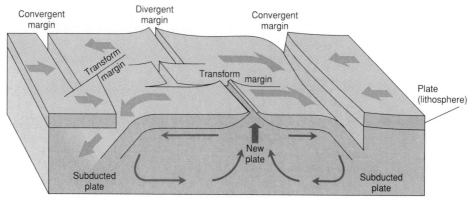

Figure 7-5   Main features of plates and types of margin. The red arrows show directions of plate movement; the blue arrows depict possible convection currents.

floors. Continents, with their sialic rocks, form on top of plates and are passive "passengers" on the denser underlying sima and upper mantle.

Seven major plates and a number of minor plates cover Earth's surface (Figure 7-6). Plates move relative to each other at variable rates, with some sections remaining locked together for millions of years before moving. Typical rates of movement vary from 0 to 100 mm (less than 4 in) per year. Such rates are slow, but 10 mm per year becomes 10 km (6.5 mi) over a million years, and 1000 km (625 mi) over 100 million years.

Earth's major relief features cluster along the plate margins. Three types of plate margin interactions produce different sets of major relief features. At **divergent plate margins,** plates move apart and new plate material is formed by molten rock rising to fill the gap created. Divergent margins commonly occur in ocean areas. At **convergent plate margins,** plates smash into each other destroying plate material and crushing continental crust to form mountain systems. At convergent margins, one plate often dives beneath another, a process known as *subduction. Transform plate margins* occur where two plates slide alongside each other. Plate material is not formed or destroyed in large quantities at transform margins and their main effect is to give a stepped appearance to maps of ocean ridges, as shown in Figure 7.5.

Although Earth's major surface features mostly form along plate margins, especially convergent margins, the rest of Earth's surface is also active, but less intensely. Interior plate areas move as one piece and so movement is less obvious than at plate margins, but such movements may cause the surface to rise or fall.

The causes of plate tectonics are not fully understood. It is believed that Earth's interior is the heat source and that movements in the upper mantle drive the plate movements. The precise mechanism, however, is still open to discussion. Some of the clues to a fuller understanding come from the study of earthquakes and volcanic eruptions.

PLATE TECTONICS, EARTHQUAKES, AND VOLCANIC ERUPTIONS

Earthquakes and volcanic eruptions nearly all occur along plate margins. They are short-lived events, taking periods of time from a few seconds to several years, but are expressions of long-term internal activity. Both may cause considerable damage and loss of life when they affect settled areas (see Environmental Issue: Earthquake Hazards—Can Humans Cope? p. 212).

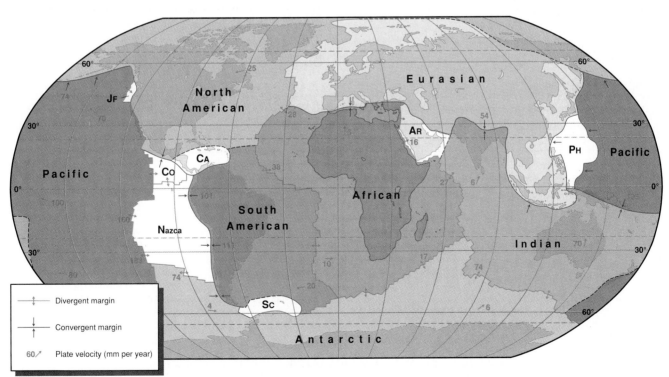

Figure 7-6 World map of major and minor plates. The minor plates (white) include Nazca, Cocos (Co), Caribbean (Ca), Juan de Fuca (Jf), Arabia (Ar), Philippines (Ph), and Scotia (Sc).

Source: Plate velocity data from NASA.

Earthquake Activity   Plate movements build up stresses in the lithosphere to the point where something has to give. The stresses slowly bend huge sections of rock until the rock snaps. The sudden failure of the rock produces shock waves known as *seismic waves*. An **earthquake** is a shaking or series of shocks generated by sudden movements in Earth's crust or upper mantle. The strongest seismic waves are close to the *focus* of the shock—the point of rock failure—and tend to affect human activities most at the *epicenter*—the surface point directly above the focus (Figure 7-7).

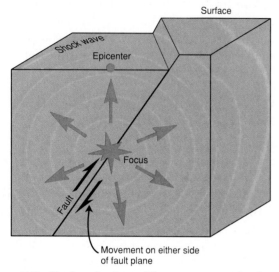

Figure 7-7   Earthquake focus. The movement of rocks at the focus inside Earth sends shock waves that reach the epicenter and then continue through Earth's rocks.

Earthquakes vary in size, or *magnitude,* as measured by instruments known as *seismographs*. In 1935 Charles F. Richter established a scale of earthquake magnitude in which each value is a tenfold increase over the value one unit of magnitude lower (Table 7-2). Large earthquakes, such as that at San Francisco in 1989 (Richter 7.1) or those in Armenia and Iran (over Richter 8.0), cause extensive surface damage within seconds of the initial shock.

The numbers on the Richter Scale reflect the magnitude of the earthquake shocks, but local conditions of relief and rock materials produce a range of effects for earthquakes of the same seismic-wave intensity. The Modified Mercalli Scale (Table 7-3) is an alternative scale for describing earthquake intensity, based on the effects on people and buildings.

TABLE 7-2   RICHTER'S SCALE OF EARTHQUAKE MAGNITUDE

Richter Magnitude	Average Number Per Year	Equivalent Energy
3	over 50,000	Smallest commonly felt
4	6,000	Detonation of 1000 tons of explosives
5	800	First atom bombs
6	120	Sufficient to launch 2 million NASA space shuttles
7	18	Niagara Falls in 4 months
8 and over	1 or less	1906 San Francisco earthquake

TABLE 7-3   MODIFIED MERCALLI SCALE OF EARTHQUAKE INTENSITY (ABBREVIATED)

Scale	Effects
I	Not felt.
II	Felt by persons at rest, on upper floors of buildings, or favorably placed.
III	Felt indoors. Hanging objects swing. Vibrations like passing of light trucks. May not be recognized as earthquake. Duration estimated.
IV	Hanging objects swing. Vibrations like passing of heavy trucks, or jolt like heavy ball striking wall. Standing autos rock. Windows, doors rattle, crockery clashes.
V	Felt outdoors. Direction estimated. Sleepers wakened. Liquids disturbed or spilled. Small unstable objects displaced or upset. Doors swing. Shutters, pictures move. Pendulum clocks stop, start, change rate.
VI	Felt by all. Many frightened and run outdoors. Persons walk unsteadily. Windows, dishes, glassware broken. Books fall off shelves, pictures off walls. Furniture moved or overturned. Weak plaster and masonry cracked. Small bells ring. Trees, bushes shake.
VII	Difficult to stand. Noticed by auto drivers. Weak masonry damaged. Weak chimneys broken at roof line. Fall of plaster, bricks, tiles. Waves on ponds, mud stirred up. Large bells ring. Concrete irrigation ditches damaged.
VIII	Steering of autos affected. Damage to ordinary masonry, but not high quality or reinforced masonry. Chimneys twist and fall; also monuments, towers, elevated tanks. Frame houses move on foundations if not bolted down. Branches broken from trees. Changes in flow or temperature of springs and wells.
IX	General panic. Poor masonry destroyed, ordinary damaged, good masonry cracked. General damage to foundations. Serious damage to reservoirs. Underground pipes broken. Conspicuous cracks in ground. In alluvial areas sand and mud ejected.
X	Most masonry and frame structures destroyed with their foundations. Some well-built wooden structures and bridges destroyed. Serious damage to dams, dikes, embankments. Large landslides. Sand and mud shifted horizontally on beaches and flat land. Rails bent slightly.
XI	Rails bent greatly. Underground pipelines completely out of service.
XII	Damage nearly total. Large rock masses displaced. Lines of sight and level distorted. Objects thrown into the air.

The world distribution of earthquakes since 1960 shows that nearly all earthquakes, especially the very large ones, occur along plate margins (Figure 7-8). The greatest concentrations are around the margins of the Pacific Ocean and in the mountainous areas of southern Europe and Asia. These are convergent plate margins, where the collision and subduction of plates causes earthquakes of high magnitude up to 700 km (450 mi) beneath the surface. Lower magnitude earthquake activity takes place along the divergent plate margins in the oceans. Few earthquakes occur away from plate margins but a few have been of high magnitude. In 1812, for example, the New Madrid (Missouri) earthquake caused widespread damage, including the collapse of river banks along the Mississippi River. This may have resulted from a jerky movement of the North American plate.

Volcanic Activity   Volcanic activity provides further evidence for the mobility and internal energy of the planet. It includes all the ways in which molten rock, or **magma,** erupts at Earth's surface. Magma is a variable and mobile mixture of liquids, gases, and solid rock that moves slowly and has the consistency of thick oatmeal. The magma erupts through pipelike vents or elongated fissures to produce a variety of volcanic features. A **volcano** is a landform produced above a pipelike vent. Surface flows of molten rock issuing from fissures or volcanoes are called **lava.**

The world distribution of volcanic activity, like that of earthquakes, is concentrated along plate margins (see Figure 7-8). Ocean basins and some continental margins are the sites of most volcanic activity. Sixty percent of the eruptions in historic time occurred in and around the Pacific Ocean, an area often called the Ring of Fire. There are some 50,000 volcanoes on the floor of the Pacific Ocean. Some of these are *active,* having erupted in historic time; others are *dormant* and likely to erupt in the future or *extinct* and unlikely to erupt again. Most of the Pacific Ocean volcanoes form along convergent plate margins around the edges of the ocean or at the divergent plate margins within the ocean basin.

Some volcanic activity also occurs in the interior parts of plates, particularly in the Pacific Ocean and on the African plate. Volcanoes in midplate positions erupt above localized zones of melting rock within the mantle known as *hot spots.* The expansion of the heated and melting rock material causes thinning and cracking of the overlying crust, letting the magma pass to the surface. Some volcanoes, such as the Canary Islands, occur in clusters over a hot spot where the plate moves slowly. Other groups of volcanoes, such as the active ones on the

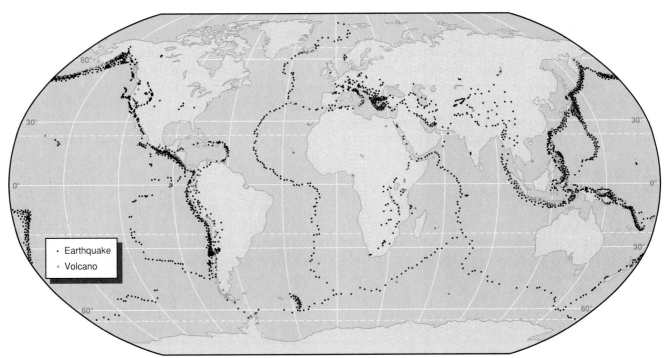

Figure 7-8   World map of earthquakes and active volcanoes. Compare their distribution to the map of plates (Figure 7-6).

Source: U.S. Geological Survey.

Hawaiian Islands and an associated group of extinct volcanoes to the northwest, form a chain carried "downstream" from a hotspot (Figure 7-9) by the plate. On the island of Hawaii, at the southeastern end of this chain, magma from the hot spot erupts and forms active volcanoes. The Pacific plate carries older volcanic peaks to the northwest, cutting them off from the magma source and rendering them extinct. The peaks moved in this way in the past sank and formed submarine remnants as distance increased from the hot spot.

Volcanic eruptions vary in their explosiveness and this helps determine the shape and size of the features produced. The greatest explosions can blow away the top of a volcano, leaving a huge crater known as a *caldera*. In 1883 the island of Krakatoa in Indonesia erupted 10 km^3 (2.4 mi^3) of rock into the atmosphere in the largest historic explosive eruption. Some 75,000 years ago, Mount Toba in the same area erupted about 20,000 km^3 (5000 mi^3) of rock fragments and lava. Such explosions occur most commonly where one plate dives beneath another, causing melting of crustal rocks. Calderas can also form after collapse of the crustal rocks around a vent when the magma has erupted.

In the least explosive volcanoes, such as those on Hawaii, liquid lava wells up steadily and flows out on the surface with only an occasional "fireworks display" of red-hot clots of lava thrown into the air by escaping gas (Figure 7-10). Such quieter volcanic activity occurs in ocean basins above divergent plate margins and hot spots.

The explosiveness of a volcanic eruption depends on the nature of the magma involved. Magma containing a high proportion of felsic minerals and originating from the melting of crustal (mainly sial) rocks erupts in an almost solid state and flows slowly and stiffly, if at all. The slow-flowing lava traps gases, and pressure inside the lava builds until the mass bursts apart in an explosive eruption. The clots of ejected lava solidify in the atmosphere and fall to the ground as particles and lumps called *tephra*. Tephra range from *fine ash* (less than 2 mm, 0.1 in, across) to *volcanic bombs* (over 64 mm, 2.5 in). At the other end of the spectrum of lava types, very fluid magma rises from the melting of mantle rock along divergent plate margins and at hot spots. The resultant lavas contain a high proportion of mafic minerals at high temperatures and are very fluid, spreading over wide areas of the surface. The gases in the lava bubble up and escape into the atmosphere but hardly disrupt the surface of the flow.

The sizes and shapes of volcanoes depend on the mobility and explosiveness of the erupted igneous material (Figure 7-11). Where eruptions of very stiff lava occur from a single pipelike vent, the landform will be a steep-sided *lava dome;* explosive eruptions of similar magma may produce a *cinder cone* formed from piles of tephra (Figure 7-11a). Extreme volcanic explosions cause rock to disintegrate in a glowing cloud of hot gas and fluid lava called a *nuée ardente,* such as that which destroyed St. Pierre, Martinique in 1902 (Figure 7-11b). Fluid magma erupted at a vent builds shallow-sloped layers into huge *shield volcanoes,* such as the Hawaiian Islands, which rise over 10 km (6 mi) from the ocean floor and cover several thousand square kilometers (Figure 7-11c). Most continental volcanoes combine layers of tephra and lava in a *composite volcano* (Figure 7-11d). The volcanic peaks of the Andes, such as Cotopaxi, are of this type. They typically have slopes that become less steep with distance from the vent since the coarse tephra fall close to the vent and fine ash particles fall farther away.

Some of the largest volcanic eruptions pour out huge masses of lava along fissures. Over time, hundreds of flows may pile up to form a *lava plateau* (Figure 7-11e). A *plateau* is high land with a flat top. The necessary quantities of magma to produce such large features are generated at plate margins. The Columbia Plateau, for instance, covers 130,000 sq km (50,000 sq mi) of the northwestern United States and has a total thickness of 2000 m (6000 ft). These flows occurred between 10 and 25 million years ago. Other lava plateaus include the Deccan Plateau of India covering half a million square kilometers (200,000 sq mi) and the Parana Plateau of Uruguay and southern Brazil covering three quarters of a million square kilometers (300,000 sq mi).

The recent history of Mount St. Helens in Washington illustrates how volcanic activity is related to plate margins and how volcanic activity impacts human affairs. It also demonstrates the extent to which humans can predict volcanic eruptions. The eruption of Mount St. Helens in 1980 affected the lives of thousands of people, remolded the local landscape, and had a short-lived but widespread effect on the atmosphere.

Figure 7-9

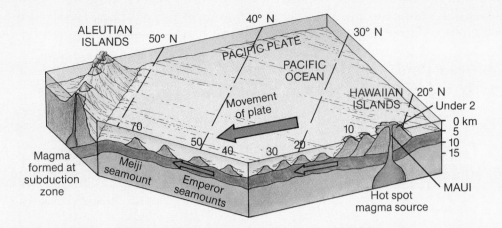

Figure 7-9   The Hawaiian hot spot. This hot spot has been a lava source for 70 million years, and at present is forming the island of Hawaii. The Pacific plate carries older volcanic peaks northwestward, the oldest peaks being at the far end of the sequence. The Aleutian Islands, by contrast, are formed along a convergent plate margin above a subduction zone. The figures next to volcanic peaks in the Hawaiian chain show the age of the oldest lava in millions of years.

Figure 7-10   Lava erupted from Kilauea, Hawaii.

Figure 7-10

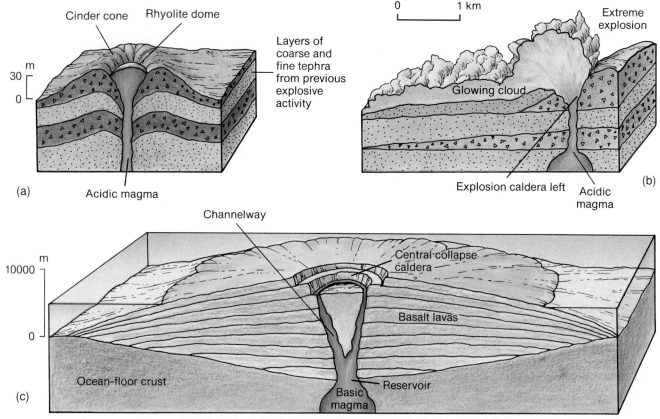

Figure 7-11   Volcanic forms. (a) A cinder cone with a dome of acidic lava resulting from an explosive eruption. (b) A nuée ardente following an extreme explosion. (c) A shield volcano formed of many layers of fluid basalt lava. (d) A composite cone with alternating layers of lava and tephra. (e) A lava plateau formed by flood basalts.

The main eruption of May 18, 1980, was not an isolated or totally unexpected event. Mount St. Helens is one of several volcanoes along the Cascade Range just inland from the Pacific coast, as shown on Figure 7-12. These volcanoes form part of the Ring of Fire around the Pacific Ocean and lie above a subduction zone at a convergent plate margin. The peak of Mount St. Helens built up over the last 2500 years by the accumulation of layers of molten lava and tephra on the remains of an older volcano. Quiet periods with little volcanic activity alternated with great explosions. Until A.D. 1400 the quiet periods lasted up to 500 years, but since then only 200 years at most. A large eruption in A.D. 1800 was witnessed by local Native Americans and a few early settlers. Smaller eruptions continued until 1857 when activity ceased. Increasing study of Mount St. Helens led geologists in 1975 to predict another large eruption before the end of the century.

From the middle of 1979 to early 1980, a bulge 150 m (470 ft) high formed on the northern flank of Mount St. Helens. On March 20, 1980 a moderate earthquake marked the start of a series of stronger and more frequent events. A small eruption occurred on March 27, with a gush of steam and ash that rose 2100 m (7000 ft) and left a summit crater. People living within 25 km (16 mi) of the mountain were advised to leave. Further eruptions opened a larger central crater 450 m (1500 ft) deep, and the constant shaking of the ground alerted geologists that molten rock was moving underneath. Scientists, reporters, and sightseers gathered, and access to the immediate area of the mountain was controlled.

On May 18, 1980, the main explosion occurred (Figure 7-13). Following a strong earthquake, 4.5 km^3 (1.2 mi^3) of rock on the bulged northern flank of the mountain collapsed. A column of finer material shot 20 km (12 mi) into the atmosphere. The collapse of the northern flank was accompanied by a powerful blast of gas and ash fragments at temperatures up to 800° C (1400° F) and moving at 160 kph (100 mph). This blast devastated an arc-shaped area extending 16 to 25 km (10 to 16 mi) to the north,

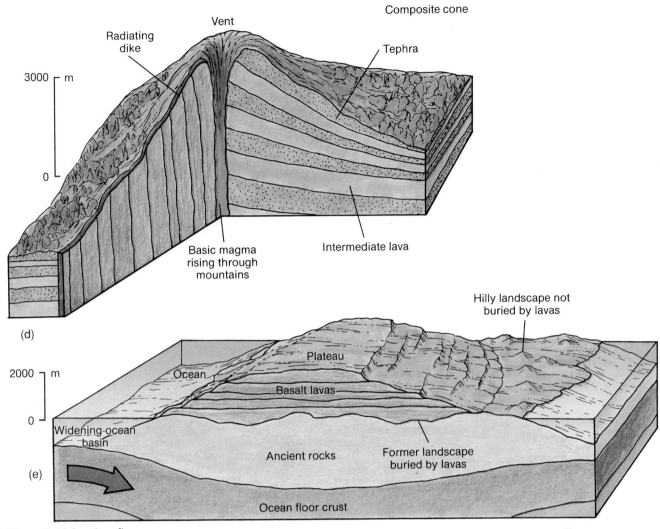

Composite cone

Radiating
dike

Vent

Tephra

3000 ⌐ m

0 ⌐

Basic magma
rising through
mountains

Intermediate lava

(d)

Hilly landscape not
buried by lavas

2000 ⌐ m

0 ⌐

Ocean

Plateau

Basalt lavas

Widening-ocean
basin

(e)

Ancient rocks

Former landscape
buried by lavas

Ocean floor crust

Figure 7-11 (continued)

felling and charring trees and destroying trucks, bridges, and homes. Melted snow and water spilled from lakes, mixed with ash, and cascaded down the valleys in a mudflow that swept away logging equipment and killed many fish in the Toutle River. The mud and ash dropped in a swath over 1000 km (600 mi) to the northeast, blocking the navigation channel of the Columbia River. Ash carried eastward in the upper troposphere was blamed for cool, wet June and July weather in Europe.

A smaller eruption occurred on May 25, and others continued through the 1980s. A dome-shaped mound of lava built up in the great hollow formed in the central part of the mountain (Figure 7-14). Since then, Mount St. Helens has continued to be a focus of geologic study, and minor volcanic activity has continued from time to time.

People reacted to the eruption in varied ways. Some were overawed by the experience, some resisted the official warnings and advice before moving out at the last moment, and some never left. Many worried about the harm the eruption might do to the regional economy or about the possibility of eruptions continuing for several years. Some of the worries over long-term effects were not realized in practice, however, since tourists came in greater numbers and the volcanic activity soon died down.

The Mount St. Helens eruption demonstrates scientists' partial knowledge about volcanic activity. Although by the mid 1970s geologists knew that an eruption was likely to take place within the next few years and posted more urgent warnings just before the event, they could not be sure of the timing and size of the eruption. Certainly, they could do nothing

Figure 7-12   Map of the area around Mount St. Helens showing other volcanoes in the area.

Figure 7-13   The eruption of May 18, 1980, and its effect on the immediate area of Mount St. Helens.

Figure 7-14   Mount St. Helens, before (left) and after the eruption of May 18, 1980.

Figure 7-12

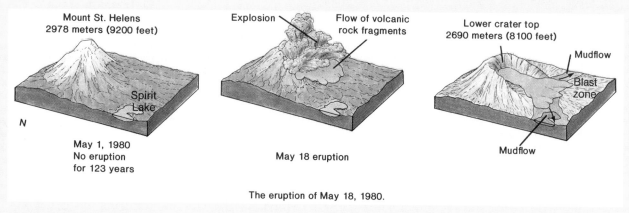

The eruption of May 18, 1980.

Figure 7-13

Figure 7-14

to modify or prevent the eruption. After 123 years of no eruptions, people ignored the warnings of imminent activity and built their habitations closer to the peak. No specific emergency procedure was devised to deal with the potential catastrophe. Even though the Mount St. Helens eruption occurred in a sparsely inhabited part of the United States, 58 people died.

Earth's surface features change continuously as the result of internal solid-earth processes that power plate tectonics. Plates are almost rigid sections of Earth's lithosphere, which converge and diverge, colliding and pulling apart. New lithosphere is produced by the cooling of rising magma along diverging plate margins and is destroyed by subduction at convergent margins. Earthquake and volcanic activity are concentrated along plate margins.

Volcanic activity has significant impacts on other earth environments. The gases erupted into the atmosphere and oceans from the **solid-earth environment** are partly responsible for the changing composition of the **atmosphere-ocean environment.** Fine tephra may remain in the atmosphere for some time, causing extra brilliant sunsets and affecting atmospheric heating by reflecting insolation back to space. Volcanic eruptions produce landforms that are molded in the **surface-relief environment.** They produce materials that affect the composition and fertility of soils formed from them (**living-organism environment**).

ROCK CYCLE

Plate tectonics processes cause movements that lead to the formation, destruction, and reformation of rocks in the **rock cycle** (Figure 7-15). Three major rock groups in Earth's crust are defined by their mode and place of origin. Rocks in Earth's interior form in different ways from those at the surface.

Igneous rocks form as magma migrates upward in the upper mantle and crust and then solidifies on cooling. New rock solidifies either within the crust or at the surface. **Metamorphic rocks** form in the crust when extreme heat or pressure changes, or "metamorphoses," minerals in older rocks. During metamorphosis the rocks do not completely melt and so there is no migration. **Sedimentary rocks** form at Earth's surface from the products of the breakup of other rocks—clay particles, sand grains, dissolved chemicals, and fragments up to boulder size—sometimes mixed with the remains of living organisms.

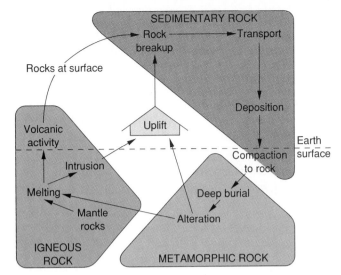

Figure 7-15 Rock cycle. How are the different types of rock involved in the circulation of materials in Earth's crust and upper mantle?

The processes that create rocks act in a continuous cycle. All rock types can be made of materials that were part of older rocks. Some igneous rocks bring material from the mantle up to the crust, but others form from the melting of crustal rocks. Metamorphic and sedimentary rocks form from previous igneous, metamorphic, or sedimentary rocks.

Igneous Rocks Igneous rocks form when magma solidifies from the molten state so that the minerals crystallize in an interlocking mass. The environments from which the magma originates and in which it solidifies determine the nature of the igneous rock. Magma forming from the melting of upper mantle rocks is rich in dense, dark-colored mafic minerals such as olivine. Magma forming from the melting of continental crust contains a wider variety of minerals, including more of the less-dense and lighter-colored felsic varieties such as quartz. If the final solidification occurs inside the crust, the rock is said to be *intrusive.* Intrusive igneous rocks cool slowly, giving large crystals time to form. When magma erupts on the surface, it produces *extrusive* rocks. Extrusive rocks cool rapidly and contain very small crystals.

Basalt and granite are the two main types of igneous rock. *Basalt* is a black, fine-grained rock formed of mafic minerals. It results from magma that originates in the upper mantle or sima and solidifies in extrusive lava flows on Earth's surface in contact with air or water (Figure 7-16). *Granite* is a lighter-colored rock with crystals of felsic minerals large enough to be seen with the naked eye. Although its magma is produced by melting crustal rocks, granite solidifies in large intrusions deep below the surface.

Igneous rocks take up distinctive forms in and on Earth's crust (Figure 7-17). Intrusive igneous rocks occur in bodies known as *plutons,* which have a variety of shapes and sizes. Smaller, sheetlike plutons, normally between a few meters and several hundred meters thick, may extend for hundreds of kilometers.

Sheets of igneous rock that solidify between layers of older rocks are called *sills.* Sheets of igneous rock that cut across older layers are called *dikes.* The largest plutons are masses up to 1000 km (600 mi) across called *batholiths.* Granite is the main rock of batholiths, and they commonly form the cores of mountain ranges (Figure 7-18).

Extrusive igneous rocks form at or near Earth's surface in contact with the ocean or atmosphere. Molten magma is either fluid and solidifies in layers of lava, or it is stiff and explodes into millions of tiny

Figure 7-16   Lava flow, Hawaii. This basalt lava covered the road in 1990.
© J. P. Lenney

Figure 7-17   A range of igneous plutons. The relationships of igneous intrusions to the older rocks and surface volcanic features (extrusive igneous rocks). Compare the sizes and shapes of batholiths, sills, dikes, and volcanic necks.

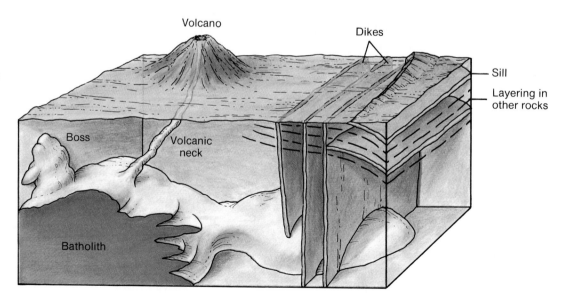

tephra fragments. Lava erupting from the ocean floor often forms rounded masses called *pillow lavas*. They form when the edges of the hot lava react with cold, salty water.

Metamorphic Rocks   The environment of rocks buried beneath Earth's surface is marked by high temperatures and pressures. These conditions do not always melt the rocks but commonly alter their minerals and physical structures. Such alterations transform, or metamorphose, the rocks.

A mineral within a heated rock may melt and recrystallize in place so that its rearranged atoms or molecules produce a new mineral. High pressures acting on a rock may flatten the weaker minerals, such as clay particles, to produce a layering effect known as *foliation* that is characteristic of many metamorphic rocks.

Several levels of metamorphic intensity and geographic extent of metamorphism result from combinations of heating and pressure. Figure 7-19 shows the effects of increasing levels of metamorphism on a clay-based rock.

Low levels or smaller geographic extents of metamorphism occur where high pressure combines with low temperature, or where low pressure combines with high temperature. High pressures without a temperature increase result in rocks being crushed. The crush zone may be several hundred meters wide and the rocks in it are a mixture of fine and coarse fragments, sometimes smashed into a solid mass by compression. High temperatures without high pressures affect the zone a few meters to a few kilometers wide around an igneous

Figure 7-18   Batholiths. Some of the largest batholiths in the world are in western North America.

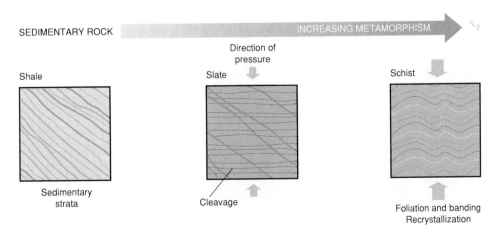

Figure 7-19   Metamorphic rocks. From an original sedimentary rock, increasing pressure and heat produce slate and then schist.

batholith intrusion without crushing the rocks. While the magma is still molten at temperatures of 500 to 1000° C (900 to 1800° F), it heats the surrounding rocks by conduction. The heat creates temperatures higher or lower than those at which the surrounding rock formed and the minerals recrystallize. The process is called *contact metamorphism* (Figure 7-20).

Moderate levels of both heat and pressure affect rocks that are buried a kilometer or so below Earth's surface. Clay minerals are particularly susceptible in such conditions; the pressures are sufficient to flatten them, forming *slate*. Slate can be split apart, or *cleaved*, into thin sheets that form perpendicular to the pressures that metamorphosed the rock.

Very high temperatures and pressures occur when rocks are buried several tens of kilometers deep. Under these conditions the minerals change in both chemical structure and physical orientation. Clay minerals alter to shiny mica, and rocks with high proportions of such altered minerals have a distinctive shiny foliation. Such rocks are called *schists*. Metamorphism of granite under these conditions

produces a *gneiss,* a rock in which bands of quartz and other felsic minerals alternate with dark mica bands (Figure 7-21).

The environment that produces slates, schists, and gneisses often covers thousands of square kilometers; such a widespread effect is called *regional metamorphism*. The common environment of deep burial, great heat, and intense pressure totally transforms rock. Regional

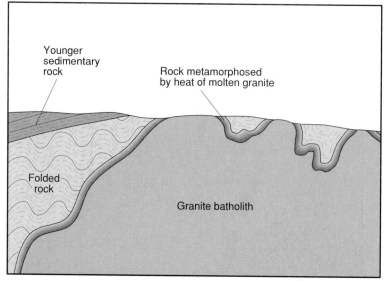

Figure 7-20   Contact metamorphism around a granite batholith. How does the intrusion affect older rocks?

Figure 7-21   Metamorphic rock, Scotland. This highly metamorphosed and contorted rock occurs in northwest Scotland. The total width of rock shown is approximately one meter.

metamorphism is often associated with huge batholiths in the cores of mountain ranges such as the Rockies and Alps. These environments are typical of convergent plate margins where continents collide.

Sedimentary Rocks   Sedimentary rocks form at Earth's surface. The raw materials may be accumulations of broken rock material called *sediment;* the debris of living organisms, such as fossil shells, bones, and plant remains; or chemical deposits. Conversion to hard layers of rock takes place over hundreds of thousands of years by a combination of compaction and the addition of a cementing mineral that glues the particles together. Layering is a common feature of sedimentary rocks and the layers are called *beds,* or strata (Figure 7-22). Strata are bounded by surfaces known as bedding planes, which mark the change from one phase of deposition or composition to the next.

Most constituents of sedimentary rocks are removed from the places where older rocks disintegrate or where the animals or plants live to an area of deposition where the material accumulates. The process of transport may mix the sediments and organic remains. Movement of the particles by water may sort them into groups of similar size. The constituents of sedimentary rocks are sometimes well-sorted and sometimes unsorted mixtures of sediment, organic remains, and chemical precipitates. Some constituents of sedimentary rocks, however, such as peat or coral reefs, are not transported but remain at the place where they later become part of a sedimentary rock.

Sites of deposition include lakes, seas, and ocean floors. A layer of one material will be buried as new layers form on top. The new layers may be similar or different in composition; sand may cover boulders, and clay may later cover the sand. The growing weight of overlying layers squeezes out water from the lower layers, reducing their thickness. The expelled water may leave behind a cement of calcium carbonate or silica that binds the particles together. The combination of compaction and cementation changes the soft, unconsolidated sediment into a hard sedimentary rock.

Sedimentary rocks are named for their main constituents. Rocks formed mainly of clay minerals are called *shale;* those formed of sand grains, *sandstone;* and those formed of larger fragments, such as gravel or boulders, *conglomerate.* Sedimentary rocks with a high proportion of calcium carbonate, be it in the form of shells or a chemical precipitate, are called *limestones.* Peat layers harden to form *coal. Rock salt* and *gypsum* form in evaporating lakes.

CONTORTED AND BROKEN ROCKS

Cliffs and road cuts make it possible to see rock strata in cross section. These rocks may occur in the original horizontal layers of sedimentary strata or lava flows. Stresses within Earth's crust caused by plate tectonics, however, often deform the rocks after they form (Figure 7-23). The resultant structures are called **folds** when the rock layers are bent but not broken and **faults** when the layers are broken and displaced on either side of the break. *Joints* are cracks where the rocks are broken but not displaced. Folds, faults, and joints vary in scale from a few centimeters to tens of kilometers in size.

Folds   It is difficult to believe that a hardened and rigid rock layer could be folded in the same way that a tablecloth folds when it is pushed. Rock folding is possible when the rock is pliable, either before it has finally hardened or when it is subjected to heat and pressure deep underground over a long time. When pressure is not very intense, simple folds form *anticlines* (upfolds) and *synclines* (downfolds). Increasing pressure produces more complex folds that are characteristic of many high mountain ranges such as Himalayas, Alps, and Andes (Figure 7-24). In describing the orientation of rock layers in folds, geologists refer to the angle between the steepest surface of a bedding plane and the horizontal as the *dip angle* (Figure 7-25). The direction of a horizontal line on this surface is the *strike.*

Folded rocks are common in the Appalachian ridge-and-valley region, extending from central Pennsylvania southeast to Tennessee (see map and photo, page 190, and Figure 7-26). The folds affect layers of sedimentary rocks including shales, sandstones, and limestones. In wearing down the folded rocks, surface processes etched out the less resistant shales and left the more resistant sandstones and limestones to form the ridges.

Faults   Rigid rocks fracture under stress, and are displaced by faults. Faults often occur in groups along a *fault zone,* as in the San Andreas fault zone of California.

Some faults displace rocks up or down; others involve sideways movements. Faults with horizontal displacements are termed *strike-slip faults* (Figure 7-27a) since movement is along the strike of the fault plane. The San Andreas fault is a strike-slip fault and movements along it occur in short, sharp jerks that cause earthquakes and displace the rocks a few meters at a time (Figure 7-28).

Faults involving mainly vertical displacement up or down the fault plane are *dip-slip faults* (Figure

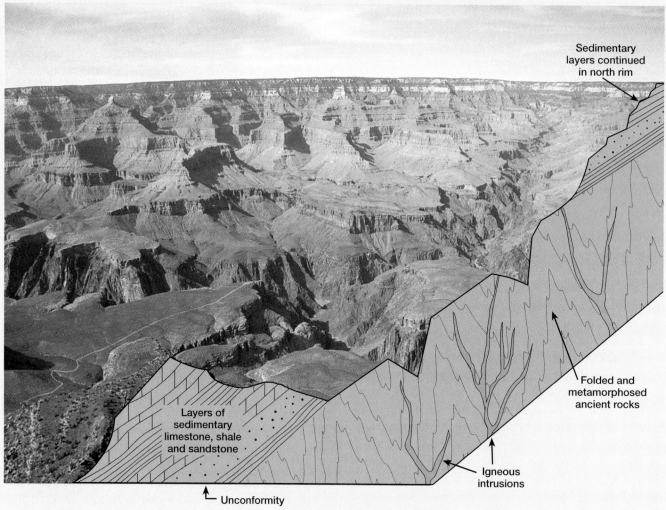

Figure 7-22   Grand Canyon, Arizona. The photo and diagram show the layered beds of sedimentary rock and the underlying metamorphic and igneous rocks.

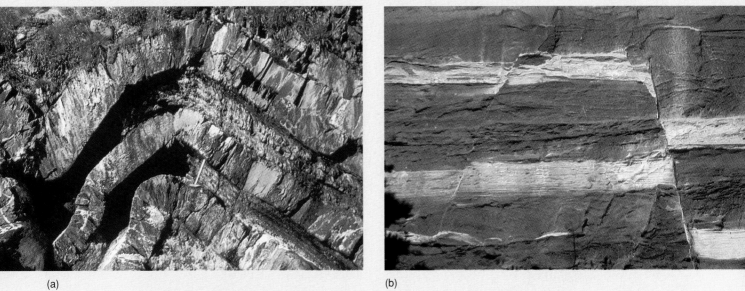

(a)   (b)

Figure 7-23   Fold and fault. (a) An anticline fold in sandstone rock, Wales. (b) A normal fault in sandstone rock near Liverpool, England.

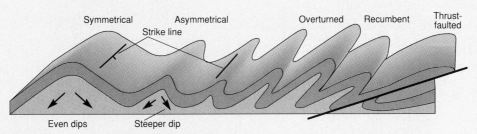

Figure 7-24   Types of fold with intensity of folding increasing from left to right.

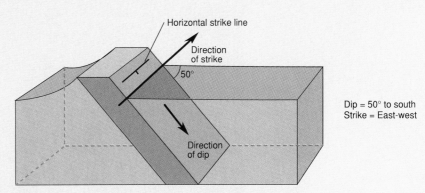

Figure 7-25   Measuring strike and dip. For strike, a horizontal line is drawn across a sloping rock surface and its direction determined by a compass. The dip angle is then calculated by drawing a line at right angles to the strike and measuring the angle between this new line and a horizontal plane. A clinometer, composed of a protractor combined with a compass and level, is used for these measurements.

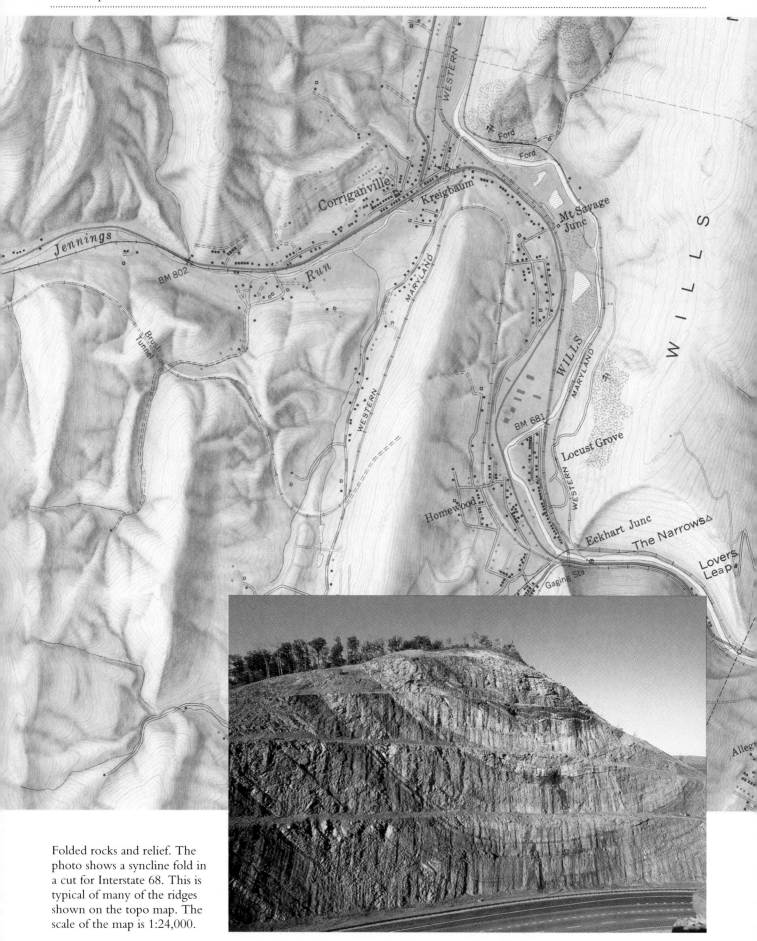

Folded rocks and relief. The photo shows a syncline fold in a cut for Interstate 68. This is typical of many of the ridges shown on the topo map. The scale of the map is 1:24,000.

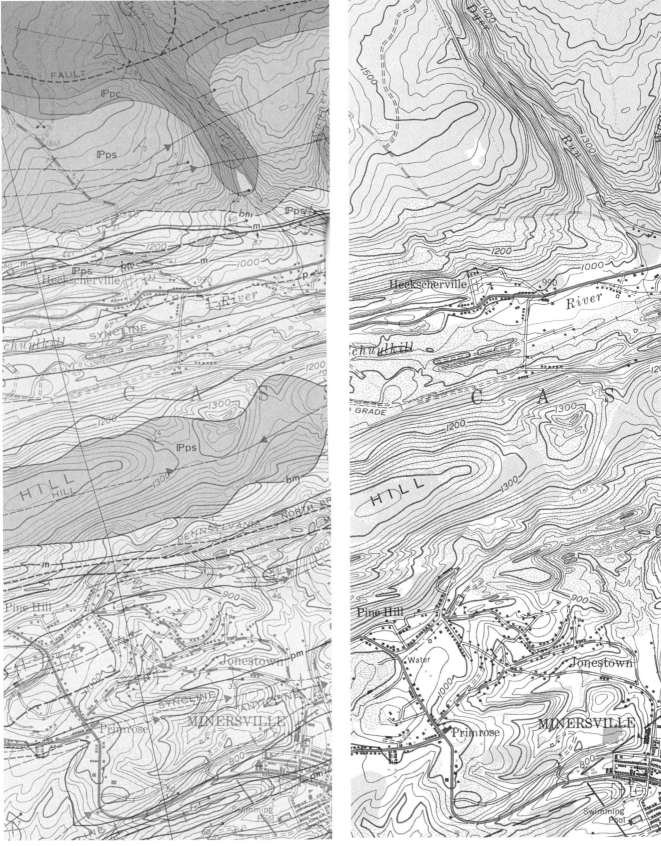

Figure 7-26   Geologic (a) and topo (b) maps of part of Schuylkill County, Pennsylvania at a scale of 1:24,000. On the geologic map the rocks are of Pennsylvanian (Upper Carboniferous) age and include sandstones (blues) and shales with coal seams (grey with black lines). Some of the rock structures are shown on the map by red lines labeled "anticline" and "syncline". The Schuylkill valley is a syncline of shales and coal seams. To the south of the valley is (Mine) Hill on an anticline that brings resistant sandstone to the surface. The topo map shows large areas of mine waste (pink stipple).

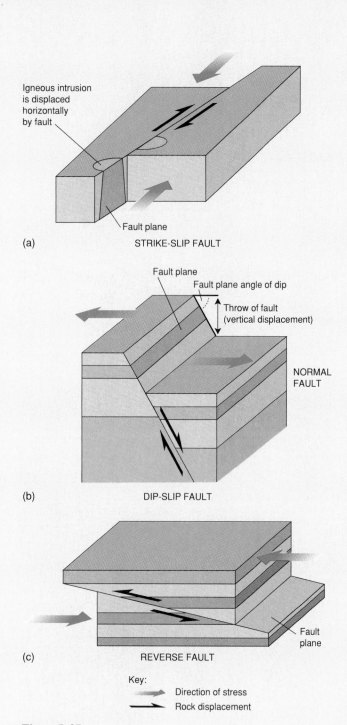

Igneous intrusion
is displaced
horizontally
by fault

Fault plane

(a)   STRIKE-SLIP FAULT

Fault plane

Fault plane angle of dip

Throw of fault
(vertical displacement)

NORMAL
FAULT

(b)   DIP-SLIP FAULT

Fault
plane

(c)   REVERSE FAULT

Key:

→ Direction of stress

→ Rock displacement

Figure 7-27

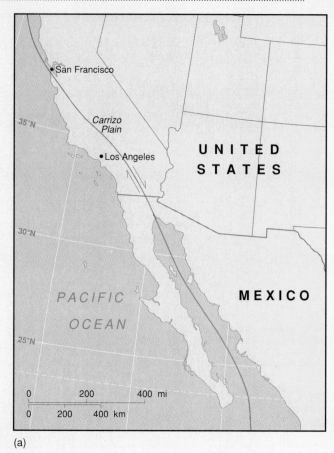

(a)

Figure 7-28

Figure 7-27   Faults. The thick blue arrows show directions of pressure that cause the rock to fracture; the black half arrows indicate the relative displacement of rock on opposite sides of the fault. (a) A strike–slip fault, where the movement is horizontal on either side of the fault. (b) A normal fault, where tension causes fractures in rocks that are being pulled apart. The rock above the fault plane moves down relative to the rock beneath the fault plane. (c) A reverse fault, where compression causes the upper block of rock to be pushed over the lower.

Figure 7-28   San Andreas fault. (a) A large slice of land on the ocean side of the fault is moving sporadically northwest. (b) The Carrizo Plain area in California illustrates the effects of the fault on surface features. The fault line is marked by the deep gash in the rocks.

(a) Source: U.S. Geological Survey.

(b)

Figure 7-28 (continued)

7-27b,c). When rocks on each side of a fault are pulled away in opposite directions, the rock above a sloping fault plane moves down the plane, producing a *normal fault*. When the fault is caused by compression, the upper block moves over the lower block in a *reverse fault*. Very intense pressures may shear off overturned fold structures (see Figure 7-24a), so that the top part moves forward over the lower and produces an almost horizontal *thrust fault*.

Faulting also produces uplifted blocks of crustal rocks and results in deep downfaulted valleys. Faulting and uplift combine in *fault-block mountains* forming plateaus. The Rhine Highlands of central Europe (Figure 7-29) are faulted on both sides,

forming a *horst*. The sinking of a valley between two horsts produces a *graben*, as in the middle Rhine valley and along the upper Rio Grande valley at Albuquerque, New Mexico.

Joints Cracks where the rocks on either side are not displaced occur in all types of rocks. In igneous rocks joints form during cooling and contraction. In coarse-grained granite, sets of joints at right angles are common (Figure 7-30), while in fine-grained basalt, the joints often produce six-sided columns. Joints form in sedimentary rocks as they contract following the expulsion of water. Folding that involves the stretching and flexing of rock layers also produces joints in rocks.

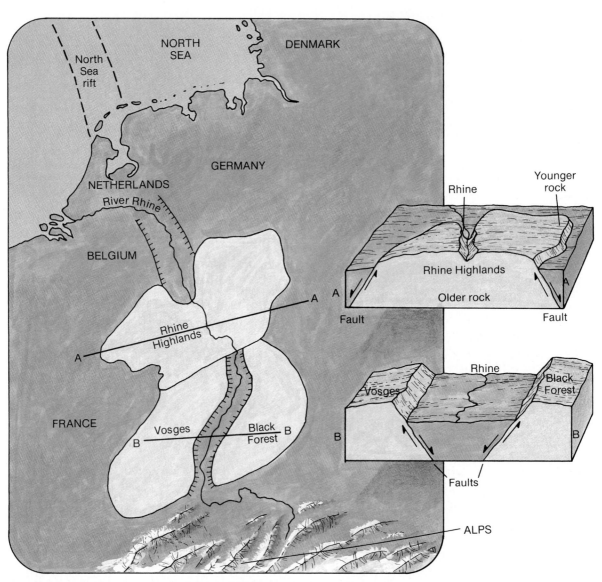

Figure 7-29 Fault-block features, Rhine valley, Europe. The uplifted mountain areas, bounded by faults, are called horsts: the rifts between, grabens. The graben extends beneath the North Sea.

Figure 7-30   Joints. Vertical cooling joints in granite, Yosemite National Park, California.

© J. P. Lenney

ROCKS AND TIME

Geologic time is the time over which rocks form. It makes the length of the human occupation of Earth appear insignificant. The oldest known "modern human" lived around 30,000 years ago, and 5000-year-old civilizations have been discovered. Geologic time, however, deals in millions or billions of years. Geologic time (Figure 7-31) has been divided into four major *eras:* the Cenozoic, "recent life"; the Mesozoic, "middle life"; the Palaeozoic, "ancient life"; and the Precambrian, rocks formed before sufficient, suitable fossils for dating existed. Within each era there are a series of *periods* of geologic time, some of which are divided into shorter *epochs.* All of human history took place after the rocks of the Grand Canyon formed and the Colorado River cut its valley into them several million years ago.

Rocks contain evidence for their own time of formation. The order of the layers and the relationships of intrusions to layered rocks make it possible to reconstruct a sequence of events. The Grand Canyon cut by the Colorado River exposes one such sequence (see Figure 7-22). At the bottom are highly metamorphosed and contorted rocks separated from the overlying piles of sedimentary rocks by a surface that forms a junction between contrasting rock groups known as an *unconformity.* The junction between the metamorphic and sedimentary rocks represents a major time gap, during which the rocks beneath the unconformity were raised into mountains and worn down. The ocean then rose to cover the land and deposited layer upon layer of sedimentary rocks on top of the worn-down mountains.

Another guide to the relative age, or sequence, of rocks is the fossils they contain. Fossils are the remains of ancient living organisms, and they display definite trends that change over time. Rock layers of a similar age contain similar groups of fossils; those of different ages contain different groups. Scientists use the fossil record to define the eras and periods of geologic time.

Some rocks do not contain fossils but contain minerals with radioactive elements. The rates of decay, or change, in these minerals is constant and is not affected by heat, pressure, or rock disintegration. Using such minerals in *radiometric dating* makes it possible to estimate the age in years of the rock containing them. Igneous rocks are particularly good for this purpose since minerals containing radioactive elements form in them as the magma cools. Radiometric dating shows that the oldest rocks with a plentiful fossil record are 570 million years old, and the oldest known rocks on Earth are 3.8 billion years old. These dates confirm the long periods of time in which solid-earth environment processes occur.

In many cases several techniques are combined in geologic time dating. For example, the studies of human origins in eastern Africa use a mixture of fossils, human tools, and radiometric dates.

A variety of rocks compose Earth's crust and their materials circulate through the rock cycle. The major rock types are igneous, metamorphic, and sedimentary. The layers and bodies of rock deform into folds and faults and are also cut by joints. The rocks contain evidence for constructing a time sequence of their formation.

The formation of sedimentary rocks involves interactions between Earth environments, whereas igneous and metamorphic rocks are products of the **solid-earth environment.** The initial rock materials are brought to the surface by solid-earth environment processes; the **atmosphere-ocean environment** interacts with these rocks in the **surface-relief environment** to produce sediment and dissolved minerals, the constituents of sedimentary rocks; the **living-organism environment** produces peat, coral reefs, and shell-based materials that become a part of some sedimentary rocks.

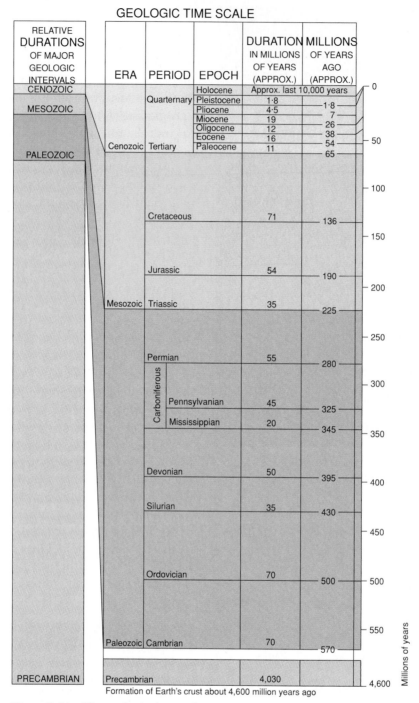

GEOLOGIC TIME SCALE

RELATIVE DURATIONS OF MAJOR GEOLOGIC INTERVALS	ERA	PERIOD	EPOCH	DURATION IN MILLIONS OF YEARS (APPROX.)	MILLIONS OF YEARS AGO (APPROX.)
CENOZOIC		Quarternary	Holocene	Approx. last 10,000 years	0
			Pleistocene	1·8	1·8
MESOZOIC			Pliocene	4·5	7
			Miocene	19	26
			Oligocene	12	38
	Cenozoic	Tertiary	Eocene	16	54
PALEOZOIC			Paleocene	11	65
		Cretaceous		71	136
		Jurassic		54	190
	Mesozoic	Triassic		35	225
		Permian		55	280
		Carboniferous	Pennsylvanian	45	325
			Mississippian	20	345
		Devonian		50	395
		Silurian		35	430
		Ordovician		70	500
PRECAMBRIAN	Paleozoic	Cambrian		70	570
		Precambrian		4,030	4,600

Formation of Earth's crust about 4,600 million years ago

Millions of years

Figure 7-31 The geologic time scale.

CONTINENTS, OCEAN BASINS, AND MOUNTAIN SYSTEMS

The major features of Earth's surface, such as continents, ocean basins, and mountain systems, are built and destroyed by plate motions. Oceans grow and shrink, continents move over Earth's surface, mountain systems rise and are worn away, and new rocks form. Such processes take hundreds of millions of years, with the progress showing from time to time in earthquakes or volcanic eruptions.

GROWTH AND DECLINE IN OCEAN BASINS

Ocean basins are large depressions in Earth's surface filled with water. They are the most fundamental result of plate tectonic processes, and their expansion and shrinkage determines movements of the continents and the formation of mountain systems. Ocean-floor plates form along divergent plate margins and move toward convergent margins where subduction destroys them in a massive recycling process. The oldest rocks in ocean basins are only 250 million years old.

The major relief features of ocean basins include ocean ridges, ocean floor plains, submarine trenches, and volcanic peaks (Figure 7-32). Each of these relief features is a product of plate tectonics.

Ocean Ridges   The creation of new ocean floor occurs along divergent plate margins as expansion of molten rock in the upper mantle lifts these margins to form ocean ridges. A connected system of such ridges, 60,000 km (40,000 mi) long, winds around the globe. **Ocean ridges** have a central depression where magma wells up through individual vents or fissures and solidifies. The rocks formed are mafic basalts and related rock types. On either side of the central depression, parallel ridges form as new magma rises and pushes aside the solidified rocks. Ocean ridges typically rise above the ocean floor to within 1500 to 3000 m (5000 to 10,000 ft) of the ocean surface but rarely break the surface. The Mid-Atlantic Ridge is an exception, rising above sea level in Iceland. Ocean ridges are often offset in a zigzag pattern by transform plate boundaries. The world map of ocean basins shows how such transform boundaries cut across ocean ridges almost at right angles.

Most ocean ridges are distant from continents. In the Atlantic Ocean, the ridge is central, winding between the American continents to the west and Europe and Africa to the east. The ridge in the Pacific Ocean lies to the east of center and is known as the East Pacific Rise. In places an ocean ridge passes beneath a continent and produces special features in the continental relief above. For instance, the Indian Ocean ridge passes into the southern end of the Red Sea, which is also connected to the rift valleys of eastern Africa to the south and the Jordan Valley to the north. The East Pacific Rise passes beneath the Gulf of California and the San Andreas strike-slip fault system (see Figure 7-28).

Ocean-Floor Plains   Wide, relatively flat areas occur in all oceans at depths of 3500 to 4000 m (12,000 to 13,000 ft) and are called **abyssal plains.** The accumulation of a blanket of sediment up to a kilometer thick on top of the ocean-floor crust results in their monotonous relief (Figure 7-33).

At the edges of the continents, clouds of fine sand and clay flow down undersea canyons at the continental margins and deposit sediment in sloping

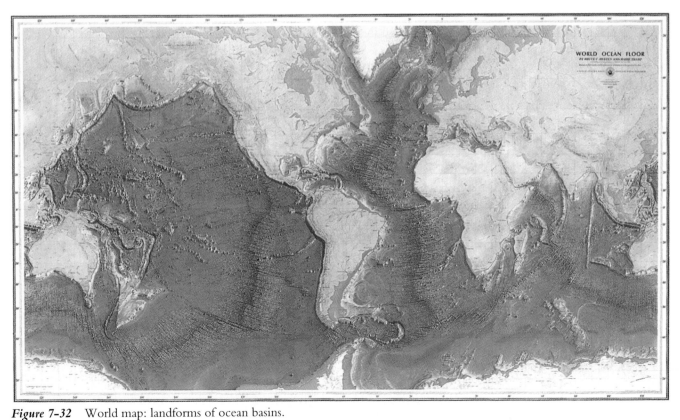

Figure 7-32   World map: landforms of ocean basins.
World Ocean Floor, Bruce C. Heezen and Marie Tharp, 1977. © by Marie Tharp 1977. Reproduced by permission of Marie Tharp, 1 Washington Ave., South Nyack, NY 10960.

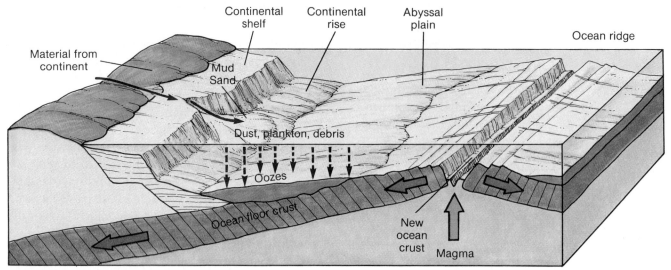

Figure 7-33   Ocean basin landforms. Abyssal plains formed of ooze dominate the basin floors; thickness of ooze is linked to the age of the ocean crust beneath. The ocean ridge and continental rise are other features of ocean basins.

continental rises along the edges of the ocean floor. Continental rise deposits may be up to 5 km (3 mi) thick on their continental margin but thin and merge into the ocean-floor plains.

Away from the continental sources of river-born sand and clay, the ocean floor sediments accumulate more slowly. A "rain" of tiny skeletons from ocean surface organisms along with some atmospheric dust sinks to form deposits called *oozes*. The uniformity of these fine deposits is interrupted by large boulders dropped from melting icebergs and by manganese nodules formed in chemical reactions on the ocean floor. Ocean floor oozes thin toward the ocean ridges. The thickest deposits are at the ocean margins and contain the longest sequence, including oozes formed up to 250 million years ago. Close to an ocean ridge, the oozes are only a few million years old because the ocean floor is youngest there.

Ocean Trenches   An **ocean trench** is an elongated trough in the ocean floor and contains the deepest parts of the oceans. A typical trench is hundreds of kilometers across and thousands of kilometers long but contains little sediment—indicating an origin in the last few tens of millions of years or a lack of time for sediment to accumulate. Nearly all the world's ocean trenches occur around the margins of the Pacific Ocean close to convergent plate margins.

Volcanic Peaks   Thousands of volcanic peaks occur in ocean basins, but few rise above sea level, and most are dormant or extinct. They occur either as single, widely separated peaks or in groups such as those formed over the Hawaiian hot spot. Where volcanoes form a distinctive arc-shaped series along the side of an ocean trench above a subduction zone (see Figure 7-9) they are known as an **island arc.** The Aleutian Islands of Alaska are an island arc. As the Pacific plate was subducted beneath the North American plate, the downward limb melted and magma rose through the overlying crust to form volcanoes. Volcanic lava and ash also erupted and accumulated to form islands that coalesced into the larger land mass of Japan (Figure 7-34).

Ocean Basin History   Ocean basins open and close through time (Figure 7-35). The stages in this sequence can be identified among the present ocean basins, providing an understanding of what happened to them in the past and what may happen in the future.

The creation of an ocean basin begins when an upward convection current heats the lithosphere. The heated rocks expand, arching the surface until the top cracks open and a rift forms, like those in eastern Africa (Figure 7-35a). As the rift widens, volcanic activity increases along its length. The zone of rifting becomes a divergent plate margin. As the rift deepens, ocean water floods in, initiating an ocean basin. The Red Sea, linked to the eastern African rifts, demonstrates this stage in which an ocean ridge forms the whole width of the new "ocean" (Figure 7-35b).

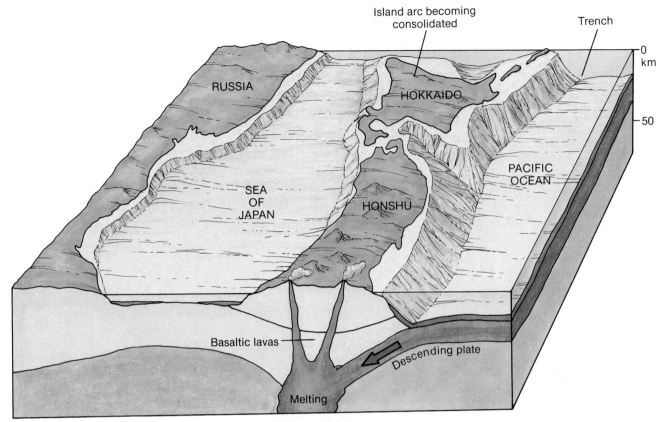

Figure 7-34 Island arc. The example of Japan shows how ocean plate is subducted and melts at 100 km (60 mi) deep; the molten magma rises through the overlying rocks, producing volcanoes of basalt lava.

The formation of new lithosphere at the divergent margin causes the ocean to widen. The ocean ridge remains central and parallel to the ocean margins as the ocean expands equally on either side of the ridge (Figure 7-35c). Ocean floor plains and a number of volcanic peaks form on either side of the ridge. The Atlantic Ocean formed in this manner and represents this stage.

Once subduction of the ocean floor begins at the ocean margins, trenches form (Figure 7-35d). The ocean basin reaches its maximum size and then begins to contract as the ocean trenches consume more ocean floor than is produced at the ocean ridge. The Pacific Ocean represents this stage. As one ocean contracts, others expand. The expanding Atlantic Ocean has pushed the American continents westward over the eastern parts of the Pacific Ocean plates, including the East Pacific Rise in places.

In the final stage of ocean basin history (Figure 7-35e), the ocean basin is reduced to a small inland sea. The Mediterranean Sea is the remnant of a former ocean basin that extended between Europe and Asia to the north and Africa and India to the south.

MOBILE CONTINENTS

As the ocean basins open and close, the continents move around Earth's surface alternately splitting apart and joining. The maps of Figure 7-36 show that most of the continents were in the southern hemisphere 540 million years ago. They moved together to form a single continent about 240 million years ago and then split apart again. The map predicting the positions of the continents in 50 million years emphasizes the continuing nature of the movements.

The largest proportions of continental masses stand above sea level although their margins may be covered by a shallow sea. As was mentioned earlier, continents stand high in Earth's relief because they have a thicker layer of low-density sial rocks. These rocks remain on the surface and accumulate there instead of being recycled in subduction. Continents

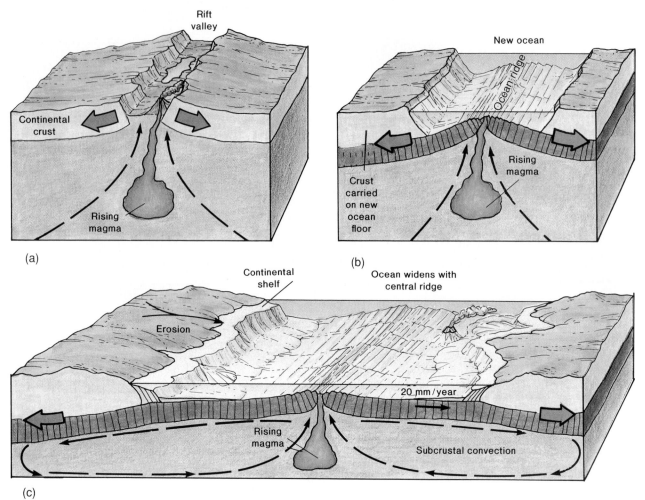

Figure 7-35 History of an ocean. Ocean basins open and close, affecting the margins of continents. Five stages are shown. (a) A rift valley. (b) New ocean forms as sides of rift move apart. (c) A mature ocean basin with broad central ridge and continental shelves on either side. (d) Ocean crust is subducted at edges, producing mountain ranges on the continents and beginning the closing of the ocean. (e) The ocean closes and mountain ranges on either side fuse.

preserve Earth's oldest rocks and have grown in areas over time. Because they are exposed above sea level, however, continents are subject to wearing down by running water, glaciers, and wind.

Continental areas contain a variety of relief features which are distinct from those in the ocean basins. High mountain systems, plateaus, and low-lying plains dominate the relief (Figure 7-37).

Young Folded Mountains   The highest mountain systems of the present day are, in order beginning with the highest, the Himalayas of northern India, the Andes of South America, and the Alps of southern Europe. They generally exceed heights of 3000 m (10,000 ft) above sea level and the highest points rise above 8000 m (25,000 ft). They occupy elongated areas up to a few hundred kilometers wide.

The highest mountain systems are called **young folded mountains** because most of their rocks formed during the last 250 million years and were folded by compression at convergent plate margins over the last 50 million years. The main belts of young folded mountains occur on the western margins of the American continents, where ocean plates collide with a continent-carrying plate, and in a zone

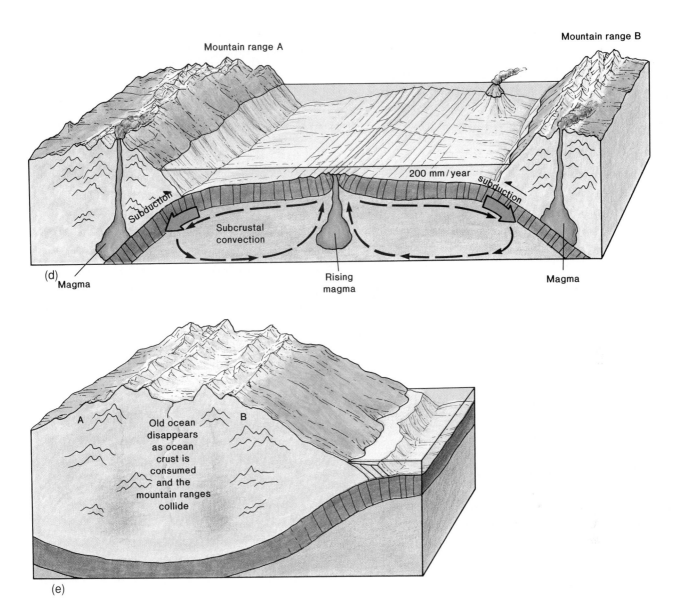

Figure 7-35 (continued)

from southern Europe to northern India, where two continent-carrying plates collide with another.

The young folded mountains consist of great thicknesses of low-density continental sial, including granites, gneisses, and sedimentary rocks. The Alps, for instance, have a sial thickness of 10 km (6 mi) compared to 2 km (1.2 mi) to the north and south. The complex structures into which the rocks have been contorted include large overturned folds several hundred kilometers across and thrust faults, demonstrating the intense compression that has occurred. The rocks in the cores of these mountain ranges, exposed by erosion, include highly metamorphosed rocks and huge granite batholiths aligned with the trend of the mountain ranges (see Figure 7-17). These rocks indicate deep burial of large sections of the crust at the time of mountain formation.

FAULT-BLOCK MOUNTAINS

A **fault-block mountain** is an uplifted section of crust with one or more faulted boundaries. Examples of fault-block mountains include the horsts of central Europe (see Figure 7-29), the Sierra Nevada of California, and the ranges of Nevada and Utah. Some

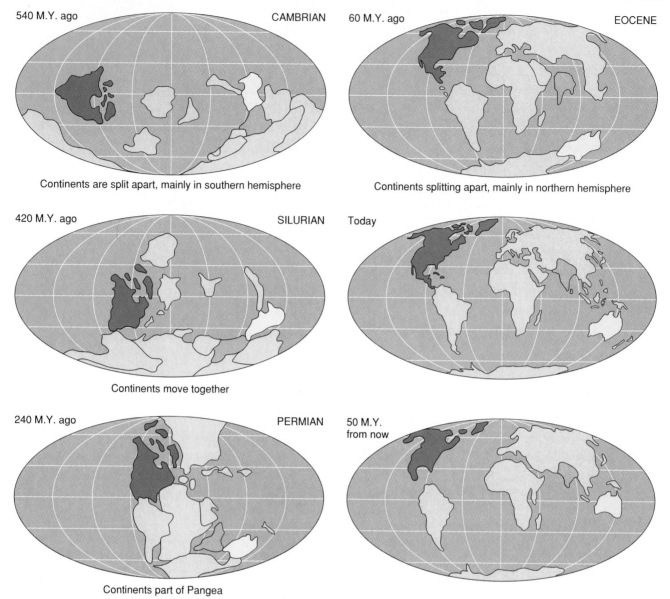

Figure 7-36 Continental drift. The changing positions of continents on Earth's surface over the last 540 million years and up to 50 million years ahead. Follow the histories of Australia (yellow), North America (red), and India (green).

fault-block mountains formed flat-topped plateaus when their surface was worn down to a plain before faulting and uplift.

The rocks in the fault-block mountains of central Europe formed mainly between 300 and 600 million years ago. The constituent rocks, the overturned folds, and the thrust faults that are typical of these closely resemble those observed in young folded mountains. Geologists conclude that the fault-block mountains were parts of folded mountains some 250 to 300 million years ago and were then worn down before being faulted and uplifted in the last 50 million years. Evidence of such ancient mountain systems makes it possible to reconstruct the convergent plate margins of the past.

PRECAMBRIAN SHIELDS

A **shield** area is a large section of the continental crust that has remained largely unaffected by earth movements since Precambrian times, over 600 million years ago. Their rocks have eroded during that time. If subsequently covered by other rocks, they have been stripped so that the Precambrian rocks occur at the surface. Such areas may form plateaus

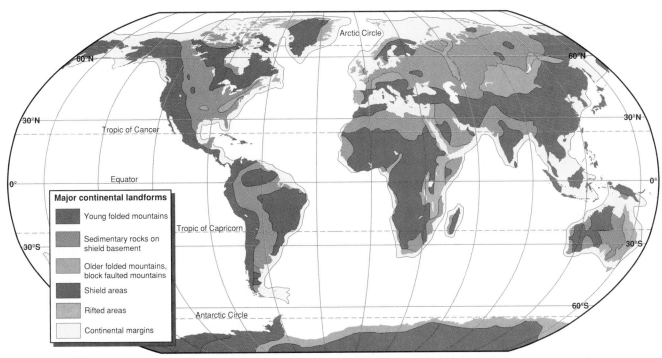

Figure 7-37 World map of continental landforms. Locate the areas where the main types of continental relief features occur. At this scale finer details are obscured.

rather like a shield lying on the floor with its front side up. Exposed shield areas make up about 25 percent of continental surfaces.

The surface relief of shield areas varies from place to place, but shows little relationship to the internal fold or fault structures formed in long-gone folded mountain systems. The knobbly relief of northern Canada, the plateaus of Africa, and the submerged rocks beneath the Baltic Sea are all parts of shield areas.

Internally, shields consist of a series of elongated zones of highly folded and metamorphosed rocks bearing close resemblances to those of the inner sections of young folded mountains. Each shield area commonly contains evidence of several phases of mountain building that took place between 3.8 billion and around 600 million years ago (Figure 7-38). The repetition of similar rocks and structures in continental crust of different ages demonstrates the basic importance of continuing cycles of mountain and rock formation and destruction since the first crustal rocks formed. The exposure of rocks that formed deep in the core of fold mountains also shows the effectiveness of surface processes that wear down mountains.

Precambrian rocks form the structural cores of the continents, often covered by younger rocks. In such cases, the Precambrian shield rocks form the continental *basement* within the sial. Shield areas covered by thin layers of sedimentary rock include the Midwest and Gulf and Atlantic coast plains in the United States, the Paris Basin of France, and large parts of Africa. The sedimentary rocks are often tilted or simply folded as a result of warping in the shield on which they lie. In the Deccan plateau of southern India, the Precambrian shield rocks are covered by little-disturbed layers of basalt.

Continental Margins   Continental margins are composed of continental crust and are designated active or passive according to their relationship to a plate margin. An *active* continental margin is where a convergent plate margin produces earthquakes, volcanic activity, and vertical movements of crust. The western coast of North America is an example. A passive continental margin, such as the eastern coast of North America, is a margin where there is no plate boundary and where offshore deposition of sediment occurs. Ocean water covers most continental margins today.

Passive margins often have broad continental shelves and continental slopes (Figure 7-39). A **continental shelf** is a gently sloping (gradient under 1 degree) offshore extension of a continent, above which the sea depth seldom exceeds 130 m (400 ft).

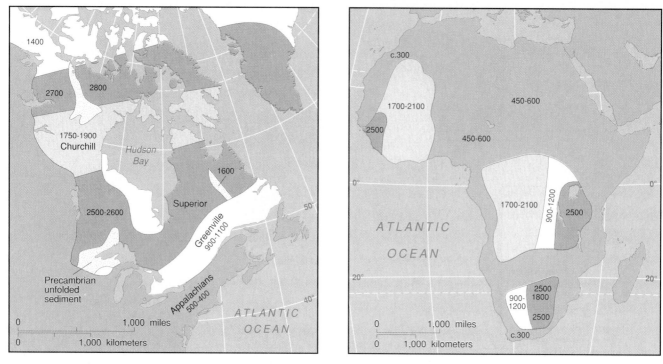

Figure 7-38 Shield areas. The shield areas of North America and Africa are composed of ancient belts of folded mountains. The numbers refer to the age of mountain formation in millions of years. Rocks younger than Precambrian are shown in green.

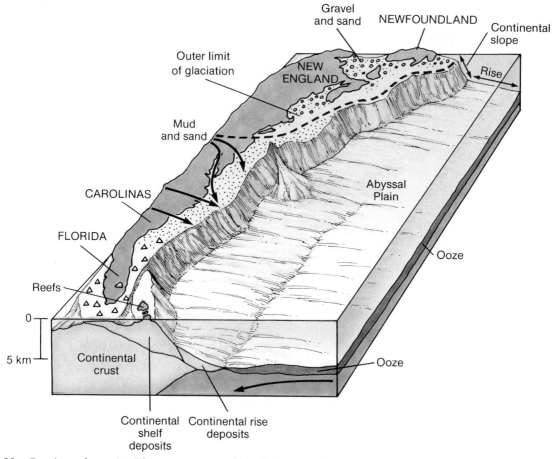

Figure 7-39 Continental margin. The eastern coast of North America has a wide continental shelf and slope leading down to the ocean floor. This passive margin is composed of a wedge of rocks formed over the last 200 million years.

Most continental shelves are areas of sediment deposition. The continental shelf along the eastern coast of the United States is a good example; the continental shelf marks the upper surface of a wedge of sedimentary rock that is 5 km (3 mi) thick at its outer edge. The wedge formed as the Atlantic Ocean opened during the last 200 million years. Streams and glaciers eroded the eastern mountains of North America depositing sediment offshore. During the Pleistocene Ice Age there were times when the surface of the continental shelf lay above sea level and was covered by ice in the north and river plains in the south.

At its outer edge, the continental shelf gives way to the *continental slope,* which has a gradient of up to 5 degrees and marks the true boundary between continent and ocean. At the base of the continental slope, the edge of the continental rise marks the beginning of ocean floor features.

The rocks beneath the continental shelf and continental rise off eastern North America resemble closely the rock types and total thicknesses of sedimentary rocks that occur in young folded mountains and the worn-down remnants of older mountain systems. The rocks accumulating in this continental margin zone will probably become part of new folded mountains when the ocean closes and continental collisions occur in a few tens of millions of years' time.

BUILDING MOUNTAIN SYSTEMS, AND OTHER CRUSTAL MOVEMENTS

Mountain systems are the main building blocks of continental crust. The process of mountain building is called **orogenesis,** a word meaning "mountain forming," and is responsible for the most dramatic movements in the continental crust. Broader warping movements of the crust uplift plateaus or cause subsidence of large areas, and are called **epeirogenesis,** a word meaning "continent forming."

Environments of Mountain Formation   Mountain ranges form along convergent plate margins where collisions of different plate types cause different results. Where two ocean floor plates collide, an island arc forms through volcanic activity, as in Japan or the Aleutian Islands (see Figures 7-9 and 7-34).

When an ocean floor plate and a continent-carrying plate collide, the denser ocean floor crust is subducted forming an ocean trench, producing earthquakes, compressing the continental margins, and uplifting the mass of thickened crust at the edge of the continent. The Andes of South America are an example of this type of mountain building, known as the *cordilleran* type (Figure 7-40). Plate convergence in this area began some 180 million years ago shortening and thickening the continental crust. It also added igneous rocks to the crust as a result of subduction and melting of the ocean floor plate. An island arc formed off the coast. By 70 million years ago, further movement of the Nazca plate crushed the island arc against the continent to form the western cordillera. This cordillera is built mainly of granite batholiths and topped by many volcanic peaks. Continuing deep earthquakes along this zone demonstrate that the Nazca plate is still advancing from the west and is being subducted beneath the South American plate. These movements also folded, compressed, and uplifted the eastern cordillera to produce the two parallel ranges of the central Andes in Peru and Bolivia.

When two continent-carrying plates collide, the masses of continental crust remain at the surface and are not subducted. This type of mountain building is known as the *Himalayan* type, after its main example (Figure 7-41). In forming the world's highest mountains, the Indian plate drove northward into Eurasia over the last 80 million years. As the Indian continental plate smashed into the Eurasian continental plate, the lower density continental rocks remained at the surface forming the huge combined mass of the Himalayan mountain system and the Tibet Plateau to the north (Figure 7-42). Major thrust faults are common, but igneous activity is less common than in the cordillera type of fold mountains.

The Mountain Systems of North America   The relief of North America is dominated by two mountain systems—the Appalachians in the east and a variety of mountain ranges and plateaus in the western third of the continent. Both result from complex events that occurred over hundreds of millions of years.

The Appalachian Mountains are a remnant of a mountain system that extended from northern Scandinavia to western Africa some 250 million years ago. These mountains formed when an earlier version of the Atlantic Ocean closed, forming a single world landmass. The earlier Atlantic Ocean resembled the present ocean in some ways, with continental shelf and rise deposits along each side. The ocean floor was then subducted and the ocean was reduced in size until the continents on either side crashed together and formed a Himalayan-type folded mountain system. After the formation of this

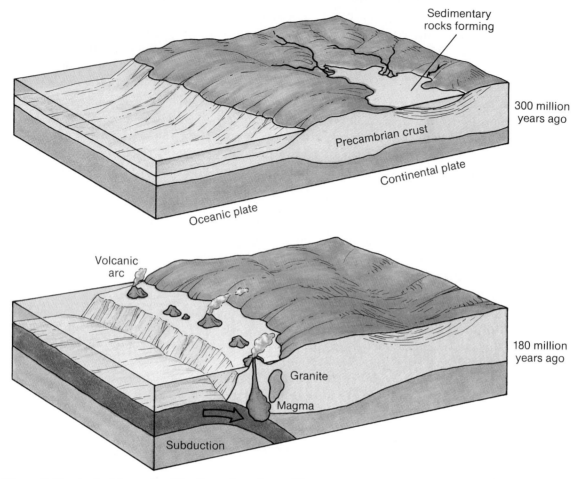

Figure 7-40 Andes Mountains. The formation of a cordilleran-type mountain system. Relate these events to plate tectonics and the operation of the geologic cycle.

huge mountain belt, the Atlantic split open again, breaking the belt into sections such as the Appalachian Mountains.

The western cordilleras, from the Rockies to the coastal ranges bordering the Pacific Ocean, began with collisions between ocean and continental plates over 100 million years ago. After a cordillera-type folded mountain system formed, some of the higher areas collapsed to form basins. Deposition of sedimentary rocks and the eruption of volcanic lavas filled the depressions, which were raised with little folding to form plateaus such as the Colorado Plateau. Faulting raised up ancient rocks to form the Sierra Nevada and sections of the Rocky Mountains. While these events affected the interior areas, subducted ocean plates continued to smash island arcs and smaller sections of continental crust against the western coastal mountains. Volcanic activity caused by this subduction, such as the Mount St. Helens eruption, continues in the Cascade Mountains of Washington and Oregon. Surface processes wore down the mountains as they rose, exposing batholiths and regionally metamorphosed rocks. The rocks of some of the highest points today formed deep in the crust and reached their present position following erosion of the overlying rocks and uplift by faulting.

Warping the Crust   While the most dramatic effects of plate movements result from horizontal collision or pulling apart at convergent and divergent plate margins, broad vertical movements may affect large areas in plate interiors. Such epeirogenesis uplifted the huge shield plateaus of most of Canada, Africa, and Australia and caused the subsidence of areas such as the North Sea basin of Europe. The mechanics of these vertical movements are poorly understood, but may be linked to variations of temperature in the weak asthenosphere layer. Such variations may result in the sagging or uplift of the overlying lithosphere.

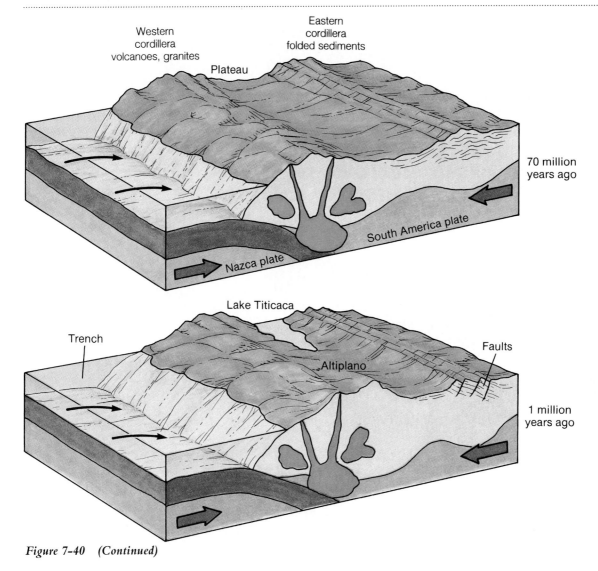

Western
cordillera
volcanoes, granites

Eastern
cordillera
folded sediments

Plateau

70 million
years ago

South America plate

Nazca plate

Trench

Lake Titicaca

Faults

Altiplano

1 million
years ago

Figure 7-40   (Continued)

Besides epeirogenic movements caused by internal changes, the crust may also sag or rise as a result of extra loads added on the surface. One example occurred during the Pleistocene Ice Age, when large ice masses accumulated on northern North America and Europe. The extra weight of ice caused the crust beneath to sink (Figure 7-43a, b). When the ice melted, the sea level rose because of the amount of water returned to the oceans, but the crust returned more gradually to its former level (Figure 7-43c, d). For instance, when the ice melted over northern Europe between 15,000 and 8000 years ago, the Baltic Sea formed by flooding of the area that had been depressed beneath ice. The crust is now rising at 8 to 10 mm (0.3 to 0.4 in) per year and will take several thousand years to return to its preglacial level.

The process by which Earth's surface adjusts its height to variations in the combined weight of crustal rocks and other loads is known as **isostasy,** which is a state of balance between the density differences of sial and sima rocks. Sialic rocks are relatively low in density and additions to them cause an adjustment in balance that has been likened to an iceberg floating in the ocean. The compression or thickening of sialic rocks during mountain building along convergent plate margins leads to the concentration of great thicknesses of low-density rocks. The extra weight of such concentrations causes the lithosphere rocks to sag, but the additional low-density mass of rocks also results in a raising of the upper surface to restore the balance between the sialic and higher density rocks. The highest mountains in the

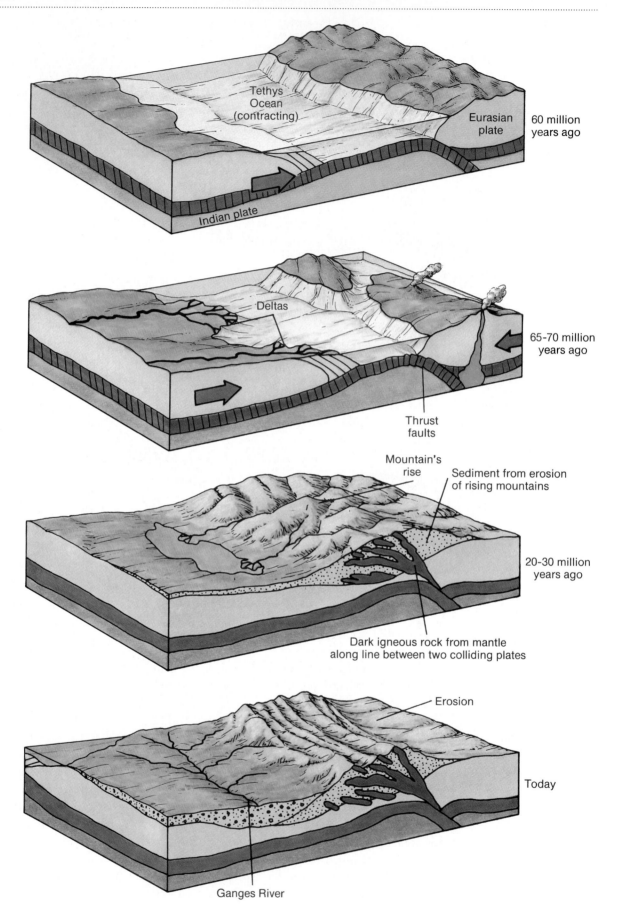

Figure 7-41 Himalayan Mountains. The formation of a mountain system where two continents collide. Compare the processes and outcomes with the formation of the Andes Mountains (Figure 7-40).

Figure 7-42 Himalayan ranges. The view to the west, taken from a space shuttle. The Tibetan plateau lies to the north (right) of the snow-capped Himalayan peaks.

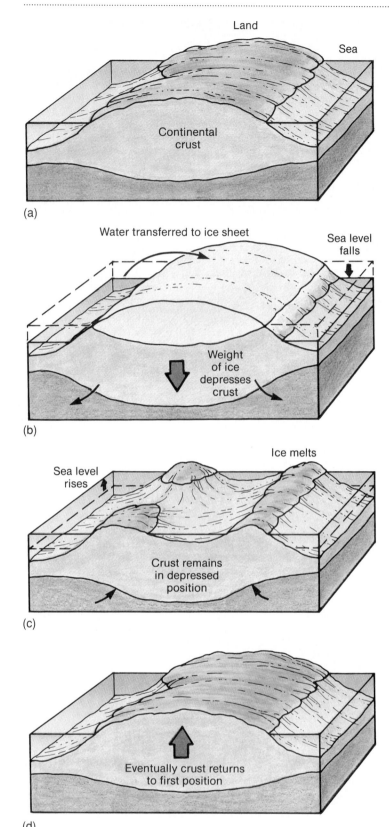

Land

Sea

Continental crust

(a)

Water transferred to ice sheet

Sea level falls

Weight of ice depresses crust

(b)

Ice melts

Sea level rises

Crust remains in depressed position

(c)

Eventually crust returns to first position

(d)

Figure 7-43   Isostasy. The impact of loading a continent with an ice sheet—during and after the event. The land mass (a) is depressed by the weight of ice (b), and adjustment occurs in the asthenosphere. When the ice melts (c), the land takes time before recoiling to its former position (d).

world today owe their altitude partly to isostasy because they are composed of great thicknesses of sialic crust.

SUPERCONTINENTS

The patterns of mountain formation, ocean basin opening and closing, and continental growth described in this chapter emerged from applying the theory of plate tectonics. More speculative ideas suggest that there have been repeated patterns of continental splitting apart and coming together over even longer periods of time. A *supercontinent* forms when continents unite in a single world continent. The formation and splitting apart of supercontinents have major impacts on all Earth environments.

A supercontinent cycle involves a distinct sequence of events (Figure 7-44). Approximately 100 million years after the formation of a supercontinent, rifting and the eruption of mantle rocks begin to tear it apart. During the next 40 million years, the continents separate as widening oceans open gaps between them. The dispersal continues for some 160 million years. Eventually subduction begins at some ocean margins. Earth is currently at the beginning of this stage in the sequence. As the rate of subduction overtakes and then exceeds the rate of ocean widening, the continents are drawn together. After another 160 million years, a new supercontinent forms. Then the cycle begins again. Supercontinents existed around 250, 650, 1100, 1650, 2100, and 2600 million years ago; each cycle lasted 400 to 550 million years.

The sequence of events in the supercontinent cycle affects many Earth environments. First, the changing shapes of ocean basins cause changes in sea level. When a supercontinent exists, the ocean basins are at their largest and deepest, and sea level falls. The continental breakup causes new ocean floor rock to be formed along ocean ridges and this takes up space in the ocean basins. Sea level rises and covers continental margins. At the time of maximum continent dispersal, the ocean ridges are less prominent and sea level falls again. The sea level rises as continents begin to move together.

Second, the diversity and numbers of marine organisms fluctuate with the changes in sea level. The greatest diversity of living organisms occurs

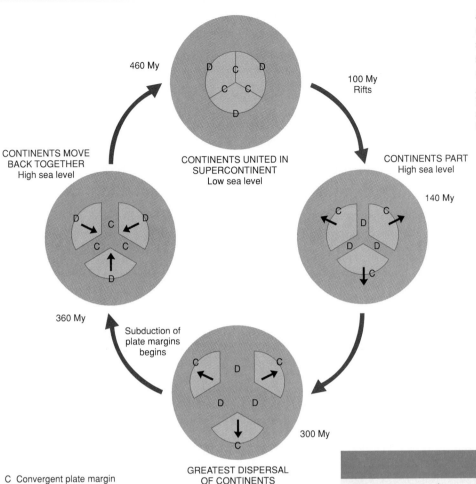

460 My

100 My
Rifts

CONTINENTS MOVE
BACK TOGETHER
High sea level

CONTINENTS UNITED IN
SUPERCONTINENT
Low sea level

CONTINENTS PART
High sea level

140 My

360 My

Subduction of
plate margins
begins

300 My

GREATEST DISPERSAL
OF CONTINENTS
Low sea level

C Convergent plate margin
D Divergent plate margin

Figure 7-44   Supercontinent cycle. The schematic diagram shows a single world continent splitting apart and reassembling over hundreds of millions of years.

Adapted from "The Supercontinent Cycle" by R. Damian Nance, Thomas R. Worsley, and Judith B. Moody. Copyright © 1988 by Scientific American, Inc. All rights reserved.

in shallow continental shelf areas that are covered by ocean waters during periods of high sea level. The falling sea level exposes continental shelves and causes major extinctions of the rich variety of life forms.

Third, the movements of continents and fluctuations in sea level cause climate changes. During periods of low sea level, a greater area of continental rocks is exposed to the atmosphere. More dissolved products are made as these rocks are worn down. The dissolved materials are carried into the oceans, where they provide additional nutrients to marine plants. Increased growth of marine plants leads to the removal of carbon dioxide from the atmosphere as it is absorbed by the growing plants and incorporated in their tissues and shells. The reduction of carbon dioxide levels leads to cooling and the formation of polar ice sheets, which lowers the sea level further. As the contrast in temperature between polar and tropical waters increases, the more turbulent ocean circulation brings more nutrients to the surface and enhances biological productivity further. The reverse conditions apply during high world sea levels, leading to lower biological productivity and atmospheric warming.

Plate tectonics produces two major environments—ocean basins and continents—in which distinctive groups of major landforms originate. The ocean basin environment is marked by ocean ridges, deep ocean plains, ocean trenches, island arcs, and isolated volcanic peaks. Ocean basins evolve from rifted sections of continental crust that widen as new lithosphere is produced along divergent plate margins. The ocean basins contract when subduction becomes dominant, and they may close altogether.

The shapes of ocean basins are the result of plate tectonics (**solid-earth environment**). They influence the circulation of ocean water and the heating of the atmosphere (**atmosphere-ocean environment**). They also affect the distribution of living organisms, both on land and in the sea (**living-organism environment**). The remains of marine plants and animals provide a large proportion of ocean-floor sediments.

Continental environments are built by folded mountains, including those that form the highest mountain systems today and the worn down remnants of older systems. Folded mountains form when plates collide. Continents join and split apart in a cycle that affects sea level, living organisms, and climate.

Just as changes in the ocean basins caused by plate tectonics (**solid-earth environment**) affect other Earth environments, so the formation and splitting of supercontinents has major influences on sea level and climate (**atmosphere-ocean environment**), on the relative roles of wearing away and deposition in continental-margin areas (**surface-relief environment**), and on the expansion and extinction of living things (**living-organism environment**).

ENVIRONMENTAL ISSUE:

Earthquake Hazards— Can Humans Cope?

Some of the issues involved with this debate include:

How much is understood about earthquakes?

Can earthquakes be forecast like hurricanes?

What can be done to mitigate the effects of an earthquake?

The largest earthquakes cause extreme damage and loss of life. An earthquake in northern China in the 1980s is thought to have killed at least 250,000 people. The Mexico City earthquake of 1985 was the largest in recent years to affect a major urban center. Earthquakes in Armenia and Iran each killed tens of thousands of people in smaller towns. The 1989 San Francisco earthquake was about one-tenth the power of these very large earthquakes and caused fewer deaths and less destruction, damaging part of a freeway, a section of bridge, and some older houses near the waterfront, as the photos show. Huge cities such as San Francisco, Los Angeles, Tokyo, Tehran, Beijing, New Delhi, and Mexico City, however, live with the daily possibility of a large earthquake that would be extremely costly in lives and property. The San Fernando Valley area of Los Angeles was struck by moderately strong earthquakes in 1971 and 1994 that destroyed homes and freeway bridges.

The violence and widespread impact of earthquakes present special hazards to human activities. Earthquakes are too powerful to modify and are very difficult to forecast. Linking earthquakes to plate tectonics has identified the zones most

likely to be affected, but determining the precise location and time of an earthquake has so far proved impossible. Moreover, some large earthquakes occur on rare occasions outside the plate margin zones, including one in Missouri and another in South Carolina in the last century. Virtually every earthquake is a surprise.

Those living near the San Andreas fault zone in California are among the best prepared for earthquakes despite the rather relaxed attitude of many Californians who have already lived through several smaller tremors. Earthquake drills include rescue exercises, and building codes for new construction are strict.

After the 1989 earthquake in the San Francisco area, the U.S. Geological Survey produced a booklet entitled *The Next Big Earthquake in the Bay Area May Come Sooner Than You Think*. It was printed in English, Spanish, Chinese, and Braille and distributed to all homes by local newspapers. The booklet includes information on the types of areas around San Francisco Bay exposed to earthquake risks and shows them on a map. Some areas are close to the fault zone and are likely to be wrenched apart by the moving rocks, some have steep slopes prone to landsliding after a shock, some have land that is liable to sink, and some have soft soils that turn to loose mud with shaking. The strongest effects of an earthquake occur around the margins of the Bay, where sands, silts, and clays were deposited after the nineteenth century hydraulic mining for gold in the Sierra Nevada. Sand was also used to fill a lagoon in the Marina district for the Panama-Pacific International Exposition of 1912, and this area suffered particular damage in 1989. The booklet emphasizes that an earthquake larger in magnitude than that of 1989 could occur in the next

decade and lists the specific localities most likely to be affected. Steps needed to reduce earthquake damage are discussed, including the need to examine building safety and to strengthen first-floor structures in older buildings. Although much has been done to strengthen freeway bridges around Los Angeles since the 1971 earthquake, a few that had not been treated were destroyed again in 1994.

A crucial question arising from the nature of earthquake hazards is how much money should be spent to improve monitoring and forecasting techniques and to develop the technology and planning required to reduce damage when an earthquake occurs. People choose to live and work in places like the coastal regions of California or the city of Tokyo where risks are high. Spending more public money on reducing the risk or forbidding construction improvements in risky areas adds to the high costs of living in California and Tokyo, yet the great majority of people and localities in these regions may never experience a major earthquake.

Linkages

Earthquake activity is caused by movements in the **solid-earth environment** but often alters features in the **surface-relief environment.**

Questions for Debate

1. Should people be prevented from living in earthquake zones?

2. Is the threat of an earthquake as great as other everyday hazards that people meet?

3. How much responsibility for protecting people from such hazards should fall on the state, on corporations, and on individuals?

(a)   (b)

(a) Aerial view shows Nimitz Freeway section where top deck collapsed onto bottom deck during rush hour, crushing scores of people in Oakland, Thursday October 18, 1989. (b) Damaged building in the Marina District, San Francisco—October 18, 1989. (c) A section of highway and cars demolished in the 1994 Los Angeles earthquake.

(c)

ENVIRONMENTAL ISSUE:

Earthquake Hazards—Can Humans Cope? (Continued)

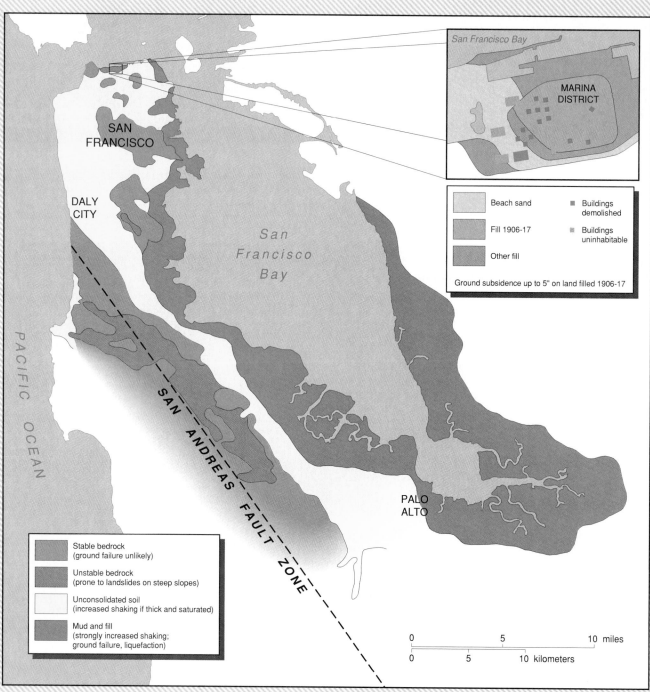

San Francisco Bay

MARINA DISTRICT

Beach sand	■	Buildings demolished
Fill 1906-17	▩	Buildings uninhabitable
Other fill		

Ground subsidence up to 5" on land filled 1906-17

SAN FRANCISCO

DALY CITY

San Francisco Bay

PACIFIC OCEAN

SAN ANDREAS FAULT ZONE

PALO ALTO

Stable bedrock
(ground failure unlikely)

Unstable bedrock
(prone to landslides on steep slopes)

Unconsolidated soil
(increased shaking if thick and saturated)

Mud and fill
(strongly increased shaking;
ground failure, liquefaction)

0   5   10 miles
0   5   10 kilometers

(d)

CHAPTER SUMMARY

1. The solid-earth environment consists of the internal layers of the planet and its rocky surface. The internal layers are the core and mantle. The crust and upper mantle include the rigid lithosphere and the underlying, weaker asthenosphere. The layers of the solid-earth environment are composed of rocks and minerals containing a small range of chemical elements.

2. Earth's surface features change continuously as the result of internal solid-earth processes that power plate tectonics. Plates are almost rigid sections of Earth's lithosphere, which converge and diverge, colliding and pulling apart. New lithosphere is produced by the cooling of rising magma along diverging plate margins and is destroyed by subduction at convergent margins. Earthquake and volcanic activity are concentrated along plate margins.

   Volcanic activity has significant impacts on other earth environments. The gases erupted into the atmosphere and oceans from the *solid-earth environment* are partly responsible for the changing composition of the *atmosphere-ocean environment.* Fine tephra may remain in the atmosphere for some time, causing extra brilliant sunsets and affecting atmospheric heating by reflecting insolation back to space. Volcanic eruptions produce landforms that are molded in the *surface-relief environment.* They produce materials that affect the composition and fertility of soils formed from them (*living-organism environment*).

3. A variety of rocks compose Earth's crust and their materials circulate through the rock cycle. The major rock types are igneous, metamorphic, and sedimentary. The layers and bodies of rock deform into folds and faults and are also cut by joints. The rocks contain evidence for constructing a time sequence of their formation.

   The formation of sedimentary rocks involves interactions between Earth environments, whereas igneous and metamorphic rocks are products of the *solid-earth environment.* The initial rock materials are brought to the surface by solid-earth environment processes; the *atmosphere-ocean environment* interacts with these rocks in the *surface-relief environment* to produce sediment and dissolved minerals, the constituents of sedimentary rocks; the *living-organism environment* produces peat, coral reefs, and shell-based materials that become a part of some sedimentary rocks.

4. Plate tectonics produces two major environments—ocean basins and continents—in which distinctive groups of major landforms originate. The ocean basin environment is marked by ocean ridges, deep ocean plains, ocean trenches, island arcs, and isolated volcanic peaks. Ocean basins evolve from rifted sections of continental crust that widen as new lithosphere is produced along divergent plate margins. The ocean basins contract when subduction becomes dominant, and they may close altogether.

   The shapes of ocean basins are the result of plate tectonics (*solid-earth environment*). They influence the circulation of ocean water and the heating of the atmosphere (*atmosphere-ocean environment*). They also affect the distribution of living organisms, both on land and in the sea (*living-organism environment*). The remains of marine plants and animals provide a large proportion of ocean-floor sediments.

5. Continental environments are built by folded mountains, including those that form the highest mountain systems today and the worn down remnants of older systems. Folded mountains form when plates collide. Continents join and split apart in a cycle that affects sea level, living organisms, and climate.

   Just as changes in the ocean basins caused by plate tectonics (*solid-earth environment*) affect other Earth environments, so the formation and splitting of supercontinents has major influences on sea level and climate (*atmosphere-ocean environment*), on the relative roles of wearing away and deposition in continental-margin areas (*surface-relief environment*), and on the expansion and extinction of living things (*living-organism environment*).

KEY TERMS

relief   *170*

landform   *170*

continent   *170*

ocean basin   *170*

crust   *172*

mantle   *172*

core   *173*

mineral   *173*

CHAPTER REVIEW QUESTIONS

1. Compare the composition and structure of the solid-earth and atmosphere-ocean environments.

2. How do Earth's core, mantle, and crust differ from each other, and how do the lithosphere and asthenosphere relate to these layers?

3. Describe the features of lithosphere plates and how plates interact with each other.

4. How do earthquakes and volcanoes support the theory of plate tectonics?

5. Relate the occurrences of earthquake and volcanic activity to a world map of plates.

6. Compare the forms and rock types of intrusive and extrusive igneous rocks.

7. What are the roles of heat and pressure in the creation of metamorphic rocks?

8. Show how sedimentary rocks are products of interactions among the solid-earth, atmosphere-ocean, and living-organism environments.

9. What are the differences among folds, faults, and joints in rocks?

10. What is the evidence for the great length of geologic time?

11. Compare the major relief features of ocean basins and continents.

12. What are the distinctive major landforms of an opening ocean and of a closing ocean?

13. How do plate tectonic processes create folded mountains?

14. What are the effects of a supercontinent cycle on sea levels, living organisms, and climates?

SUGGESTED ACTIVITIES

1. Reconstruct a geologic history outline of North America in terms of plate tectonics and mountain building; refer to the shields, the Appalachian Mountains, and the western mountains.

2. Examine some specimens of different types of rock. Describe them in terms of color, mineral size and shape, and other characteristics you notice. Compare your description with a fuller account in a manual and attempt to link the rock to the processes that formed it.

3. When an earthquake or volcanic eruption occurs, find its position on a world map and assess whether it is near a plate margin or in the center of a plate.

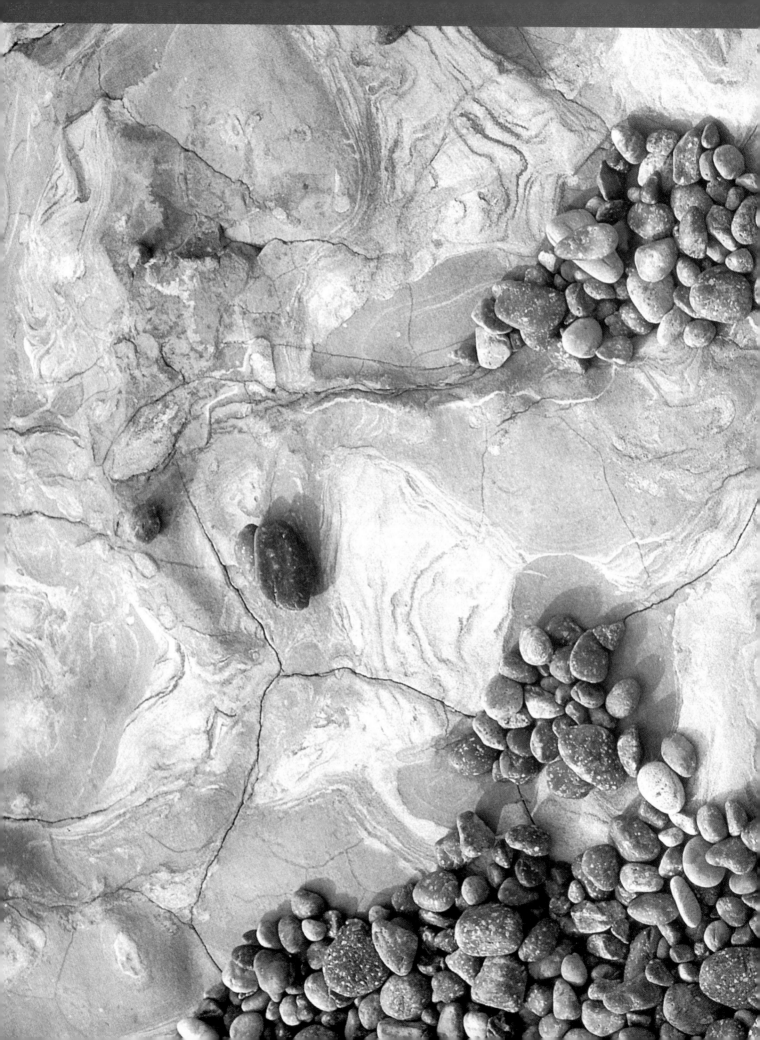

CHAPTER
8

SHAPING THE LAND

THIS CHAPTER IS ABOUT:

◆ Surface-relief environment
◆ Weathering processes
◆ Products of weathering
◆ Weathering, rocks, and climate
◆ Weathering and the human factor
◆ Mass movement transport
◆ Mass movements, rocks, and climate
◆ Human activity and mass movements
◆ Slopes
◆ Environmental Issue: Landslides—Weak Rocks and Ignorance

Rocks at Earth's surface break into fragments because frost action and chemical reactions weaken them. The fragments may then move downhill under gravity as in the catastrophic avalanches of 1962 and 1970 in the Peruvian Andes (inset), or disintegrate into smaller particles and form soil.

Yungay

Río Santo

Matacoto   Ranrahirca

Huascaran Summit

N

1970 avalanche deposit

1962 avalanche deposit

Towns

SURFACE-RELIEF ENVIRONMENT

As soon as mountains form or any dry land appears, the atmosphere works to destroy them. It breaks rocks into small pieces, which move downhill toward the oceans. Just as it took millions of years to raise the mountains, so it takes millions of years to wear them down.

Landforms are products of the interactions among the internal forces that raise or lower Earth's crust and the external atmospheric and marine processes that mold this surface. Figure 8-1 shows how these opposed forces produce the detailed features of Earth's surface relief in the narrow zone of the *surface-relief environment*. This environment is only a few meters or tens of meters deep, but it is the scene of continuous activity.

The Andes Mountains in South America demonstrate how the atmosphere-ocean and solid-earth environments create surface relief over many millions of years. The processes of plate tectonics gradually built the Andes into the world's second highest mountains,

topped by a number of active and dormant volcanoes (see Figure 7-40). As the ranges rose, surface-relief processes modified the landscape. Initially the tropical arid climate had little impact on the low land on the present Chile-Peru border (Figure 8-2a). Then tectonic uplift produced low mountain ranges, which forced air to rise and rain to fall. Increased humidity caused the rocks to break down more rapidly, while streams flowed down the slopes, carving valleys (Figure 8-2b). As millions of years passed, continuing uplift took the tops of the ranges above the clouds, where arid conditions prevailed again. The dry former river valleys filled with rock debris, worn from the mountain tops by the wind and occasional floods (Figure 8-2c). Most recently the Andes have risen to altitudes where temperatures are low enough to let snow fall, accumulate, and turn to ice. Glaciers carve distinctive deep and wide valleys as they flow slowly downhill (Figure 8-2d). Each set of surface environmental conditions—arid, rainy with flowing streams, and cold with glaciers—has left its mark on the landforms of the Andes Mountains.

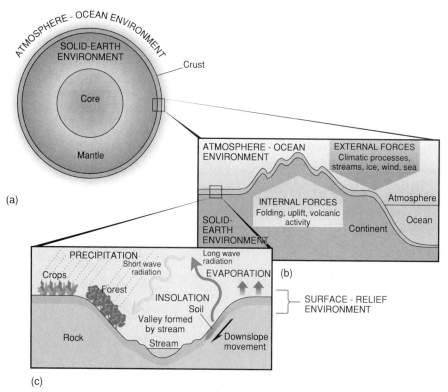

Figure 8-1   Surface-relief environment.
(a) The global scale—at the intersection between the solid-earth and atmosphere-ocean environments. (b) The continental scale—where internal and external forces interact and form major relief features. (c) The local scale—detailed features of hill slopes are formed by atmospheric processes acting on solid-earth features.

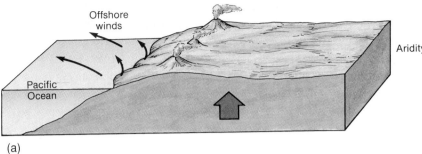

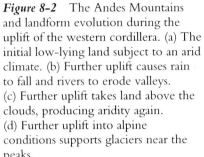

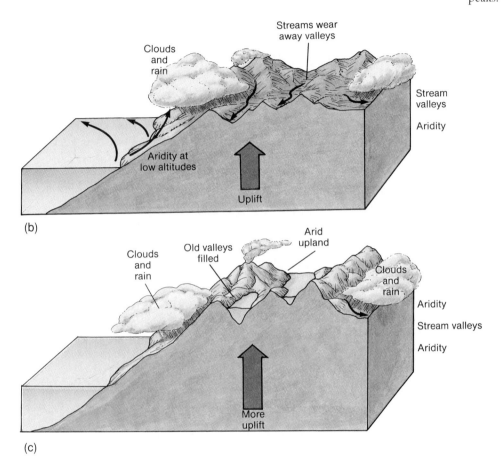

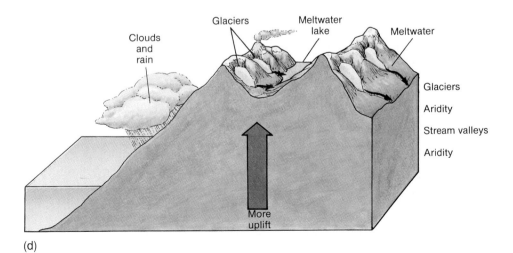

Figure 8-2 The Andes Mountains and landform evolution during the uplift of the western cordillera. (a) The initial low-lying land subject to an arid climate. (b) Further uplift causes rain to fall and rivers to erode valleys. (c) Further uplift takes land above the clouds, producing aridity again. (d) Further uplift into alpine conditions supports glaciers near the peaks.

Chapters 8 through 11 focus on the processes acting in the surface-relief environment. These processes result from the atmosphere-ocean environment's circulation of heat, air, and water. Overall, they act to wear down the land, an outcome known as **gradation.** Gradation lowers the overall land surface but includes both destructive (degradation) and constructive (aggradation) processes. Mountains are destroyed by rock disintegration, the carving of valleys by water and ice, and the planing of wider areas by combinations of rock disintegration and the action of running water and ice masses. The rock fragments produced are moved to lower areas and to the sea where they are deposited to form constructional landforms and new rocks.

Gradation combines several processes. **Weathering** is the disintegration and decomposition of solid rocks making rock materials available for movement. The products of weathering range from large blocks several meters across, through sand grains (0.02 to 2 mm across) and fine clay particles, to dissolved matter. Agents such as running water, ice, and wind **transport** rocks, rock debris, and dissolved rock materials. **Erosion** is the wearing away and removal of rock by these agents. **Deposition** is the dropping of rock particles to form sediments. Depositional features typically have distinctive compositions and shapes. Weathering and the transport accomplished by gravity, which is called **mass movement,** are the subjects of the remainder of this chapter. Chapters 9 through 11 cover erosion, transport, and deposition by running water, ice, wind, and the sea.

The gradational processes are limited in their work by a lowest level of erosion, known as the **base level.** For most continental areas, sea level is the base level since rivers and glaciers flow downhill to sea level but have little impact below sea level. In basins of interior drainage, however, the rivers do not reach the sea and local base level may be higher than sea level, as in Great Salt Lake, Utah, or lower than sea level, as in the Dead Sea between Israel and Jordan. Interior drainage basins are common in arid and semiarid regions.

WEATHERING PROCESSES

Once rock is exposed to the atmosphere-ocean environment, it is subjected to forces that start to break it into pieces and move them elsewhere. Exposure to the atmosphere is not merely at the rock surface but wherever air and water penetrate downward along joints, bedding planes, or faults in the rock. Once rock is broken into small pieces by weathering, it can be transported more easily.

Weathering includes a complex set of physical, chemical, and biologic processes. Rock fragments produced by weathering may also be subject to the continuing action of such processes disintegrating further into their component minerals, decomposing into chemical products such as clay minerals, or dissolving into ions. The physical, chemical, and biologic processes usually act in concert, but it is necessary to understand how the separate processes work (Figure 8-3). **Physical weathering** processes cause disintegration without significant chemical alteration. **Chemical weathering** processes create chemical changes that decompose rocks and their minerals; in most places chemical processes are by far the most important aspect of weathering. **Biologic weathering** processes involve living things and include both physical actions, such as plant root growth and animal burrowing, and chemical actions, such as the excretion of acids.

PHYSICAL WEATHERING

Physical weathering causes rocks to disintegrate into fragments ranging in size from large boulders to sand grains. It involves stresses imposed by the release of confining pressure, changes of temperature, the formation of ice, or the addition or removal of water. Such stresses build up over time and cause the rock to break apart. Rock fatigue, rather like metal fatigue in machinery, may be in progress for some time before failure occurs.

Physical weathering is helped by, and adds to, cracks in rocks. Initial cracks include cooling joints in igneous rocks, drying joints and bedding planes in sedimentary rocks, slaty cleavage in metamorphic rocks, and fault planes. As rocks break into smaller fragments, more surfaces become exposed to weathering of all types.

Wearing down of Earth's surface often brings to the surface rocks that were buried several kilometers deep. When at depth, such rocks are subject to strong confining pressures. The removal of the overlying rock, or *unloading,* reduces the pressure and subsequent rock expansion leads to the formation of cracks that are parallel to the ground surface and known as expansion joints. Such joints are common in granite batholiths, and they separate sheets of rock from a few centimeters to 30 m thick. The process by which rocks break into such sheets is known as **exfoliation.**

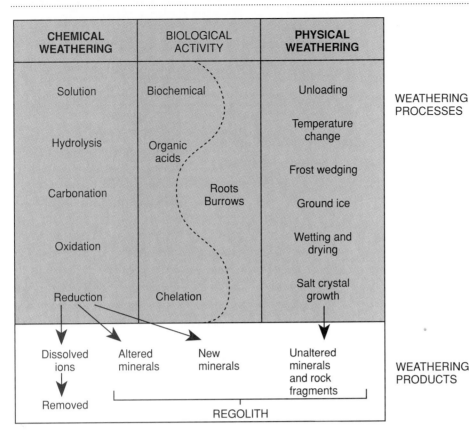

CHEMICAL WEATHERING	BIOLOGICAL ACTIVITY	PHYSICAL WEATHERING	
Solution	Biochemical	Unloading	**WEATHERING PROCESSES**
Hydrolysis	Organic acids	Temperature change	
Carbonation	Roots Burrows	Frost wedging	
		Ground ice	
Oxidation		Wetting and drying	
Reduction	Chelation	Salt crystal growth	
Dissolved ions	Altered minerals	New minerals	Unaltered minerals and rock fragments	**WEATHERING PRODUCTS**
Removed				
	REGOLITH		

Figure 8-3 Weathering processes and products. Physical and chemical processes reacting with rocks give rise to a range of weathered products from dissolved ions to large rock fragments.

Temperature changes also result in rocks breaking apart. Intense insolation heats bare rock surfaces, causing different minerals to expand at different rates. Colder temperatures can cause the minerals to contract. Day-to-night temperature fluctuations might be expected to provide repeated cycles of expansion and contraction that cause rock to disintegrate by flaking into thin layers or by the forcing apart of individual grains. Laboratory experiments, however, show that rock seldom disintegrates unless water is present to first weaken the rock by chemical action. The use of fire by traditional cultures to crack rock is most successful when water is added. It is unlikely that day-to-night expansion and contraction of rocks alone is responsible for rocks breaking apart.

Temperature fluctuations are more significant when low temperatures cause ice to form from water trapped in rock joints and pores. When water freezes, its volume increases by 9 percent exerting pressure on the surrounding rocks. Ice continues to increase in volume as the temperature falls below freezing, with maximum pressure being exerted at −22° C (5° F). This increase in pressure splits rocks apart. The process is known as **frost wedging.** Although angular rock fragments that might have been broken off by frost wedging are common in arctic and alpine areas, it is not clear how the process works. Tests have attempted to determine whether the crucial factor is the speed of freezing and thawing, the presence of sufficient water to saturate all cracks and pores in the rock, or a temperature drop to below −5° C (23° F). They have not identified the main cause, however. Frost wedging is a significant consideration in planning civil engineering projects such as road building and in the choice of bricks for construction. Engineers use information on frost activity and winter precipitation to divide the United States into weathering regions. They apply this information to determine how weathering will affect types of brick at different locations.

The growth of *ground ice* in very cold areas exerts another physical stress that breaks rocks apart. Ground ice includes both the fillings of tiny pore spaces in soil or rock and also large continuous masses of ice several tens of meters across. Ground ice and the landforms associated with it are discussed more fully in Chapter 10.

Water movements cause changes in the volume of rock. Some minerals, especially clays, swell after absorbing water and shrink when drying out. Alternate *wetting and drying* produces cycles of expansion and shrinking that disintegrate some rock surfaces. On coasts, rocks in and just above the tidal zone are wetted and dried twice a day and such weathering can be significant (see Chapter 11). In deserts and along coasts, *salt crystal growth* occurs as evaporating water precipitates dissolved minerals. Continuing evaporation causes these crystals to expand and exert a pressure against the surrounding materials that may be sufficient to fracture a rock or cause heaving of the ground.

Plants and animals exert some physical stresses on rocks. Plant roots of tiny lichens cause rock surfaces to crumble, and larger tree roots penetrate cracks and force blocks of rock apart. Worms, termites, and burrowing animals churn the weathered debris and reduce particle sizes. Such biologic activity may be significant locally but plays a minor role compared to other forms of weathering.

CHEMICAL WEATHERING

Chemical reactions are generally much more important than physical stresses in the breakdown of rocks. Important constituents of rocks, such as the cement that holds the grains together or a major mineral, may be susceptible to chemical changes. Such chemical changes cause a rock to decompose by removing important constituents in solution and detaching less affected minerals.

Several types of chemical reaction occur in decomposing rocks. Of these, simply dissolving minerals in water is one of the least common. **Solution** occurs when a solid or gas joins the chemical composition of a liquid. Put solid salt (sodium chloride) into water and it becomes part of the water as separate sodium and chlorine ions. Minerals that dissolve so easily, however, are not common in rocks and this reduces the significance of the process.

Most chemical weathering involves exchanges of materials between rock minerals and water. The most important form of chemical weathering, **hydrolysis,** involves the exchange of the ions in rock minerals with those in water. In hydrolysis the hydrogen ions from water penetrate the chemical structure of rock minerals and displace ions such as potassium, sodium, calcium, and magnesium. The displaced ions are removed in solution. In the chemical weathering of granite, for instance, hydrolysis acts on the feldspar and mica minerals, reducing them to dissolved matter and clay minerals. The quartz in the rock changes little, but so much of the rock is affected that it disintegrates, leaving quartz-sand and clay behind.

Water containing dissolved chemicals also reacts with rock minerals. One example is when, after flowing through soil, a weak solution of water and carbon dioxide acts on limestone rock. In this process known as **carbonation,** the calcium carbonate minerals that compose the limestone break down to calcium and bicarbonate ions and much of the rock is carried away in solution. Water with a high carbon dioxide content and little bicarbonate in solution is said to be aggressive since it attacks and removes the calcium carbonate. This produces a distinctive landscape known as karst on some limestone rocks (see Chapter 9). An excess of bicarbonate ions in water with little carbon dioxide causes the deposition of calcium carbonate, filling joints or pore spaces or creating columns in underground caverns (Figure 8-4).

Water that contains dissolved oxygen affects surface soil and rocks after rain or when rain percolates through them. In this process called **oxidation,** dissolved oxygen reacts readily with minerals that contain iron to form insoluble red oxides of iron (as in rust), or yellow-brown varieties that crumble when they dry out. Such minerals often provide color to soils, and red soils are common in the southeastern United States and in many parts of the tropics. The reverse chemical changes occur in places where oxygen is in short supply or absent, by the reaction called **reduction.** In reduction the weathered material loses oxygen from its chemical structure. This occurs most commonly where rock or soil is totally waterlogged by still water containing little or no dissolved oxygen, as in wetland areas. The red oxides of iron change to more soluble green oxides. Soils subject to alternate waterlogging and drying out are often mottled green-orange because they have a mixture of red and green oxides of iron.

Biologic activity enhances chemical weathering. Plants and animals live, die, and decay in the surface-relief environment and create their own chemical conditions. As plant roots extract nutrients from soil, they replace them with hydrogen ions that increase soil acidity. Fungi secrete acids that attack silicate minerals. Termites and other insects are abundant in tropical soils and release acids in their excreta, increasing the effectiveness of chemical weathering. Water passing through decaying vegetation on the soil or rock surface picks up acidic organic extracts. These combine readily with iron and aluminum in soil or rock minerals and then remove those elements in solution. This process is known as **chelation** and is particularly effective beneath the carpets of pine needles in northern midlatitude forests.

The extent and rapidity of chemical weathering depend partly on the availability of water and its rate of circulation through a rock. Chemical weathering is very effective when water circulates through a rock at moderate speeds, giving time for reactions with susceptible minerals but not allowing the water to become saturated with dissolved matter. If water circulation is sluggish, the water may become saturated, leading to precipitation of dissolved materials and blocking of the water movement. Temperature is another factor controlling the rate of chemical weathering. Chemical weathering is particularly active in the humid tropics, where high temperatures combine with water availability and intense organic activity.

Since chemical weathering attacks a rock from the outside, it often creates a sequence of layers that show different degrees of weathering. Some rocks, such as basalts, clearly show the depth of weathering in a series of discolored layers or where the outer layers flake off. This feature of weathered rocks is termed *spheroidal,* or *onion-layer, weathering.*

PRODUCTS OF WEATHERING

The broken rock products of weathering often form a layer of rock debris overlying the unaltered parent rock (Figure 8-5). This layer, **regolith,** comprises rock fragments, resistant quartz grains, and altered clay minerals. The flow of water through the regolith removes the more soluble matter.

Regolith may take hundreds or thousands of years to develop because weathering is generally a slow process. In places where downslope movement, the action of streams, or wind removes the weathered material, the regolith accumulates even more slowly, if at all. If removal rates exceed the renewal of regolith by weathering, all the material is transported.

Plants and other organisms often become established on and in the regolith. They combine with the passage of water and air through the weathered material to convert the regolith into a soil (see Chapter 13). Plants with deep root systems may also bring long-term stability to regolith in areas where streams or the wind might otherwise remove the loose material. The removal of trees and grass for farming, housing, or highway construction may, however, expose the regolith to the possibility of removal. In Italy, Greece, and other Mediterranean lands, the removal of vegetation by farmers led to the loss of soil and regolith and the exposure of bare rock outcrops.

WEATHERING, ROCKS, AND CLIMATE

Weathering varies greatly on a worldwide basis. Take some examples:

> By 1928, several centimeters of crude soil had formed on new volcanic ashes and lavas erupted on the island of Krakatoa (Indonesia) in 1883.
>
> A 45 centimeter layer of decomposed granite formed beneath building foundations in Rio de Janeiro (Brazil) within 20 years of construction.
>
> No soil or weathered layer has developed on 1200-year-old mud flows around Mount Shasta in northern California.
>
> Rock surfaces bared by the action of glaciers during the last glacial advance in Britain show little evidence of 10,000 years of postglacial weathering.

These examples show that weathering intensity varies from one part of the world to another and in some places can be very significant in rock destruction. In general, weathering is more rapid in the humid tropical areas such as Indonesia and Brazil than in midlatitude areas such as northern California and Britain (Figure 8-6). Weathering rates also depend on the nature of the rocks being attacked and on the local relief.

Weathering is a battle, the outcome of which depends on the forces attacking the rocks and the resistance that the rocks make to the attack. A rock and its minerals are resistant to weathering if they are slow in decomposing chemically or in disintegrating physically. Rocks that resist weathering stand out in the landscape as ridges, since the less resistant rocks break down and are lowered more rapidly. The sandstone and limestone rocks stand out in the relief of the area

Figure 8-4   A limestone cavern at Postojna, Slovenia. Some columns grow down from the cavern roof, but others are beginning to grow upward.

Figure 8-5   Weathered rock, or regolith. Weathering has caused the rock to disintegrate. Weathering processes attacked along the joints and bedding planes in the rock, breaking it into blocks, and those have been broken into smaller rock fragments.

Figure 8-4

Figure 8-5

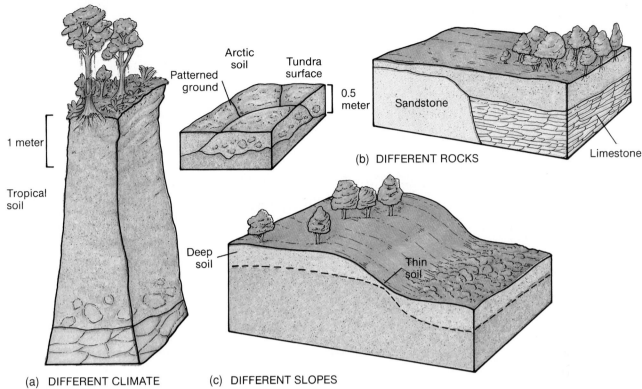

Figure 8-6   Weathering, climate, and rocks. What differences in the thickness of soil occur in (a) tropical and arctic soils, (b) above sandstone and limestone, and (c) on flat and steep slopes?

shown in Figure 7–26 because they are more resistant to the combined attack of weathering and surface water flow than the clay rocks of the lowlands.

The susceptibility of minerals to chemical weathering is controlled by the strength of the bonds between the mineral ions. Sometimes these are weak, as in olivine, and chemical weathering carries out exchanges that rapidly decompose the mineral. In other minerals, such as quartz, the bonding is strong and the mineral has a strong resistance to chemical weathering. Sandstones formed of quartz grains cemented by silica are among the most chemically resistant rocks.

Physical characteristics also influence a rock's resistance to weathering. The sizes and mixture of grains that compose a rock determine its texture. Fine-grained and even-grained rocks tend to be more resistant to both physical and chemical weathering. Texture also affects the way in which rocks hold water or allow it to pass through. **Porosity** is the proportion of space in a rock not occupied by mineral matter; this space is usually filled by air or water. Rocks with a low porosity or very fine texture do not store water or let it pass through, while those with a high porosity and medium-to-coarse grains let water flow through easily. Both types are more resistant to weathering than rocks that hold water. Rocks that hold water are liable to frost wedging in midlatitude and polar climates or to chemical weathering in warmer lands.

Climatic conditions influence the type and rate of weathering processes at a place since temperature and the availability of water are two controlling factors in weathering processes. The amount and rate of weathering affect the depth of regolith produced (Figure 8-7). Physical weathering dominates the *very cold regions* of polar climates or high altitudes. Near midlatitude margins where there are frequent freeze-thaw cycles, the processes include the action of ground ice and frost wedging. Chemical attack is generally less significant than in other climates because chemical reactions are slow and because organisms are less active at low temperatures.

Physical weathering and slow change also dominate many *arid regions*. Salt crystal growth and granular disintegration occur in such areas, but the low relative humidity in the air and the small amount of water flowing through soil and rock reduce the effects

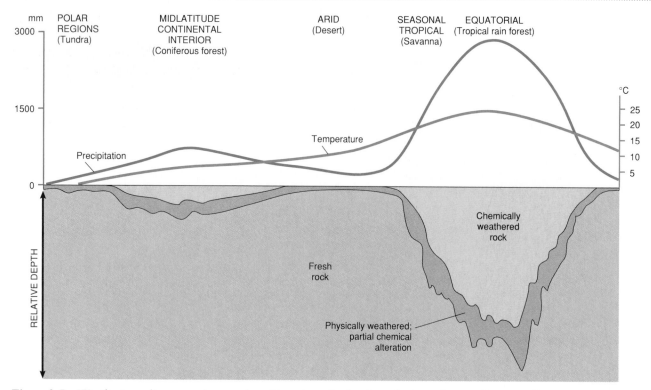

Figure 8-7 Weathering, climate, and vegetation. How does depth of weathering relate to climate and vegetation? This diagram assumes that there is little relief.

of chemical processes. The Sphinx and pyramids of Egypt demonstrate, however, that weathering in deserts can be significant despite the slowness of chemical reactions. The disfigured face of the Sphinx, carved some 4000 years ago, is a limestone rock that holds water. Evaporation draws water to the rock surface, precipitates calcium carbonate crystals there, and hardens the surface stone. This layer becomes heavy, flakes, and falls off. The pyramids had a protective layer of resistant limestone blocks when built, but most of these blocks were removed about 1000 years ago by people who used them to construct new buildings. This left a variety of underlying blocks exposed that have different resistances to weathering. Where the resistant limestone remains, it shows little alteration, but softer limestones are scored by pits 1 to 2 cm deep, and a poorly cemented sandstone has been reduced to rubble.

In *midlatitude humid regions,* chemical weathering is more significant than physical weathering, although both occur. Frost wedging takes place in winter in places where freeze-thaw cycles are common. Sufficient water is available to make chemical decomposition continuous, but its intensity and depth of impact are limited by the moderate average temperatures. Biochemical activity including chelation is important beneath evergreen forests.

Chemical weathering dominates *tropical humid regions* where the warmer and wetter climates support large numbers of organisms. At depths of several meters in the regolith, temperatures remain high all year and water is continuously available. Chemical weathering decomposes most minerals, leaving a clay residue rich in iron and aluminum.

WEATHERING AND THE HUMAN FACTOR

Human activities have an increasing impact on many gradation processes, but have little direct effect on weathering rates. Indirectly, however, human actions may be increasing rates of chemical weathering in particular.

Burning fossil fuels makes the atmosphere more acidic, especially over and downwind of urban-industrial areas (see Environmental Issue: Acid Rain—Who Is to Blame? p. 94). Although it is difficult to assess the impact of increased atmospheric acidity in the natural landscape, exposed building stones in urban areas show the effects. Major buildings in London, England, such as the Houses of Parliament and St. Paul's Cathedral have been exposed to an urban-industrial atmosphere with high levels of sulfur dioxide for nearly 200 years. The limestone used in building the Houses of Parliament is particularly susceptible to chemical weathering and much of the building now has new stone. The limestone used in St. Paul's Cathedral is more resistant, but even so, the surface has been reduced by 1 cm (nearly half an inch).

The drainage of wetland areas, as in the northern Everglades of Florida, changes the chemical environment by letting oxygen into formerly waterlogged land. Soils reclaimed from the former swamp for farming were high in plant debris and reduced minerals. Once water began to flow through the soils, oxidation altered the minerals and destroyed the organic matter.

Weathering is the physical disintegration or chemical decomposition of solid rock into fragments, minerals, or dissolved chemical ions. Physical weathering is caused by mechanical stresses that fracture or disintegrate rocks. Chemical weathering involves chemical reactions that weaken and decompose rocks or their constituent minerals. Living organisms assist both physical and chemical weathering.

The broken rock material produced by weathering is regolith, a basic component of soil. Differences in the resistance characteristics of rocks and in their climatic and surface-relief environments produce differences in the effectiveness of weathering from place to place. Human actions often enhance weathering indirectly by making the atmosphere more acidic, by altering water regimes, and by bringing solid rock into direct contact with the atmosphere.

MASS MOVEMENT TRANSPORT

Mass movement is the downhill transfer of sections of solid rock or masses of weathered debris under the influence of gravity. In many cases, the presence of water or ice in rock or weathered material reduces strength or friction and so enhances the effect of the pull of gravity.

The main feature of mass movements is that a large volume of rock or regolith moves downhill as a unit. This unit may be a rigid whole in the form of a huge block of rock or a mass of individual fragments, but its constituents are in close contact with each other. In streams or glaciers, individual particles of rock are carried separately by a large quantity of water or ice. Mass movements, on the other hand, contain much rock and little water or ice.

Mass movements may act rapidly or slowly. There are three main types of movement, classified by a combination of the speed of movement and the amount of water or ice present. These movement types are flows, heaves, and slides and falls (Figure 8-8). Lastly, subsidence is a form of mass movement that may happen after material is removed or consolidation occurs below the surface.

FLOWS

Where there is sufficient water or ice for lubrication, regolith may flow rapidly. Mass movement **flows** have greater proportions of rock debris than of water or ice. Material with less water or ice in it requires a steeper slope before movement will begin. The most rapid flows occur where the slope is steepest and where the water or ice content is greatest.

Mudflows consist of clay and silt with pebbles and even boulders mixed with a relatively high water content, and they flow rapidly. They are common on the valley floors of semi-arid regions where occasional rainstorms cause flash floods. The sudden onset of water sweeps up large quantities of loose rock debris. If rock boulders are present, the movement will be slower and shorter-lived. Following the Mount St. Helens eruption, huge quantities of volcanic ash mixed with water from overflowing lakes and formed a massive mudflow (Figure 8-9).

Earth flows are usually slower than mud flows because of a lower water content, and they are more localized in extent. They are common on hill slopes of humid regions in places where there is so much soil water that the regolith swells and bursts through a cover of grass or other vegetation (Figure 8-10a). Over a period of many years, earth flows may create extensive spreads of debris at the foot of a slope.

Avalanches are high-speed flows of ice and wet snow containing varied proportions of rock. A snow avalanche may be almost purely snow; a debris avalanche consists almost entirely of rock (Figure 8-10b). Avalanches occur in high mountains with steep slopes and, besides transporting snow and rock downhill, may carve depressions in the surface, called avalanche chutes, as they travel downward.

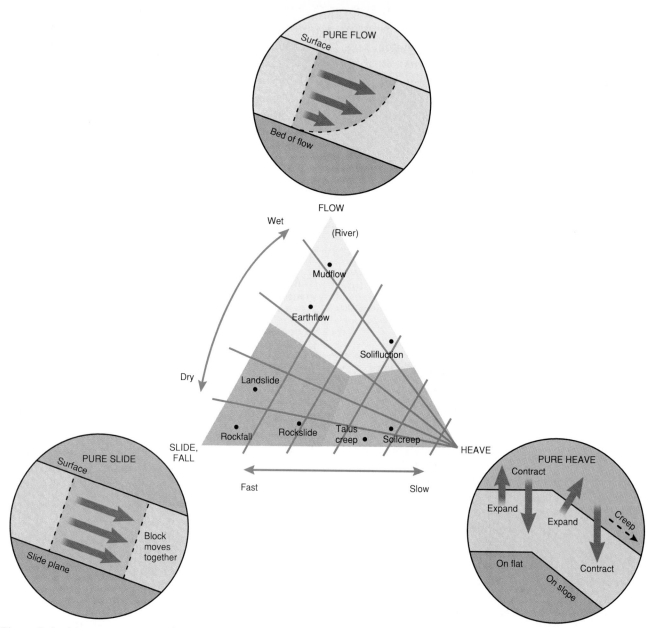

Figure 8-8   Mass movements. The importance of water content and speed of movement in defining types of mass movement. Use the inset diagrams to write your own definition of flow, slide, and heave.

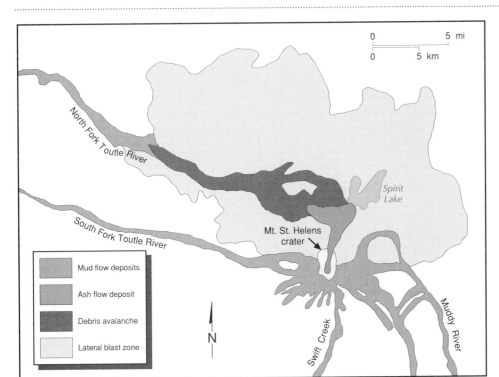

Figure 8-9 The Mount St. Helens mudflow was formed by a mixture of volcanic ash from the eruption and water from Spirit Lake.

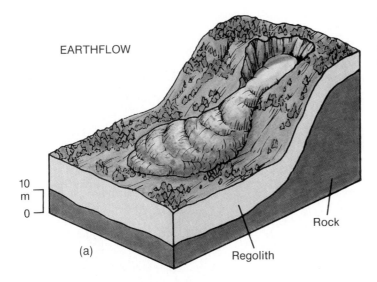

EARTHFLOW

(a)

Regolith

Rock

10 m
0

Figure 8-10 (a) An earthflow, in which a mass of regolith flows down a slope. (b) A debris avalanche that begins as a mixture of rock with snow and ice.

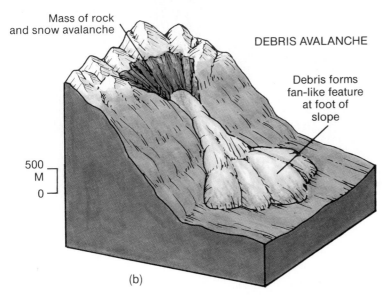

Mass of rock and snow avalanche

DEBRIS AVALANCHE

Debris forms fan-like feature at foot of slope

500 M
0

(b)

Solifluction is a slow downhill flow movement that may occur in cold climates. In summer, melting ground ice traps water in a surface zone 50 to 100 cm (20 to 40 in.) deep above a permanently frozen layer. On slopes between 6 and 20 degrees, the weight of meltwater causes the soil material to collapse and flow occurs beneath the vegetation cover. Such flow produces lobelike features where water is localized or steplike terraces where water is more abundant (Figure 8-11).

SLIDES AND FALLS

Slides, or landslides, move masses of rock from a meter to several hundred meters across. The sliding rock remains in contact with the slope, but requires little or no lubricating water or ice. A **fall** is also fast, and since detached blocks or boulders fall freely, it does not involve a lubricant.

Slides can move along planes inclined at a few degrees. These are often lines of weakness such as a bedding plane or a weak rock layer such as rock salt or clay (Figure 8-12a). Movement may be rapid or slow depending on how far the moving rock mass becomes detached from its foundations. A *slump* (Figure 8-12b) is a distinctive type of slide in which a mass of rock collapses along a curved fracture surface and moves downward and outward. This is common where clay lies beneath a layer of stronger rock such as sandstone or limestone. The clay loses its strength when it becomes saturated with water, and the weight of the overlying rock causes the collapse.

Rockfalls occur on steep slopes with rocky outcrops (Figure 8-12c) and especially along coastal cliffs and river bluffs. On such slopes sections of rock become unstable and fall. If the fallen material is not removed from the base of the slope, it forms a pile of broken rock fragments known as *talus,* or scree. The surface slope of such deposits has a maximum angle called the *angle of repose,* which is determined by the size and

Figure 8-11 Solifluction lobes, Spitzbergen. Soil saturated with thawed water flows down a low-angle slope beneath the vegetation.

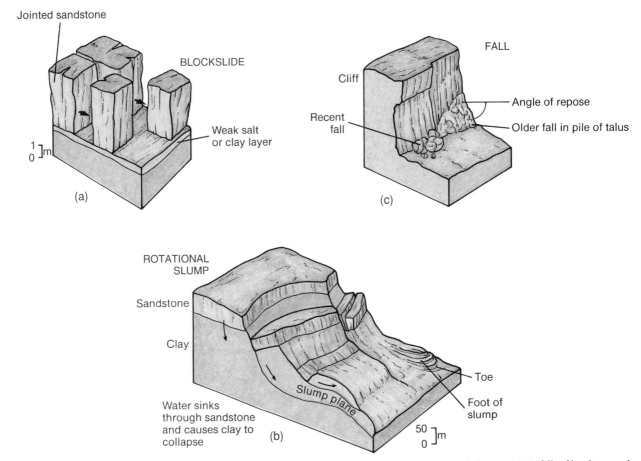

Figure 8-12   Slides. (a) A block slide in massive rock overlying a weak layer. (b) A rotational slump. (c) A fall of broken rock on a vertical cliff face, producing a pile of talus.

shape of the materials. Talus composed of larger fragments has a steeper angle of repose than one formed of smaller fragments. If the slope angle is greater than the angle of repose for a particular size fragment, the mass may become unstable and move downslope.

Frost action is often significant in starting rockfalls. In northern Sweden, researchers found that few rockfalls occur when temperatures remain below freezing point, but they occur frequently when warmer days alternate with frosty nights in spring and later summer (Figure 8-13). Such conditions produce intense physical weathering, including frost wedging, but the ice holds the rock pieces together and has to melt before rockfalls occur in great numbers.

HEAVES

Cycles of freezing and thawing or of wetting and drying cause the regolith on flat or low-angle slopes to heave. **Heaving** is the combination of vertical expansion caused by wetting or freezing, and contraction caused by drying or thawing. If there is a regular cycle of such movements, the surface will repeatedly rise and fall a few millimeters or centimeters.

On flat ground, heaving movements are simply up and down. In cold climates, alternate freezing and thawing pushes rock fragments toward the surface, where they accumulate in mounds and are distributed in a distinctive arrangement called *patterned ground*. This process is known as *frost heaving* (Figure 8-14).

On a slope expansion takes place at right angles to the slope surface, but contraction causes the regolith to settle vertically under gravity (see Figure 8-8). As a result, the surface material moves slowly downslope in a process known as *creep*. Soil creep is active on most slopes, and the process is assisted by relatively small amounts of water acting as a lubricant, and by frost heaving. Such creep is often

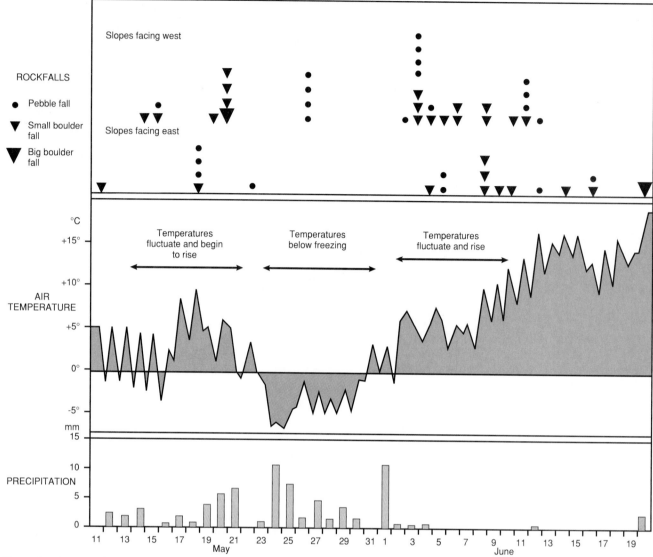

Figure 8-13 Rockfalls, northern Sweden. Under what temperature conditions did the main periods of rockfall activity take place?

apparent in the landscape because it affects surface features by tilting fences and telephone poles. *Rock creep* affects masses of larger fragments such as talus deposits and results in a slow shift of the surface particles toward lower ground.

SUBSIDENCE

Subsidence is the lowering of the ground surface when subsurface material is removed or consolidated. The coastal area near Los Angeles subsided up to 9 m (30 ft) following oil extraction (Figure 8-15). Similar subsidence is common where oil or water are extracted from rocks or when rock salt or coal are mined underground. The removal of oil, water, rock salt and coal cause settling and consolidation of

the rock and surface subsidence. The ground surface in subsided areas often becomes filled with water.

Collapse depressions, or *sinkholes,* are a natural characteristic of limestone regions and the distinctive landscape of these regions is discussed in Chapter 9. In Florida, where sinkholes are common, houses may fall into sinkholes after a masking layer of soil collapses.(Figure 8-16).

In very cold regions such as Alaska, attempts by early settlers to plow fields and the removal of vegetation for roads and airfields deepened the surface layer of summer ground-ice melting. Where the water could flow out from the base of the melted layer, the soil subsided or collapsed, creating sinkholes that are called *thermokarst* (Figure 8-17).

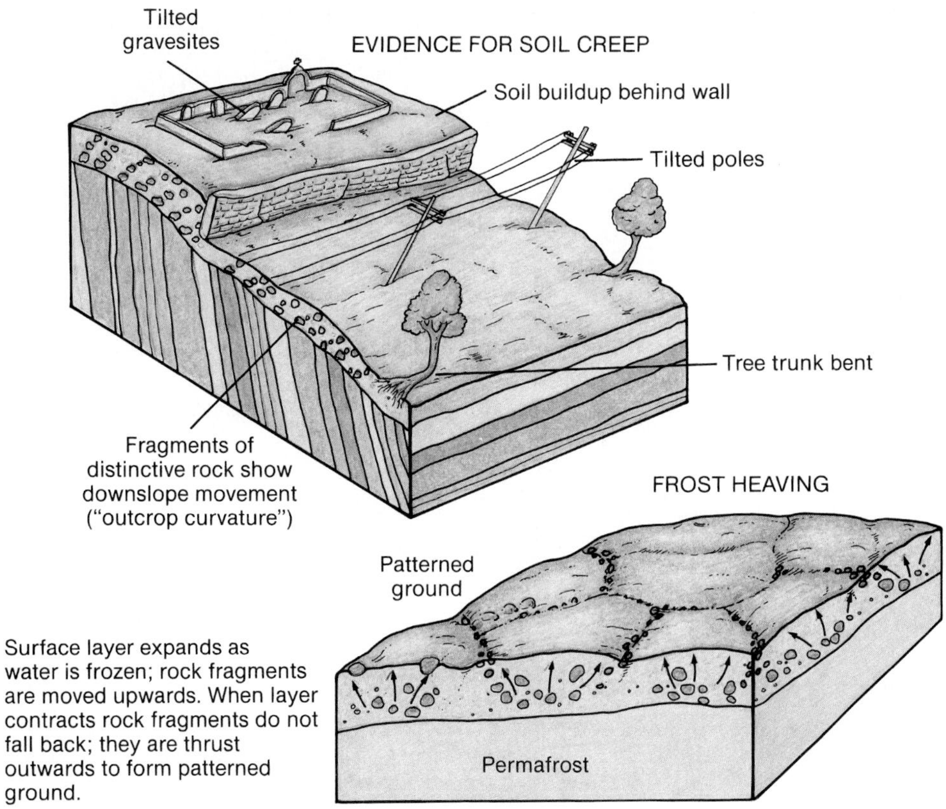

Tilted
gravesites

EVIDENCE FOR SOIL CREEP

Soil buildup behind wall

Tilted poles

Tree trunk bent

Fragments of
distinctive rock show
downslope movement
("outcrop curvature")

Figure 8-14 Heaves. Soil
creep and frost heaving.

FROST HEAVING

Patterned
ground

Surface layer expands as
water is frozen; rock fragments
are moved upwards. When layer
contracts rock fragments do not
fall back; they are thrust
outwards to form patterned
ground.

Permafrost

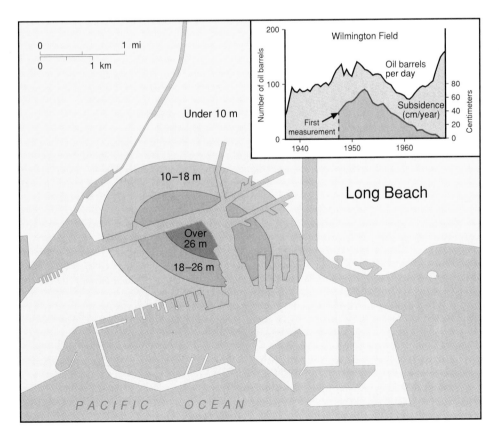

0   1 mi

0   1 km

Under 10 m

200

Wilmington Field

Number of oil barrels

100

Oil barrels
per day

First
measurement

80

60

40

20

0

Centimeters

Subsidence
(cm/year)

0

1940   1950   1960

10–18 m

Long Beach

Over
26 m

18–26 m

PACIFIC   OCEAN

Figure 8-15 Subsidence
beneath Long Beach,
California. The map shows the
area affected, and the inset
graphs show how subsidence
was related to oil extraction
until water was pumped
underground to replace the oil.

Figure 8-16   Sinkhole in Winter Park, Florida. What damage was caused by the subsidence?

Figure 8-17   Thermokarst subsidence over permafrost. The surface layer that melts in summer may comprise up to 50 percent water. Removal of surface vegetation for farming increases the depth of the melting layer; and if water flows out from the base, the surface will collapse.

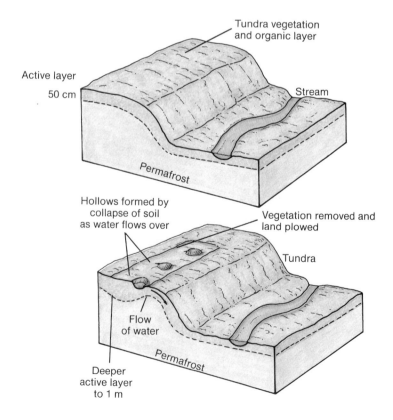

MASS MOVEMENTS, ROCKS, AND CLIMATE

Mass movements take place on all slopes, but the most rapid and dramatic forms occur in areas of high relief and those subject to earthquake and volcanic activity. Certain types of rock and rock structures also favor mass movements. Rock salt, shale, and clay, for example, are likely to fail under moderate stress and collapse even sooner when water builds up inside them or additional weight is placed on top.

Mass movements are often the dominant surface processes in high mountain areas such as the Alps and Andes. Individual mass movements are commonly triggered by an event such as an earthquake shock or sudden melting of snow. In 1962 earthquakes in the Peruvian Andes produced huge avalanches that moved 3 million tons of ice mixed with 9 million tons of rock at speeds of over 100 kph (60 mph) for up to 20 km (12 mi), killing 3500 people (Figure 8-18). In 1970 another earthquake triggered an even larger avalanche in the same region. It traveled at 300 kph (180 mph), jumping over a 300 m (1000 ft) ridge to bury the town of Yungay and some 40,000 people.

Climatic conditions affect the distribution of mass movement types in both space and time. Mass movements are prevalent in climates with saturated rock or regolith near the surface and in those with limited surface vegetation cover. *Cold climates* in high latitudes or at high altitudes experience rockfalls, frost heaving, and solifluction. Slides, slumps, and earthflows may be common where water builds up in subsurface materials or if there is little surface vegetation; piles of talus build up at the bottom of steep slopes.

In *arid climates* the combined results of weathering and mass movements produce angular landforms with little or no regolith cover. Rock falls are frequent on the common bare, steep slopes, and mudflows often follow flash floods. Wet-dry heave accompanied by salt crystal growth affects smaller relief forms.

In *humid climates* soil water and groundwater can lower resistance to regolith and rock movement on slopes, and help to lubricate movement once it begins. The vegetation cover, however, often reduces the size and frequency of mass movements; slow creep is the main form of mass movement. In combination chemical weathering and creep create a landscape of gentle slopes grading into each other without sharp angles. Where intense rain falls on steep slopes, as in the Appalachian Mountains or the coastal ranges of Washington state, the water content of soil may be so great that ground collapses in earthflows even though it is covered by protective forest.

Many parts of the world that today have few mass movements contain landforms produced by such activity in the past. In Britain and other midlatitude countries, landslides and talus slopes that formed during cold phases of the Pleistocene Ice Age are now grassed over and stabilized. In Arizona there are slump blocks that formed when the climate was rainier tens of thousands of years ago.

HUMAN ACTIVITY AND MASS MOVEMENTS

Human activities often cause, accelerate, or increase the likelihood of mass movements. Steep cuttings or embankments resulting from highway construction in weak materials are subject to slope failure (see Environmental Issue: Landslides—Weak Rocks and

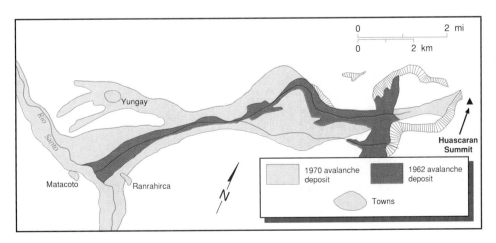

Figure 8-18 Avalanche deposits in the Peruvian Andes that resulted from earthquakes in 1962 and 1970.

Ignorance, p. 244). Removing vegetation for farming increases the likelihood of mass movements in humid climates. In particular, the intense rains and deep soils of humid tropical regions cause landslides when the natural forest is removed for farming or for building housing or transportation facilities.

Human activities also produce spoil heaps of waste rock material near mines and open pits. These are often poorly constructed and are easily penetrated by water that may cause a spoil heap to flow. Since many mines occur in hilly areas, there is seldom space for adequately built spoil heaps. In the Appalachian coalfield, spoil heaps fill valleys; some have collapsed in concentrated mud flows, burying valley-floor communities. Other spoil heaps are sited in unstable positions on slopes above the busy valley floor. Water from an underground spring flowing into the Aberfan coal spoil heap in South Wales caused the heap to collapse in 1966. The mass collapsed, burying a school in which over 100 children were killed. As a result, other spoil heaps in South Wales were leveled and planted with grasses, but some are still liable to collapse.

SLOPES

All landscapes are formed of slopes. A **slope** is a unit of Earth's surface relief that has common characteristics of angle and orientation. The form of each slope reflects the interactions between the slope materials and the surface processes acting on them. Weathering and mass movements are particularly significant processes in the formation of slopes, but other processes also mold them. A hillside can be mapped to show the arrangements of slopes (Figure 8-19), and it is often possible to link the processes at work on the slopes to a pattern of evolution for the hillside.

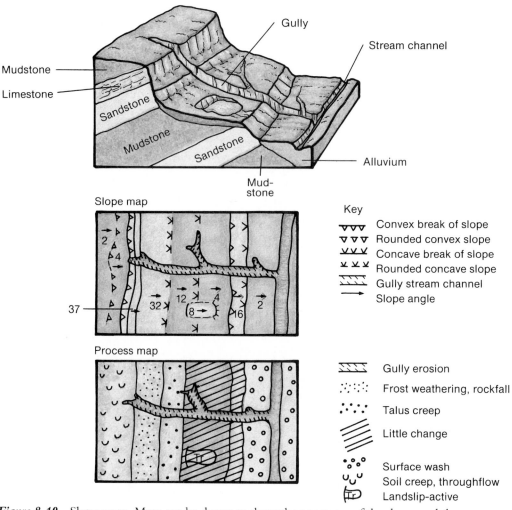

Figure 8-19   Slope maps. Maps can be drawn to show the geometry of the slopes and the processes acting on each segment.

SLOPE FORMS

Slopes are three-dimensional landforms but are often studied as two-dimensional profiles. A typical slope profile has three types of unit—a convex (getting steeper downslope) top, a straight middle section, and a concave (getting less steep downslope) lower part (Figure 8-20a). The three-dimensional form can also be described by these three unit types (Figure 8-20b) and has important applications in tracing routes taken by water, rock particles, and solutes.

A straight slope unit with rock at the surface is known as a *free face,* and its slope angle is controlled by the rock strength. Free face slopes are most common in arid, coastal, and glacial environments and are least common in humid inland areas, where regolith covers virtually all slopes. Straight units on regolith-covered slopes form when mass movement and other forms of slope transport produce a constant depth of regolith. The angle of slope depends on the regolith material and is related to the angle of repose of talus fragments. For example, clay regolith has slopes of 8 to 11 degrees, sandy regolith slopes are 19 to 21 degrees, and slopes on larger rock fragments may be 25 to 45 degrees, depending on how much water the regolith and talus contain and the degree of packing in the rock fragments.

Convex slope units are common at the tops of hillsides and are characterized by increasing rates of soil creep as the slope steepens below the crest. Concave slope units generally occur at the base of a hillside and form from materials washed down the slope. On a concave slope, the grain size of particles decreases as the slope gradient gets less down-slope; the finer particles are transported farther across the lower slope gradient than larger ones.

SLOPE MATERIALS AND PROCESSES

The forms of slopes and the rates of change that affect them result from the interaction between surface materials and the forces acting on them. The crucial balancing factors in this interaction are the strength of the materials and the aggressiveness of the surface processes. Stable slopes, subject to little change, occur where the strength of the rock or regolith is greater than the effect of the processes. Unstable slopes, liable to failure, occur where the reverse is true. Slopes that

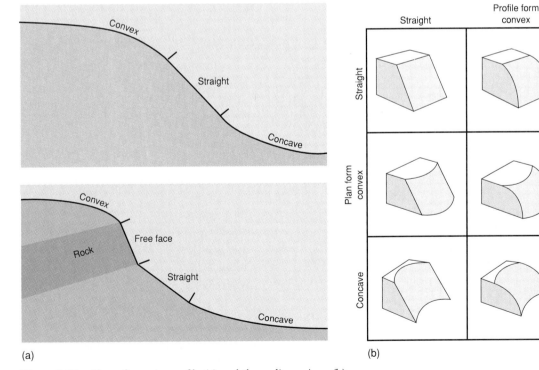

(a)

(b)

Figure 8-20 Slope forms in profile (a) and three dimensions (b).

are initially stable may become rapidly unstable if the strength of the materials is reduced or the processes become more active.

Rocks are hard, coherent, and basically strong and are not greatly weakened by saturation with water. Joints and other fractures crossing rocks provide potential lines of weakness. Regolith, including soil, is inherently weak, however, because it is formed of unconsolidated particles and is further weakened when saturated with water.

A variety of processes act on slope rocks and regolith. Weathering weakens rock resistance by breaking solid rock into particles, by removing solu-ble minerals in solution, and by reducing the size of regolith particles. Mass movements transport rock and regolith down slopes. Raindrops dislodge surface particles, the overland flow of water moves particles up to sand-grain size, and water flowing through soil removes solutes and fine particles. Surface ice flow, ground ice, the wind, and the sea also affect the for-mation of slopes (see Chapters 9 to 11).

The rates at which slope processes modify a slope depend on the slope angle, slope materials, and cli-mate (Figure 8-21). Faster rates occur on steeper slopes, in clay and loose materials, and where freeze-thaw fluctuations are frequent.

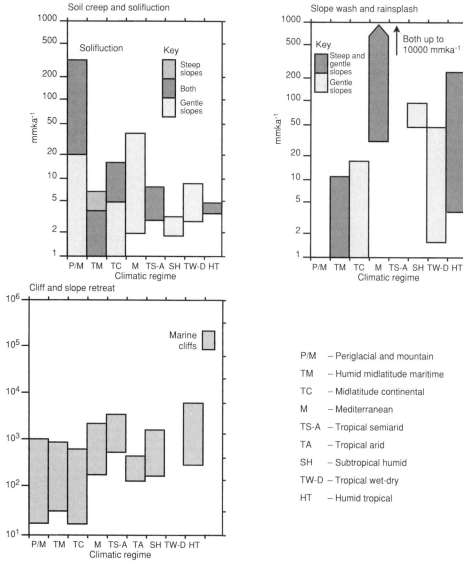

Figure 8-21   Rates of slope processes in different climates (millimeters per thousand years).

(a)

(b)

Figure 8-22   Contrasting slopes with varied origins. (a) Island of Mull, western Scotland. The free face near the top of the slope is formed in basalt lava and there is a narrow convex segment above. Below the rock outcrop, the slope is formed of fragments weathered from the rock face and moved downslope by gravity and surface wash. This lower slope has a concave profile.
(b) Masvingo Province, Zimbabwe. The granite blocks called kopjes have been rounded by weathering, and there is a sharp difference in slope between the rock face and the regolith at the base.

Figure 8-22 (Continued)
(c) Yellowstone River gorge, Wyoming. The powerful river has eroded into rocks of low resistance. Steep, unvegetated straight-segment slopes reflect this rapid erosion. (d) Western Scotland. A formerly glaciated area has slopes that have changed little since the ice melted around 10,000 years ago. Water action has begun to affect the details of the valley floor.

(c)

(d)

SLOPE EVOLUTION

Slopes change their shapes and angles of slope over time in response both to solid-earth processes that cause uplift or subsidence and to surface-relief processes. The nature of the underlying rocks and regolith is a further important factor. Examples of slopes with varied histories are shown in Figure 8-22.

Some slopes maintain the same angle of slope as they retreat. Such parallel slope retreat is common in free faces, when the rock strength remains constant and rock fragments produced by weathering are removed from the base of the slope. Over time, however, particularly in humid climates, regolith covers exposed rock and the slope forms are controlled increasingly by the transport processes at work on the regolith. This usually causes the slope angle to decrease as regolith accumulates at the slope base, while the upper part of a hillside becomes rounded.

Mass movements are the downslope transport of rocks, regolith, or soil. They occur as a flow, a slide, or by heaving. The amount of rock material in the moving mass almost always exceeds the water, ice, or air content. Subsidence is the downward vertical movement of a section of regolith or surface rock resulting from withdrawal of rock or fluids beneath the surface. Mass movements such as flows, falls, slides, creep, and subsidence often occur together.

Weathering processes and mass movements involve many linkages among Earth environments. Rock minerals form in the **solid-earth environment.** Rocks and minerals react with the **atmosphere-ocean environment** through processes such as temperature changes and precipitation, and with animal and plant activity in the **living-organism environment.**

Geologic conditions provide a variety of resistances to mass movement; climatic conditions and the impact of human activities, on the other hand, promote mass movement by placing distinctive stresses on rock and regolith that vary from region to region. The **atmosphere-ocean environment** produces conditions of temperature and moisture supply that may influence slope failure.

Slopes are basic units of landscapes that have concave, straight, or convex segments. They reflect the interactions between slope materials and surface processes, in which weathering and mass movements play significant roles. The **solid-earth environment** causes the uplift that forms steep slopes. The **living-organism environment** helps stabilize slopes with vegetation growth.

ENVIRONMENTAL
ISSUE:

Landslides—Weak Rocks and Ignorance

Some of the issues in this debate include:

What are the causes of landslides?
Do human activities increase the likelihood of landslides?
What actions are needed to prevent landslides?

Pittsburgh, Pennsylvania, is built in and around deep valleys carved into the Appalachian Plateau. Landslides are a major hazard in the area, destroying homes, blocking roads and railroads, and breaching reservoirs. Damages from landslides in Allegheny County cost millions of dollars each year. Most of these landslides are slumps; but earth flows, rock falls, and debris slides also occur. Fortunately many of the mass movements are slow moving and personal injuries are rare.

The area is susceptible to landslides because of its rock structure. The plateau surface is capped by thick layers of sandstone or limestone. Water flows downward through cracks in these layers, weakening the shales and clay rocks below. Over thousands of years, however, the valley slopes have been stabilized by the accumulation of rock debris at their base. Once the angle of slope was reduced in this way, vegetation anchored the slope surfaces. In the natural state, few landslides would occur today.

Human activities, the main cause of the increasing number of landslides, promote slope instability and failure in several ways. Buildings constructed at the top of a slope exert downward pressure on deep unconsolidated regolith beneath. Slag and other industrial wastes dumped on slopes add more weight. Removal of vegetation eliminates its binding effect on steep slopes. The widening of narrow valleys for road construction involved the excavation of material at the base of slopes, causing the slopes to steepen. All of these actions make slopes unstable or affect the movement of water through the slope materials. Former landslide areas once stabilized by vegetation are particularly vulnerable to being mobilized again. In backyards along the sides of the Monongahela valley, cracks up to 3 m (10 ft) wide and 10 m (30 ft) deep opened because of the extra weight of new buildings and the removal of slope vegetation.

People who buy homes in the Pittsburgh area or in other areas with a history of mass movements should look for the signs of landslide activity. Cracks in buildings, poorly fitting doors and windows, leaning fences, tilted trees or utility poles, broken underground pipes, leaking pools, and ground cracks are all significant indicators.

Actions to mitigate landslide damage include avoiding building or dumping close to the tops of steep slopes, preserving the lower portions of old landslide deposits, and maintaining a vegetation cover on slopes. Such measures are increasingly part of building codes, especially in urban jurisdictions.

Linkages

Landslides occur where slopes underlain by susceptible rock sequences *(solid-earth environment)* become subject to surface processes *(surface-relief environment),* often enhanced by the removal of vegetation cover *(living-organism environment).*

Questions for Debate

1. Is human ignorance a major cause of landslide hazards, or do humans hope that actions that might lead to landsliding will not be detected?

2. How important is it to try and prevent landslides?

A valley side slides into a road in western Pennsylvania.

CHAPTER SUMMARY

1. Weathering is the physical disintegration or chemical decomposition of solid rock into fragments, minerals, or dissolved chemical ions. Physical weathering is caused by mechanical stresses that fracture or disintegrate rocks. Chemical weathering involves chemical reactions that weaken and decompose rocks or their constituent minerals. Living organisms assist both physical and chemical weathering.

2. The broken rock material produced by weathering is regolith, a basic component of soil. Differences in the resistance characteristics of rocks and in their climatic and surface-relief environments produce differences in the effectiveness of weathering from place to place. Human actions often enhance weathering indirectly by making the atmosphere more acidic, by altering water regimes, and by bringing solid rock into direct contact with the atmosphere.

3. Mass movements are the downslope transport of rocks, regolith, or soil. They occur as a flow, a slide, or by heaving. The amount of rock material in the moving mass almost always exceeds the water, ice, or air content. Subsidence is the downward vertical movement of a section of regolith or surface rock resulting from withdrawal of rock or fluids beneath the surface. Mass movements such as flows, falls, slides, creep, and subsidence often occur together.

   Weathering processes and mass movements involve many linkages among Earth environments. Rock minerals form in the **solid-earth environment.** Rocks and minerals react with the **atmosphere-ocean environment** through processes such as temperature changes and precipitation, and with animal and plant activity in the **living-organism environment.**

4. Geologic conditions provide a variety of resistances to mass movement; climatic conditions and the impact of human activities, on the other hand, promote mass movement by placing distinctive stresses on rock and regolith that vary from region to region. The **atmosphere-ocean environment** produces conditions of temperature and moisture supply that may influence slope failure.

5. Slopes are basic units of landscapes that have concave, straight, or convex segments. They reflect the interactions between slope materials and surface processes, in which weathering and mass movements play significant roles.

   The **solid earth-environment** causes the uplift that forms steep slopes. The **living-organism environment** helps stabilize slopes with vegetation growth.

KEY TERMS

gradation   222
weathering   222
transport   222
erosion   222
deposition   222
mass movement   222
base level   222
physical weathering   222
chemical weathering   222
biologic weathering   222
exfoliation   222
frost wedging   223
solution   224
hydrolysis   224
carbonation   224
oxidation   224
reduction   224
chelation   225
regolith   225
porosity   227
flow   229
slide   232
fall   232
heave   233
subsidence   234
slope   238

CHAPTER REVIEW QUESTIONS

1. Define gradation, weathering, transport, erosion, and deposition.
2. What are the main processes of physical and chemical weathering?
3. What factors affect the depth of regolith?
4. Distinguish between flows, slides, falls, and heaves of rock and regolith.
5. How do weathering and mass movements affect the degradation of land surfaces?
6. Show how the chemical composition and physical structure of a mineral or rock affect its resistance to weathering.
7. Summarize the effects of weathering and mass movements in different climates.
8. Describe some human impacts on weathering and mass movements.

SUGGESTED ACTIVITIES

1. Find out the weathering characteristics of building stones and construction materials in local use. Has the use of some materials been more successful than others?
2. Investigate an example of mass movement that has caused damage to humans or property. Write a short account of its supposed causes and effects, suggesting how disasters of this kind may be avoided.
3. Study the slopes on a hillside. You may use Figure 8-22 or a local example. Measure or draw a profile of the component units and suggest the relationship between the strengths of the slope materials and the processes acting on them.

CHAPTER

RIVERS AND LANDFORMS

THIS CHAPTER IS ABOUT:

◆ Running water and landscapes
◆ Watersheds
◆ Water flow at the surface and underground
◆ Stream transport
◆ Stream erosion and deposition
◆ Stream channel landforms
◆ Flood plain and alluvial fan landforms
◆ Climate, rocks, and stream landforms
◆ Human impact on stream processes
◆ Environmental Issue: Flood Hazards

Rivers wear down Earth's surface in some places, picking up particles of sand and clay and sometimes moving boulders. Rivers deposit the clay and sand they carry in other places where the river flow slows. In this way rivers shape the landforms that provide a stage for human activities. The rapids on the river in Point Lobis Park, California, are lowering the river bed. The meanders on Muddy Creek, Wyoming, form in previous river deposits.

RUNNING WATER AND LANDSCAPES

unning water is the most important agent of gradation in the surface-relief environment; rivers move some 90 percent of the sediment worn from the continents. **Rivers** are streams of water flowing from higher to lower ground. They are a dominant agent in wearing down the landscape in all parts of the world except beneath glaciers and ice sheets. Running water, however, can be an effective agent even beneath the margins of ice masses. Rivers also flow from time to time in deserts and have important impacts in these areas where other erosion agents work much slower.

Rivers are part of the hydrologic cycle. They complete the cycle studied in Chapter 4 by concentrating the flow of water precipitated on the continents and taking it to the ocean (see Figure 4-1). When rain or snow falls on the continents, water may be stored initially in rocks or ice masses, but it eventually flows into rivers and then downslope to the ocean, possibly passing through a lake on the way.

This chapter shows how rivers affect surface landforms. The first section focuses on how rivers work, what causes their flow to vary from place to place and from time to time, and how they carry rock debris. Understanding river processes makes it possible to examine the origins of river landforms, such as river channels, flood plains, and alluvial fans. The chapter concludes by describing the effects of climate, rocks, and human activities on river flow and related landforms in different parts of the world.

WATERSHEDS

The area drained by a river and its tributary streams is known as a **watershed,** or *drainage basin* (Figure 9-1). A watershed is bounded by a **drainage divide**—the line that determines whether precipitated water flows toward one river or another. Within a watershed, water flows down and through slopes into a network of stream channels.

Each watershed is a discrete unit in the surface-relief environment. Water supplied by precipitation flows within each watershed, moving weathered particles with it. The water is concentrated into stream channels that are shaped by the flow of water and sediment. Sediment is deposited where the flow slows or stops.

The stream channels within a watershed form a connected network. The network begins in the upper parts of the watershed in small streams that have no tributaries and are called *headwaters.* Tributary streams join each other at *confluences* and add water along the length of the river, normally increasing flow until the river enters a lake or the ocean at its *mouth.*

In general terms a watershed consists of three sections. Water and sediment are concentrated into stream channels and erosion is dominant in the upper section; sediment transport is dominant in the middle section; and sediment is deposited in the lower section. This division conveniently describes the main aspects of stream activities within watersheds. In different proportions, however, erosion, transport, and deposition are present along the whole length of a river.

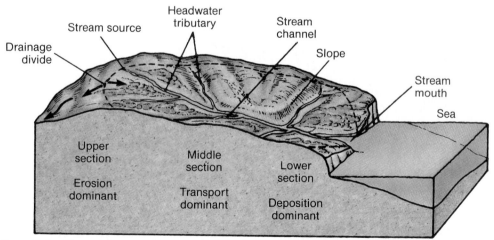

Figure 9-1   General features of a watershed, including the network of stream channels, drainage divide, and mouth.

Watersheds exist at a variety of geographic scales. For instance, the Mississippi-Missouri watershed (Figure 9-2) covers over 40 percent of the continental United States (3,322,000 sq km or 1,244,000 sq mi). This huge watershed comprises several major tributary watersheds including the Missouri and Ohio, which in turn are composed of smaller watersheds. The smallest watersheds contain only a single headwater tributary and commonly drain areas of up to 2 sq km (up to half a square mile).

WATER FLOW AT THE SURFACE AND UNDERGROUND

Rain and snow supply watersheds with water that travels through soil and rock and over slopes until it enters a river (Figure 9-3). The waterflow is then concentrated in channels as streamflow.

SOURCES OF WATER

The water that flows through watersheds comes from a variety of sources such as rain, snowmelt, and melting glacier ice. Each of these sources has its own pattern of supply to the stream that is reflected in fluctuations of flow through the year.

Individual rainfall events vary in amount, intensity (volume per unit time), and frequency. If intense rain lasts for a long period, the ground becomes saturated. Local flooding occurs when the supply of water exceeds the capacity of the stream channel to carry it away within its banks (see Environmental Issue: Flood Hazards, p. 281). Many parts of the world have rainy and dry seasons for part of each year resulting in seasons of high and low streamflow respectively. Some regions have a fairly even distribution of rain throughout the year, and their streams flow with less variation of water

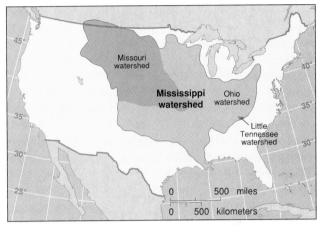

Figure 9-2 The Mississippi watershed is one of the world's largest watersheds and includes tributary watersheds such as those of the Missouri and Ohio Rivers. The Little Tennessee watershed is one of the smaller drainage areas contributing to flow in the main streams and has headwater tributaries in the Great Smoky Mountains.

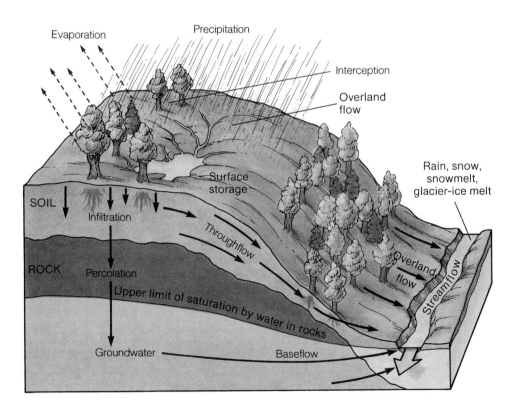

Figure 9-3 The movement of water through a watershed. Trace the passage of precipitated water over the ground surface and through vegetation cover, soil, and rock to the stream. Precipitation falls over the whole watershed but is concentrated in stream channels.

level. Other regions receive very little rain at any season, and their streams only flow for short periods of time after occasional storms.

Snowmelt flow generally peaks in spring or early summer (Figure 9-4). During winter several snowfalls may accumulate without melting. In spring the total snow cover melts rapidly, producing extra volumes of river flow that may cause flooding. Glacier melt is usually less dramatic, peaking in late summer and then diminishing until winter cold prevents melting.

WATER HITS THE SURFACE

Rain or snow falling on a watershed first encounters trees, grass, or crops. **Interception** is the percentage of rain or snow that plants prevent from directly striking the ground. Plants hold raindrops and snowflakes on their leaves, and some of that water evaporates. Snow may remain on plants until it melts. Water that is intercepted by vegetation, but not evaporated, drips off the leaves or runs down branches and trunks to the ground. Interception reduces the amount and intensity of rainfall at the ground and slows its arrival at ground level. Where there is no vegetation cover in deserts or on plowed land, the rain or snow strikes the ground directly.

Different plant covers intercept varying proportions of rain or snow. Deciduous trees intercept

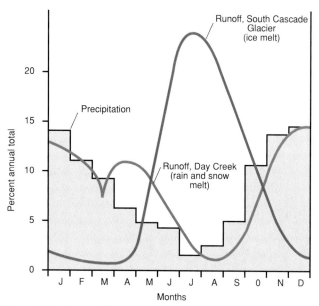

Figure 9-4 Streamflow from snowmelt and ice melt: contrasting flow patterns in the North Cascade Range, Washington. Explain the differences in flow in Day Creek and from the South Cascade glacier.
Source: U.S. Geological Survey.

some 15 percent, evergreen conifers 20 to 25 percent, and tropical rainforest up to 30 percent. A grass cover intercepts about 20 percent.

FLOW OVER AND THROUGH SOILS

Once at the soil surface, water follows one of several paths. The rainwater or snowmelt may remain on the surface as puddles, soak into the soil, or run off over the surface. Evaporation may accompany any of these, reducing the amount of water flowing toward a stream.

The process of water sinking into the soil is known as **infiltration,** and surface flow across a slope is known as **overland flow.** The proportion of water following each of these paths is determined by the *infiltration capacity* of the soil. The infiltration capacity depends on soil permeability and the amount of water already in the soil. **Permeability** is the rate at which water passes through a material. In soil it is determined by the amount of fracturing and the size and number of gaps between particles. A soil composed mostly of large sand grains has many large and interconnected pore spaces and allows water to soak in and pass through rapidly. A clay or fine silty soil may have pore spaces that are too small to allow water to pass through. Infiltration capacity is high in soils of high permeability and low water content, such as sands, and low in soils of low permeability, such as clays. If it has rained recently, the pore spaces in any soil may be full of water, saturating the soil and reducing the infiltration capacity to zero. Once precipitation exceeds the infiltration capacity, water accumulates on flat surfaces and overland flow occurs on slopes. Overland flow is particularly common on clay soils and on slopes lacking vegetation cover, but it is rare on forested land where vegetation intercepts the rain and provides trunk and root pathways into the soil.

GROUNDWATER

Water entering the soil may flow through the soil virtually parallel to the ground surface (*throughflow*), or it may penetrate deeper into the underlying rocks by **percolation.** Water movement through a rock depends on the rock's permeability. If a rock is impermeable, percolation rates are very slow or zero, and throughflow in the overlying soils will be more significant. Rates of both throughflow and percolation are slower than overland flow or flow in stream channels, but they can be quite high in soils or rocks that have large interconnected pore spaces and other voids such as fractures or joints. Where the spaces in rock are large enough, underground streams may flow at speeds equal to those on the surface.

In a permeable rock, water moves downward under gravity filling the pore spaces in the deepest section. This section is the *zone of saturation* and its upper surface is known as the **water table** (Figure 9-5). Water accumulated in the saturated parts of permeable rock below the water table is known as **groundwater.** Groundwater below the water table moves slowly. In the *zone of aeration* above the water table, water moves continuously—downward from the surface after rain and upward near the surface in dry periods to replace evaporated water. The level of the water table fluctuates during the year if precipitation has a strong seasonal pattern; it rises in the wet season and falls in the dry season. The water table level also falls when humans pump water to the surface.

Water entering the groundwater zone after rainfall is known as *recharge* when it adds to the groundwater and raises the water table. The deeper the water goes, the longer it is likely to remain in rock storage. Water in long-term groundwater storage may eventually emerge at a surface outlet, or **spring.** Springs occur where a water table reaches the surface or where fractures in the rock concentrate the flow of water and bring it to the surface. Water may also flow directly from groundwater into

streams providing the **baseflow** component of streamflow. Water leaving the groundwater store to flow into streams may lower the water table.

Rocks that allow water to move through them may also store water if the geologic structure and bounding layers are right. Rocks which store water that can be extracted for later use are known as **aquifers.** An aquifer traps water between impermeable layers above and below it. For example, a syncline beneath London, England, contains a limestone aquifer some 300 m (1000 ft) thick. Water falling on the hills surrounding the city flows into this aquifer. Land in the center of London is lower in height than the level of the water table beneath the hills, and this places the water in the aquifer beneath London under pressure. It flows out of wells dug there without a need for pumping—or it did until so much water was extracted for use in London that the water table fell. Water that flows out of wells without pumping is known as *artesian water.* Another example of artesian water occurs in eastern Australia where water falling on the coastal highlands flows westward in an aquifer beneath impermeable rocks. Water flows into wells drilled into this aquifer at the base of the hills because of the pressure acting on the water.

One of the major aquifers in the United States, the Ogallala, lies beneath the Great Plains. Water moves through this layer, which is tilted to the east, and parts have artesian conditions. Some 170,000 wells tap this aquifer and its water is used on 20 percent of U.S. irrigated land (Figure 9-6). Although this

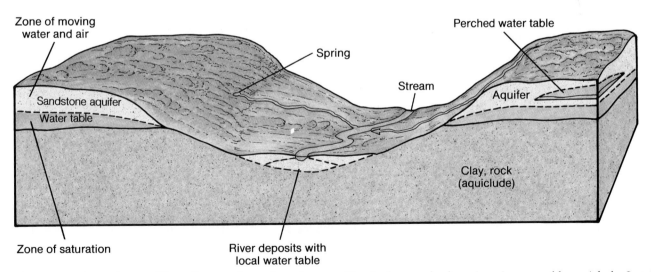

Figure 9-5 Groundwater. The yellow sandstone layer is an aquifer; the brown clay layer is an impermeable aquiclude. Local water tables occur within the aquifer and also in the valley floor deposits.

Figure 9-6 Irrigation circles in Nebraska. A well taps the underlying aquifer rock and supplies the sprinklers along the arm. The arm rotates to water the whole circle when the crop is growing.

aquifer covers such a large area and averages 65 m (200 ft) thick, the water is being extracted so rapidly that the level of the water table is falling faster than rainfall can replenish the water.

Impermeable rock layers may store some water, but water adheres to the rock particles so strongly that it is not available to supply streams, plant roots, or to be extracted for human use. Rocks that prevent water storage and extraction are known as *aquicludes.*

Groundwater is an important resource for human use. Much of the water used in farming in California and Arizona, as well as in the Great Plains, comes from groundwater sources. Groundwater also brings prosperity to dry areas in less developed countries. In humid areas, groundwater is tapped by utility companies. Groundwater extraction for human use often draws on sources that took thousands of years to accumulate. When extraction exceeds the rate of replenishment, groundwater use ceases to be a sustainable practice.

In some cases, groundwater is contaminated and becomes unusable by humans, livestock, and plants. Toxic waste materials in surface spoil heaps may seep into underlying aquifers where they are concentrated before flowing out in wells and springs. Polluting materials include sewage, tanker spills, and buried chemical waste. In Florida the excessive pumping of groundwater has allowed salt ocean water to penetrate the freshwater aquifers.

WATERFLOW IN RIVERS

Water enters a stream from four major sources. Precipitation falling on the water surface is a tiny percentage of the total. Overland flow from rainwater, melting snow, or melting ice reaches the stream quickly after rain. Throughflow that emerges from soil at the base of a slope gets to the stream more slowly. Baseflow from springs generally has the greatest delay after percolating into rock storage unless the rock is broken by wide fractures that allow water to flow rapidly through. The contributions of overland flow, throughflow, and baseflow vary from watershed to watershed and from time to time in the same watershed.

After rain the combination of direct precipitation into the stream, overland flow, and throughflow causes the stream level to rise within an hour or so of the maximum rainfall. This is the *stormflow* component of streamflow. Baseflow from groundwater is fairly constant, maintaining streamflow when it is not raining. After rain the baseflow rises somewhat and then diminishes gradually until the next fall of rain, but usually forms the main component of streamflow between rainfalls.

Not all stream channels always contain water. Those that do are termed *perennial*. *Intermittent* streams contain water during the wet season, while *ephemeral* streams are shorter-lived, flowing for only a few hours after a single fall of rain. A single river may contain all these types along its length. It is common, for instance, for headwaters to contain water only during the wet season or after a heavy storm.

Streamflow is measured by calculating the area of the waterflow cross section in the channel and multiplying this by the average velocity of waterflow at that point (Figure 9-7). At any point along the river, the total volume of flow per unit of time is known as the **discharge** and is measured in cubic meters per second (cms) or cubic feet per second (cfs). The volume of water flowing in a river may also be measured in acre feet in the United States. This measure has particular significance to irrigation farmers, since an *acre foot* is the amount of water that would cover one acre to a depth of one foot. An acre foot is equivalent to 1219 m^3.

Flow can be measured at any point along the river's course, but when regular records are required, a permanent gauging station structure is built. This structure ensures that the channel cross section does not vary and only the depth of water fluctuates in it.

As a result, only one measurement—the depth, or *stage*—of the water in the channel is needed, and the water discharge is calculated from a graph. Many automatic stream gauging stations record the stage every 15 minutes.

Stream discharge varies over time, rising after storms and falling in drier weather. The response of streamflow to a storm depends on the nature of the watershed. Where a watershed is well vegetated, has permeable soils and rock, and the slopes are gentle, much of the rain will be intercepted and flow into soil and rock. This delays the passage of water through the watershed and evens out its arrival at the river. The stream level rises slowly and has a moderate peak. Between storms baseflow maintains a high average flow in the river. By contrast, where there is no vegetation, the soils and rocks are impermeable, and the slopes are steep, most rain becomes overland flow and reaches the river quickly after a storm. The stream level rises sharply to a high peak and then falls almost as rapidly. Between storms the small quantity of baseflow maintains only a low streamflow. Most watersheds are between these two extremes and have a mixture of factors that promote small or large stormflow peaks.

Streamflow also fluctuates seasonally and over time. The proportions of overland flow and infiltration often vary from summer to winter in midlatitude humid areas. In winter there is less evaporation from the soil and transpiration from plants, so more water remains in the soil. When rain falls, the soil becomes saturated and there is a higher proportion of overland flow. In summer evapotranspiration removes water from the soil and plant cover increases

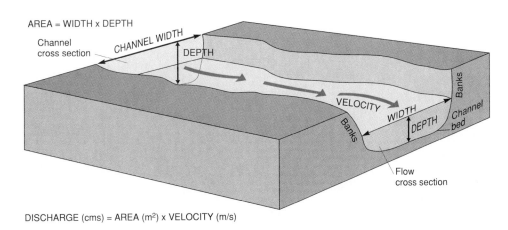

Figure 9-7 Measuring streamflow. The discharge of water is calculated by multiplying the area of waterflow cross section by the velocity of the water passing through.

interception. The soil has a higher infiltration capacity and overland flow is reduced. Climate change causes variations in flow over periods of years. The natural flow of water in the Colorado River, for example, now averages only 75 percent of the flow in the early 1920s. This reduced flow is cut further by evaporation losses from the huge reservoirs built along the river.

As water flows down a river's course, it is joined by water from tributaries increasing the volume of water in the main channel. The channel size and the water velocity both increase downstream (Figure 9-8). For instance, average velocity along the Mississippi River increases from less than 0.5 m/sec

(2 ft/sec) in the Rocky Mountain Missouri tributaries to nearly 2 m/sec (6 ft/sec) near the river mouth. Friction between the water and channel sides is lower in the larger channel downstream, allowing water to flow through at higher velocities despite the lower channel gradient.

Waterflow through a river system may be interrupted on encountering a lake or reservoir. Such bodies slow and store water for days or weeks before it continues downstream.

STREAM TRANSPORT

STREAM LOAD

The movement of rock debris and dissolved matter is one of the most important relief-forming activities of streamflow. The material carried in streams is known as **load** (Figure 9-9).

The stream's supply of rock debris and dissolved matter principally results from weathering processes that break up solid rocks into transportable sizes. Mass movements and other slope processes, such as overland flow and throughflow, bring weathered materials to the river channel. Baseflow contains dissolved matter when water that percolates into rocks reacts chemically with them. Streams move large quantities of dissolved matter and finer clay and sand particles, together with smaller amounts of larger pebbles. A few boulders may be moved short distances during high streamflows.

Different amounts and sizes of particles occur in a stream channel at different points along its course. Where steep rocky slopes end close to the channel, large blocks and boulders from weathered and well-jointed rocks may fall into the channel. When a stream flows through a wide plain, it acts on those materials that it deposited there previously; these

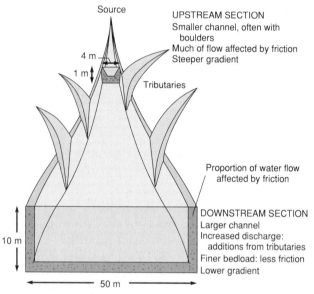

Figure 9-8   Channel size and discharge. This schematic diagram compares the upstream and downstream sections of a stream channel. How are channel slope, size, and discharge related in each place?

Figure 9-9   Watershed and stream load. Sediment and rock material in solution move from the upper to the lower part of the watershed and out to sea as part of the gradation process.

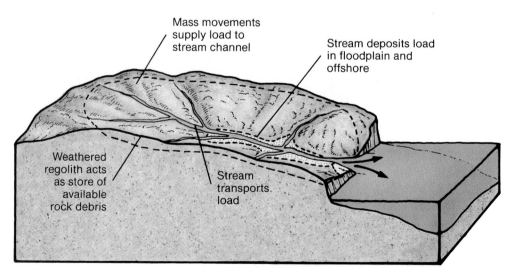

materials are often much smaller than the fragments in its upper course. In the Mississippi-Missouri River, for example, large rock fragments are common in the channels of its upper tributaries in the Rocky Mountains. Where the river enters the Great Plains, the size of load material abruptly decreases to fine sand, silt, and clay.

The total load of a stream may consist of a wide range of materials of different sizes. There are three fractions of stream load—dissolved, suspended, and bed load—each of which reacts differently to waterflow.

DISSOLVED LOAD

As water passes through soil and rock, ions produced by chemical weathering of elements or compounds are dissolved in the moving water. These dissolved ions, or *solutes,* form the **dissolved load** of a stream. Overland flow and throughflow contain few solutes because water is moving too quickly to take up dissolved matter. Most solutes come from baseflow, but in agricultural areas they are augmented by excess fertilizer spread on fields by farmers. Water officials often talk about water quality in the same breath as dissolved load because many chemicals added by human activities lower stream water quality.

The concentration of solutes in stream water varies over time and is often diluted during peak stormflow. Once in a stream, most solutes remain there until they reach an ocean or an evaporating lake such as the Great Salt Lake of Utah.

SUSPENDED LOAD

Particles picked up from the channel bed or worn from the banks may be carried along in the flow of water as **suspended load.** Particles in suspended load are usually no larger than clay, silt, or fine sand—all under 1 mm in diameter. Such fine particles may enter the streamflow directly from the surrounding slopes in stormflow. The size and quantity of particles carried in suspension increase with stream velocity. Faster flow increases the turbulence of the water and the stream's ability to lift particles and keep them moving. Some of the larger particles only stay in suspension for a short time, but the tiny clay particles remain in suspension until they reach the sea.

As streamflow peaks after a rainstorm, the quantity of suspended load rises even more rapidly. When water discharge doubles, for instance, the volume of suspended load carried may increase fourfold. High concentrations of suspended load have been measured. Before the Colorado River was dammed in so many places, it carried up to 600,000 parts of sediment per million parts of water—about one-third of the total stream was sediment. The Platte River of Colorado had a similar level of suspended load and became famous as being "too thin to plow and too thick to drink." The Yellow River of China often carries nearly one million parts of sediment to one million parts of water.

The **capacity** of a stream—the maximum load it can carry in a given time—is seldom reached in the suspended load of fine particles. The supply of such particles is often exhausted before the stream's capacity to carry them is reached.

BED LOAD

Rock fragments rolled along the stream bed or lifted by the streamflow for a few seconds at a time as they move downstream are known as the **bed load.** Bed load includes particles varying in size from coarse sand to boulders a meter or so across.

For a large particle to be moved, the force of streamflow must overcome the resistance of the particle to movement (Figure 9-10). The water exerts one force by its downstream flow and another through its turbulence, which assists lifting. The size, weight, and shape of the bed load particle determine its resistance to movement. Resistance increases when larger particles block others or form an armored river bed with interlocked rock fragments. Bed load movements are usually brief and discontinuous because the resistance factors are commonly greater than the force of river flow. The diameter of the largest particle that a river can move at a particular discharge is known as the stream's **competence.** Most channels contain more bed load than the stream can move in a single flood peak.

In high flows, however, streams move considerable quantities of bed load. During such movement, individual particles become smaller as their corners are rounded by banging against each other. Less resistant material may be completely pulverized after only a short journey. The study of a meltwater stream issuing from a Canadian glacier showed that the bed load at the point of emergence was a mixture of large and small, weak and resistant, and mainly angular rock fragments. Four kilometers (2.5 mi) downstream the average size was smaller, the particles were better sorted and more rounded, and fragments of weaker rocks were no longer part of the bed load.

Figure 9-10 Suspended load and bed load. Compare the ways in which these two fractions of the stream load are moved.

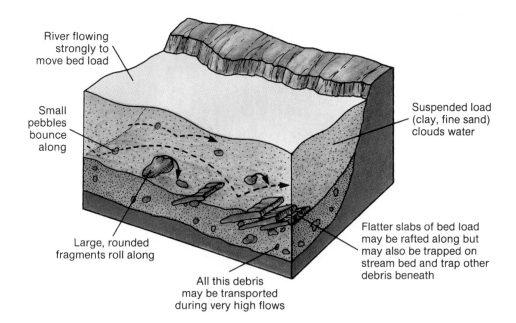

River flowing strongly to move bed load

Small pebbles bounce along

Large, rounded fragments roll along

Suspended load (clay, fine sand) clouds water

Flatter slabs of bed load may be rafted along but may also be trapped on stream bed and trap other debris beneath

All this debris may be transported during very high flows

TOTAL STREAM LOAD

The total stream load is composed of the dissolved, suspended, and bed load fractions. Bed load makes up 70 percent of total load by weight in mountain areas and 20 to 30 percent in arid environments, but elsewhere it is usually less than 10 percent.

More accurate figures are available for solute and suspended load than for bed load. For the world as a whole, the average annual amount of suspended load is 88,000 kg/sq km (251 tons/sq mi) of the area drained. The average amount of solute load is 34,000 kg/sq km (96 tons/sq mi).

The distribution of stream load fractions varies around the world. Eighty percent of the world's suspended load is carried by the rivers of southern and eastern Asia. These rivers drain the high mountains of the Himalayan ranges and their extensions, where physical and chemical weathering are rapid, mass movements are frequent, and the streams have high flows of water during the spring snow melt and summer monsoons. By contrast, streams draining many humid midlatitude watersheds in areas of low-to-moderate relief have a low suspended load. In such cases the solute load may make up over 90 percent of the total weight of stream load.

STREAM EROSION AND DEPOSITION

The flows of water, sediment, and solutes in rivers cause erosion and deposition. As a stream erodes the bed and sides of its channel, the valley in which it flows may be deepened and widened. Deposition occurs when the stream's capacity for moving sediment falls.

Stream erosion is the result of several processes acting together. Two of the most important are corrasion and corrosion. **Corrasion,** or abrasion, is the wearing of solid rock when the stream load scrapes against the stream's bed. This commonly smoothes and cuts grooves or potholes in the rock and contributes to the wearing down of the bed load particles (Figure 9-11). **Corrosion** is chemical activity in water. Corrosion by moving groundwater and underground streams is particularly effective in limestone rock where it complements the work of chemical weathering.

Erosion also occurs when the force of high floodwater flows dislodges loose rock fragments held in channel beds or banks. Such action undercuts channel banks so that they collapse when the water level falls. In underground caverns below the water table in limestone rock, water flows under pressure, increasing its erosive effectiveness and making it able to drill out almost circular tunnels.

Following transport in the stream, deposition of the load occurs when the stream discharge decreases or when it enters a body of standing water, such as a lake or the ocean. At that point, stream capacity decreases and the load is deposited. Load deposited by a stream is called *alluvium.* Deposits at river mouths include deltas and estuarine mud (see Chapter 11).

Figure 9-11 Corrasion. Circular potholes in a rocky stream bed in northern England. The flow of water in this stream circulates pebbles that enlarge the hollows.

Running water flows mainly in stream channels. Water enters watersheds in precipitation (***atmosphere-ocean environment***) and passes into stream channels. Its passage through the watershed is slowed if it encounters vegetation (***living-organism environment***), permeable soil or rock (***surface-relief environment***), or if it is stored in ice masses or lakes.

The inputs of precipitation and the pathways of water movement through the watershed control the flow in stream channels.

Stream load is the rock material carried in streamflow and includes dissolved, suspended, and bed load fractions (***solid-earth environment***). The movement of water and its load in rivers acts as an agent of gradation through the erosion, transport, and deposition of rock particles.

STREAM CHANNEL LANDFORMS

A **stream channel** is a troughlike landform along the floor of a valley that carries the main flow of water through a watershed (Figure 9-12). The size and shape of the channel adjust to the flows of water and passing load. Water flowing through the channel is in constant contact with the landform it produces.

CHANNEL SIZE

With the exception of those in karst landscapes, stream channels in humid regions almost always contain running water. The channel is large enough to cope with most flows of water supplied by precipitation on the watershed. On average, the stream fills only from one-sixth to one-third of the channel depth. Unusually heavy rainfalls or additional water resulting from the melting of several layers of snow causes many streams in humid regions to overflow their channel once or twice a year, flooding the surrounding valley floor. Once a stream channel reaches the state when its dimensions and gradient change little as the result of erosion or deposition, it is in equilibrium with the flow of water and load generated in the watershed.

Stream channels often develop from smaller flows of water that eventually erode larger features. Overland flow across a slope may occur as *sheetflow*, which is a thin layer of water covering the whole slope. Sheetflows seldom have depths of more than a few millimeters and on most slopes soon break into concentrations of flow. The concentrated flow creates small channels a few centimeters deep known as *rills*. As flow in a rill deepens, corrasion and other

(a)

(b)

Figure 9-12 Stream channels. Two sections of streams in White Mountains, N.H. (a) An upstream section with large boulders up to a meter across and a steeper channel gradient. (b) A downstream section without boulders.

erosional processes excavate small valleys up to a few meters deep called **gullies** (Figure 9-13). Rills and gullies contain water only immediately after rainfall. They form most commonly where a surface is bare or almost bare of vegetation, as in arid climates or after plowing. Gullies may become part of the stream channel network when water flows more consistently in them.

CHANNEL SHAPE

A stream channel can be viewed from three different perspectives. Imagine standing on the bed of a stream and looking either upstream or downstream. This perspective shows a channel cross section, in which both banks and the bed can be seen (see Figure 9-8). If the stream could be straightened out, and cut along the center of the channel from end to end, the longitudinal profile would be seen. This view shows the change in gradient from source to mouth. The third perspective is the plan view, which gives a bird's-eye view of the channel features; this is the perspective shown on a plan or map.

Channel Cross Section   The size and shape of the channel cross section reflects the amount of water passing through. Small channels transmit little water and large ones transmit large quantities. The cross

section increases downstream (see Figure 9-8). A comparison between the Mississippi and Amazon stream channels shows that although the Amazon watershed is not much bigger than that of the Mississippi, the Amazon channel is much larger than the Mississippi. This is because of the greater rainfall and flow of water in the equatorial climate of the Amazon basin (Figure 9-14).

Differences in shape among channel cross sections are based on comparisons of their depth and width. The *width-depth ratio* (width divided by depth) indicates whether the channel is wide and shallow (high ratio) or narrow and deep (low ratio). Wide and shallow channels often have coarse bed load, erodible banks, and fluctuating seasonal flows. The channel shape forms during the high flows when the stream's competence and erosional force are high. It moves the larger load and erodes the banks and bed. During low flows, the load drops and fills the channel so that its width does not change but its depth decreases. Stream channels that are relatively deep and narrow often have a fine load and stable banks and occur where there is little variation in flow through the year.

River Longitudinal Profile   The channel gradient is usually steeper near the source and flatter near the

Figure 9-13   A gully near Cuiaba, central Brazil. The removal of surface vegetation made it possible for a storm to wash out this channel, which was enlarged by subsequent storms. Between storms no water flows in the gully.

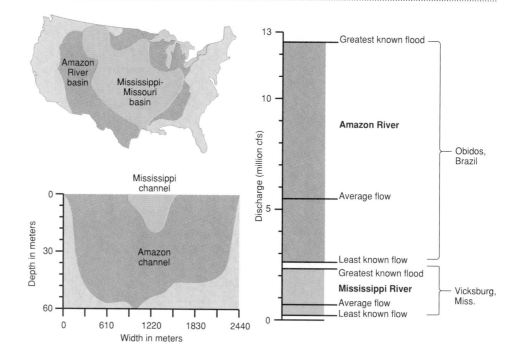

Figure 9-14 Watershed areas and stream discharge. How do the Mississippi and Amazon rivers compare in terms of watershed area and stream discharge?

mouth, giving a concave longitudinal profile. A steeper profile in a headwater tributary is typical of streams transporting coarse bed load with a small water discharge. The lower gradient downstream reflects the increase in water discharge and decrease in load particle size. The frictional effects of the channel boundary on the waterflow decrease downstream and so make it possible for a greater flow to occur across a lower gradient slope.

In some streams the concave longitudinal profile is broken by steps that have a gentle upstream gradient and a steeper downstream gradient. Such breaks are known as **knick points** and are frequently marked by waterfalls or rapids. Knick points often occur where resistant rocks come to the surface along the stream channel, slowing erosion and supplying coarser bedload to the river.

Channel Plan View Stream channels have distinctive plan-view patterns that reflect the relationship between waterflow and load. Some channel reaches are straight, others wind or meander, and another group splits into separate strands of flow. All these patterns may occur along a single river as circumstances of slope, load, and water discharge change from source to mouth.

Straight channel reaches are rare in nature and their lengths seldom exceed ten channel widths. Within a straight channel, the flow tends to wind from side to side through a series of deeper *pools* and shallower *riffles*. Riffles form in bed load deposits and produce an even depth across the stream (Figure 9-15).

Stream channels that wind in looplike patterns are said to **meander.** Meandering reaches occur where a relatively constant flow of water combines with a fine sediment load and low-to-moderate channel gradient. Such channels have small width-to-depth ratios and are common in the lowland areas of humid midlatitudes (Figure 9-16). The size of meanders depends on the amount of water flowing through the channel: the greater the streamflow, the larger the meanders.

River meanders often have pools on the outside of a bend and riffles on the straighter sections between bends. The main flow of water is concentrated toward the outside of the bend where it moves through the pools and strikes the banks with its greatest force. Material eroded at these points moves downstream and is deposited where flow is slack on the inside bend of the next meander.

Meanders vary in degree of activity and development. In active meanders erosion wears back the outsides of bends and deposits sediment on the insides of

Figure 9-15 Stream channel: riffle and pool. The deep water in the foreground of this upland area in western Scotland is a pool; at the bend in the channel the bed load forms a shallower section that is above the water level in a dry period—a riffle.

bends, leading to downstream migration of the meanders. Many stream meanders, however, do not migrate. Some meanders in southeastern England still have the same position they had in Roman times, two thousand years ago. Such meanders become inactive when vegetation protects the banks or when stream discharge is too weak to cause erosion. People often attempt to prevent meander activity to stop loss or flooding of land they own or to provide a permanent waterway for transportation. Narrow valley floors, such as those in New England, confine meanders naturally in a zone that is narrower than that needed to develop the meander loops. The cramped conditions result in Z-shaped or box-shaped meander patterns as the channel switches sharply from one side of the valley floor to the other. Figure 9-12b shows one example of such a confined meandering stream.

Braided stream reaches often have a relatively shallow and wide channel. During seasons of low water, the waterflow within the channel splits into threads that combine and split again around largely unvegetated bars formed of bed load (Figure 9-17). *Anastomosing rivers* also have a split flow but within distinct channels that divide around more permanently vegetated islands or low hills.

Braided channels are most common where sediment is coarse in mountainous, arid, or arctic environments. In these areas, spring snowmelt produces abnormally high streamflows, the channel gradient is steep, and the bed load comprises coarse sand, gravel, and boulders. In the high-flow season, water fills the whole channel. The braids and bars appear in the low-flow season when the smaller flow of water has to find its way through deposited bed load. Short braided channel sections may also occur on lower terrain where there is a steep local gradient or where a tributary brings in coarser bed load.

FLOODPLAIN AND ALLUVIAL FAN LANDFORMS

Stream deposition produces landforms in the lower parts of watersheds. These landforms are often attractive

Figure 9-16 Meandering stream channels of (a) Muddy Creek, Wyoming, flowing through fine sediment with a highly sinuous course, (b) and the Madeira River in western Brazil.

(a)

(b)

Figure 9-17   Braided stream channel. Some of the bars in this braided stream channel in Prince Rupert are covered with vegetation.

for human settlements and land use but are subject to flood hazards (see Environmental Issue: Flood Hazards, p. 281).

A **floodplain** is the part of the valley floor adjacent to the stream channel that is covered by water when flow exceeds the channel capacity. Waterflow normally occurs within channels but occasionally overflows for a few hours or days. In much of the United States, overbank floods occur in spring following snowmelt or at other seasons following prolonged heavy rain. Floods deposit part of the stream load on top of the floodplain. The floodplain is thus effectively part of the natural river system—an expanded storage area that accommodates excessive supplies of water and load.

The materials that constitute floodplain landforms are a combination of both present and past deposits, including deposits in the channel and from overbank floods. Typical floodplain features associated with a meandering stream are shown on Figure 9-18. The in-channel deposits form on the inside bends of meanders and are known as *point bars*. If a meander loop is cut off from the main channel, it forms an *oxbow* lake that gradually fills with plants and with fine clay that settles out of suspension in the still water. Oxbows are most common where a

section of cohesive clay slows meander-bank erosion compared to the rate in surrounding sandy sediment. A narrow meander neck forms as another meander limb migrating downstream catches up with the slowed one, and the stream cuts through the neck isolating the oxbow.

Overbank floods spread a shallow layer of water over the floodplain. They soon drop any coarser rock particles to form low ridges called *natural levees* along the tops of channel banks. The finer load settles out more slowly as the flood waters spread farther from the channel. Marshy areas subject to annual inundations are known as *backswamp* zones. Once a natural levee forms, subsequent floods may break through gaps and deposit the coarser load in concentrated splays on the far side of the levee.

Braided stream channels produce different floodplain landforms. Such channels often occupy the whole valley floor, and gravel bars composed of coarse bed load dominate much of the channel. Braided channel floodplains are less varied in their deposits and landforms than meandering channel floodplains.

Alluvial fans are fan-shaped deposits of alluvium that range from several hundred meters to tens of kilometers across. Watersheds in which alluvial fans occur

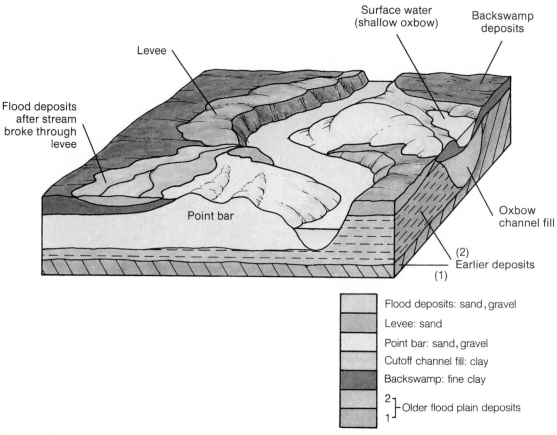

Figure 9-18   Floodplain: the distinctive deposits formed when a meandering stream flows across a wide valley floor. Which of these features were deposited by overbank floods and which by processes within the stream channel?

have two distinct halves (Figure 9-19). In the upper mountainous half, a stream with a steep channel gradient in a steep-sided valley erodes rapidly, producing an ample supply of sediment. In the lower half on flatter land, the stream channel widens and deposition is dominant. The water spreads out, slows down, and drops its load. One phase of deposition may block flows in the channel causing the stream's path to shift from side to side. Over time a series of such shifts leads to a radiating pattern of deposition. In this way, the stream builds a fan-shaped landform.

Alluvial fans are particularly common in the mountainous deserts of the southwestern United States where they cover one-third of the surface (Figure 9-20). Streamflow is fed by snowmelt or precipitation over the mountains. Fans also occur at the edges of mountainous regions in humid areas, as along the southern flank of the Himalayas.

The sizes and shapes of alluvial fans depend on the supply of water and rock fragments from the source area in the upper part of the watershed.

More water or rock particles produce larger fans. In dry regions the streams often cut trenches in the fan surface during an unusually high waterflow and carry sediment through this trench to the outer edge of the fan. Mudflows and earthflows can also move sediment after sudden floods that permeate loose and dry rock debris, causing it to flow. The higher parts of the fan between trenched sections dry out when the stream is not flowing and a surface crust may form. Such dry sites close to water tempt people to build homes on them, but a heavy storm may cause the stream to shift its course in the unconsolidated fan sediment and erode the land beneath the settlement.

CLIMATE, ROCKS, AND STREAM LANDFORMS

Streams flow in almost all climatic environments of the world although they act infrequently in very dry areas such as the central Sahara. Streams do not operate in the central parts of ice sheets but can be

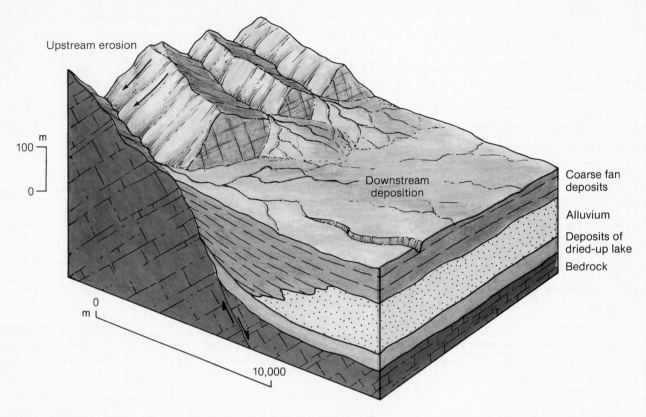

Upstream erosion

100 m

0

0
m

10,000

Downstream
deposition

Coarse fan
deposits

Alluvium

Deposits of
dried-up lake

Bedrock

Figure 9-19

Figure 9-19   The main features typical of alluvial fans in the southwestern United States. Compare the upstream valleys with the lower part of the fan beyond the mountain front.

Figure 9-20   Alluvial fan: the mouth of Copper Canyon, Death Valley, California. The erosional features of the upper part of the watershed contrast with the deposits of the fan.

Figure 9-20

important where meltwater shapes marginal areas. Distinctive landforms are produced by streams in different climate regions. Some rocks also interact with streams to produce characteristic network patterns and suites of landforms. Climatic and geologic factors produce geographic variations in stream activity and landforms.

STREAM LANDFORMS IN DIFFERENT CLIMATES

The effectiveness of streams in molding landforms depends on the level of water discharge and the amounts of rock waste available for transport. Streams are particularly effective agents of denudation where there is a combination of deep weathering, rapid mass movement, and high water discharge. Where there is slow weathering, little mass movement, or low discharge streams do not modify landforms in their watersheds to any great extent. Figure 9-21 summarizes some of the features of river-made landforms in different climates. The climatic environments described include humid areas, where streams are continuously active, and cold and dry regions, where stream action is more irregular.

In *humid midlatitude climates,* such as western Europe and the eastern United States (Figure 9-21a), a protective cover of vegetation increases interception and infiltration capacity, slowing the passage of water and preventing much overland flow. Flow in stream channels is fairly even over time, so water seldom floods over the banks. Inactive meandering channels are common. Landforms change slowly.

In the *humid seasonal tropics,* such as eastern and southern Africa (Figure 9-21b), chemical weathering breaks down rock fragments to fine clay particles in a deep layer of soil and regolith. During the dry season, streamflow is reduced to a trickle, plants die down, and the soil surface dries out, losing cohesion. At the start of the wet season, rains rapidly remove the loose debris in sheetflow and streams carry it away before the growing vegetation slows erosive activity. These processes cause wearing down of the whole landscape and not just along the line of the river. The stream's main contribution is to remove the fine debris that results from intense weathering and is available for transport after the drought.

On the *subpolar coasts and islands* around the Arctic Ocean, streams freeze for much of the year, and flow occurs only in the two weeks or so following snowmelt in early summer (Figure 9-21c). Some rain may fall later in the short summer, but most of these areas receive little precipitation because they are not in the paths of midlatitude cyclones. Such a pattern might suggest that subpolar rivers are ineffective agents of denudation because water seldom flows in them. When the rivers do flow, however, they move huge volumes of snowmelt water in a short time. Such flows are very effective in changing landforms. The valley features on the islands of northern Canada and elsewhere in the Arctic show that the combination of rapid weathering with annual high levels of streamflow provide a basis for rapid erosion. The humid subpolar islands of Spitzbergen in the northernmost reaches of the Atlantic Ocean, for instance, have a high annual precipitation, mainly as snow from midlatitude cyclones. Active physical weathering by frost-wedging and ground ice action feed large volumes of broken rock into solifluction and massive snowmelt runoff. The streams carve valleys, deepening them at rates as rapid as 100 m (300 ft) per 1000 years. Valleys only 6 km (4 mi) long have braided river channels up to 70 m (210 ft) wide. By contrast, meandering channels in humid midlatitude watersheds of the same size are only up to 2 m (6 ft) wide.

Stream processes in *arid regions* (Figure 9-21d) act infrequently because of the general lack of water, but they may still be the most effective agents of gradation at work in such environments. In arid and semi-arid regions, rivers are often ephemeral and commonly drain to a central lake, such as the Great Salt Lake, Utah, rather than to an ocean. Internal drainage of this type has a *centripetal* pattern (Figure 9-22). Infrequent rainfall results in slow erosion despite the lack of vegetation and high availability of weathered debris. Where rainfall is a little more regular, flash floods after occasional storms may produce short-lived streamflows or mudflows. In semi-arid environments, streams become extremely active where higher precipitation totals fall on sparsely vegetated land. Streams fed by snowmelt from outside arid areas can also have powerful effects on the landscape, as in the case of the Colorado River carving the Grand Canyon in northern Arizona.

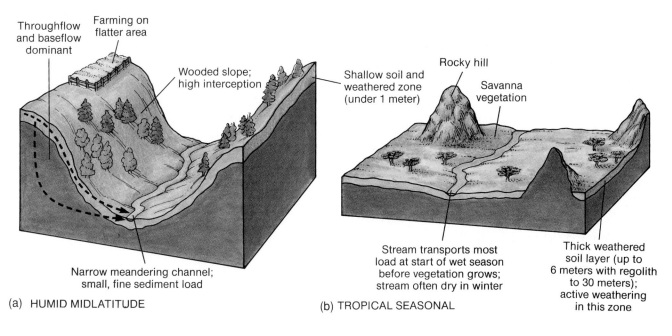

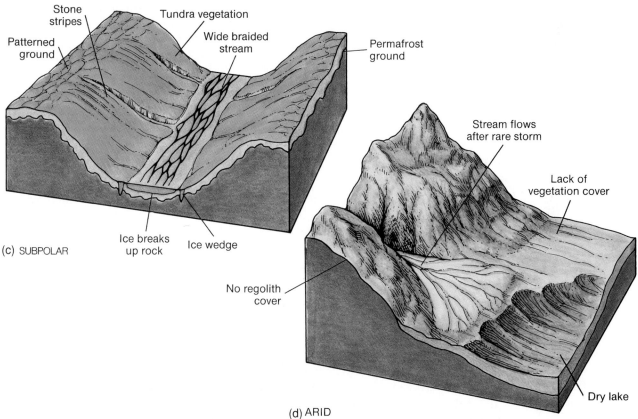

Figure 9-21   Climate and stream processes. Landforms in four contrasting climatic environments. (a) Humid midlatitude climate where rain occurs at all seasons but temperatures are not very cold in winter or very hot in summer. (b) Tropical seasonal climate, where summer rains contrast with winter drought. (c) Polar climate where long frozen winters alternate with short, cool summers. Large volumes of snowmelt supply the stream with two weeks of high flow in early summer. (d) Arid climate where streams flow for short periods.

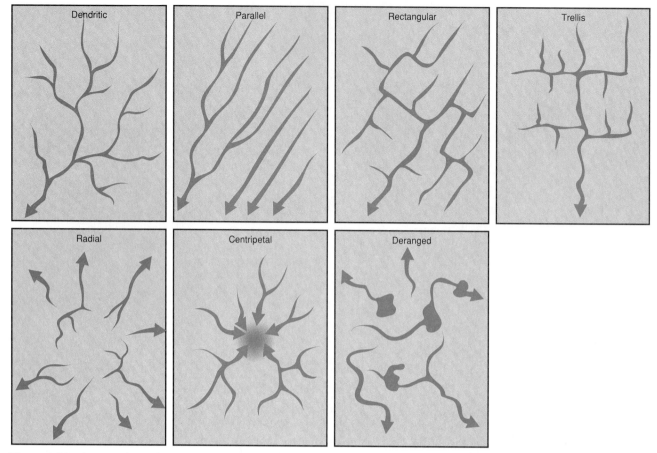

Figure 9-22   Stream channel network patterns.

CHANGING CLIMATE AND STREAM LANDFORMS

Changes in streamflow and load result from changes in climate and produce different groups of landforms. The valley of the lower Mississippi River is an example (Figure 9-23). Today the meandering channel winds across a flat floodplain between bluffs of older rocks. Features produced during the last 20,000 years are buried beneath the floodplain. Toward the end of the last glacial advance, the stream carved deep valleys to a sea level over 100 m below the present (Figure 9-23a). As the ice sheets melted and the ocean level rose, the debris-charged meltwaters flowed down the Mississippi filling the channels with coarse bed load (Figure 9-23b). The ocean level continued to rise, but waterflow in the river fell when the flow changed to a path along the

St. Lawrence River and the supply from melting ice ended. This produced braided and then meandering stream channels (Figure 9-23c,d).

More recently, stream channel changes have occurred during the twentieth century in the Cimarron River in western Kansas and northwestern Oklahoma. At the beginning of the century, the narrow meandering channel was 16 m (50 ft) wide and flowed across a grassed floodplain. After floods in 1914, the channel gradually widened until by 1940 it was nearly 400 m (1200 ft) across in places. The channel became braided. A railroad company had to rebuild its bridges three times during this period to cope with the widening. After 1940 many sections of the channel began to fill and their margins became vegetated. The channel narrowed but did not return to the 1900 state.

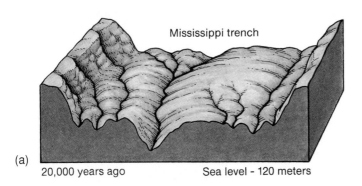

(a)

20,000 years ago   Sea level - 120 meters

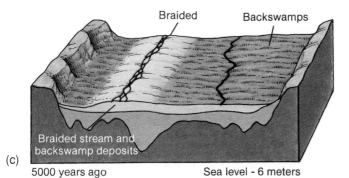

(b)

10,000 years ago   Sea level - 30 meters

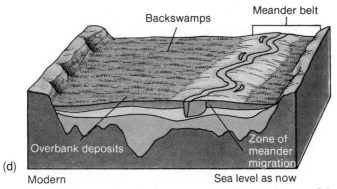

(c)

5000 years ago   Sea level - 6 meters

(d)

Modern   Sea level as now

Figure 9-23 Floodplain and climate change. The evolution of the lower Mississippi River valley since the end of the last glacial phase. The sea levels were 120, 30, and 6 meters below the present sea level.

Variations in the amount and type of rainfall caused these changes. The late 1800s and early 1900s were periods of relatively high rainfall in the watershed, but the rain was evenly spread over time in moderate individual falls and drought years were few. There was sufficient rain for the floodplain and channel banks to be covered by grass and the meandering channel coped with most flows. In 1914 catastrophic floods tore away the protective bank vegetation and were followed by twenty-five years of lower rainfall, more frequent drought years, and more intense storms. This regime prevented the bank vegetation from growing and produced high and erosive flood peaks. After 1940 the rainfall resumed its former pattern.

ROCK TYPE AND STREAM LANDFORMS

Differences in rock type produce variations among the landforms in watersheds. Resistant rocks usually create greater relief, whereas weak rocks underlie flatter, lower land with less relief because they are more easily worn away. The rocks that form the ridges of the Appalachian Mountains are more resistant than those forming the intervening lowlands.

Rock characteristics in combination with relief and precipitation produce different densities of streams and valleys. On permeable rock more water infiltrates the soil and there is less overland flow. Consequently, there is less surface drainage and the density of stream channels is lower. On impermeable rock most water runs off over the surface and drainage densities are higher. Flowing water is more effective in shaping the surface where there is a higher density of channels.

Rock Type and Stream Networks   Rock type also controls river network patterns since streams flow along lines of weakness in the rocks (see Figure 9-22). Where the underlying rock is evenly resistant, streams have a branching, or *dendritic,* pattern. When an evenly resistant rock surface slopes steeply, the branches are drawn out into a set of almost *parallel* streams. On rocks such as granite that have strongly developed, right/angle joint patterns, the stream network often has a *rectangular* pattern.

Where the rock layers of a region alternate in resistance and have been tilted or folded, another group of stream network patterns becomes important. Along the Atlantic and Gulf Coast plains of the southern and eastern United States, a *trellised* pattern occurs in which the main streams flow at right angles to the ridges, while tributaries develop parallel to the ridges along the weaker rocks (Figure 9-24). In areas of domed rock structures or volcanic peaks, a *radial* drainage pattern often evolves.

Karst Landforms   The most distinctive suite of landforms associated with a specific rock type is that which develops on limestone. Such landscapes are termed **karst,** after a region in former Yugoslavia named for its bare, stony ground. Surface sinkholes and underground caverns are common in karst areas.

Limestones include a variety of rock types, but calcium carbonate is the main constituent of all of them. It sometimes occurs in mixtures with clay, but the most distinctive karst landforms occur on rock that is more than 90 percent calcium carbonate. Massive limestones that are highly permeable, have well-spaced joints, and

have a vertical thickness of several hundred meters show the greatest development of karst features.

Chemical weathering attacks limestones as rainwater moves through pore spaces or along joints in the rock. There are three main zones of alteration in limestone according to the way in which water moves through the rock (Figure 9-25). In Zone 1 on the surface there is little or no running water because of easy percolation into the rock. Surface flow on limestone is more common in the humid tropics where rainfall is more intense than in midlatitudes. In humid midlatitudes landforms such as sinkholes result from solution and subsidence, and are more common than landforms produced by surface streams. Zone 2 extends from the surface down to the water table. In this zone of aeration, water percolates through the rocks or moves freely through joint cracks and in underground streams. It corrodes and corrades the rock in a manner similar to surface streams. In Zone 3 below the water table, groundwater moves slowly. The pressure of the weight of water within the saturated zone can force the groundwater through narrow openings and even uphill.

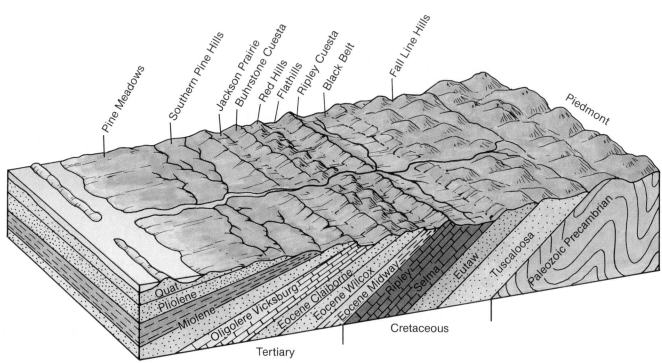

Figure 9-24   Watershed and rock structure. This west-to-east section across the coastal plain of the southeastern United States shows how tilted rocks of varying resistances produce a trellis pattern of drainage. The ridges form on more resistant sandstones and limestones, while lower areas between are on less resistant clay rocks.

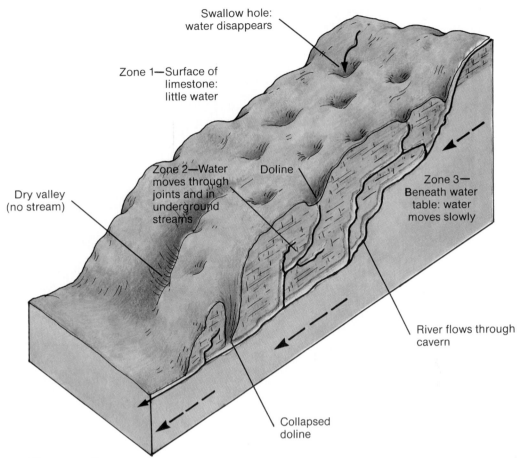

Swallow hole:
water disappears

Zone 1—Surface of
limestone:
little water

Zone 2—Water
moves through
joints and in
underground
streams

Doline

Dry valley
(no stream)

Zone 3—
Beneath water
table: water
moves slowly

River flows through
cavern

Collapsed
doline

Figure 9-25   Karst landscape. The zones of water movement.

The surface landforms of karst areas are often dominated by a chaotic set of pits and hollows known as **dolines** or sinkholes—the essential features by which karst is recognized (Figure 9-26a). In parts of former Yugoslavia they occupy over 60 percent of the surface. Dolines range in size from 2 to 100 m (6 to 300 ft) deep and 100 to 1000 m (300 to 3000 ft) across. In humid midlatitude environments in western Europe, they result from a combination of solution and subsidence. They are associated with *swallow holes* where surface water disappears underground (Figure 9-26b), and *dry valleys* that do not contain streams. Dry valleys in karst are often "blind" at one end where a cliff blocks the valley. Larger *polje* depressions up to several kilometers across form after the widespread erosion of limestone. The flat, impermeable floors of poljes are often subject to seasonal flooding as the water table in the limestone rises.

The underground features of karst include **caverns** and their connecting tunnels. The chambers and tunnels run vertically and horizontally according to rock structure. Their size and shape are determined by whether they were formed above or below the water table. In the saturated zone beneath the water table, groundwater movement under pressure produces tunnels that are bore tubes having an almost circular cross section.

Deposits of dripstones, rock piles from collapsed cavern roofs, and mud brought in by streams fill caverns and tunnels to varying extents. The *dripstones* form as calcium carbonate precipitates from solution in water dripping from a cavern roof. Some grow down from the roof (stalactites) and some grow up from the floor (stalagmites) as shown in Figure 8-4. Large caverns in the United States occur in Mississippian limestones around the margins of the Appalachians including western Virginia and Kentucky.

Different varieties of karst landforms occur in different climatic environments. Tropical karst has

274   Chapter 9

Figure 9-26 Karst landforms. (a) Solution dolines in Croatia. (b) A swallow hole. Fell Beck enters a swallow hole near Ingleborough, Yorkshire, England.

(a)

(b)

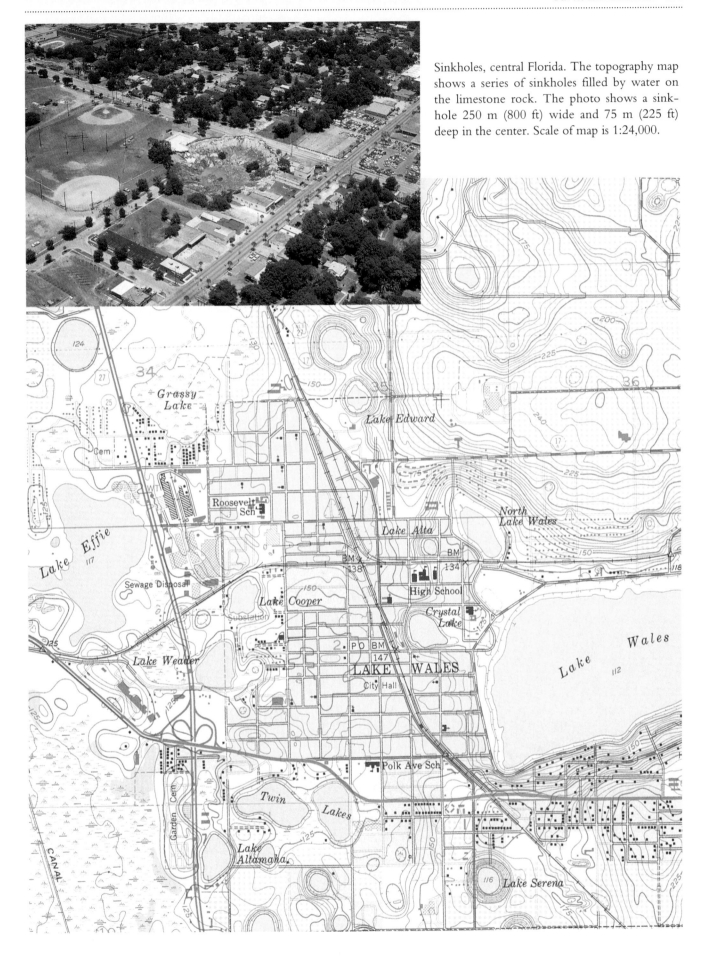

Sinkholes, central Florida. The topography map shows a series of sinkholes filled by water on the limestone rock. The photo shows a sinkhole 250 m (800 ft) wide and 75 m (225 ft) deep in the center. Scale of map is 1:24,000.

higher rates of chemical decomposition and greater quantities of surface water. In tropical cone karst there are wider depressions, or *cockpits,* carved largely by surface waterflow. In the *tower karst* of southeastern Asia, high, steep-sided hills are honeycombed by massive caverns. Karst areas that were beneath ice during the Pleistocene Ice Age, such as the Niagara escarpment of Ontario, Canada, were scraped bare of soil and exhibit widened joints in *limestone pavement.* Karst occurs in arid areas such as northern Egypt and Arabia, but the landforms were produced under former humid conditions.

HUMAN IMPACT ON STREAM PROCESSES

Human beings have long occupied the watersheds of the world. Their activities have produced many changes in the natural processes of streamflow and in the landforms typical of such environments. The human impact is great even in arid areas where the river waters were the basis of irrigation by the earliest civilizations. In humid midlatitude climates, where natural landform change is slow and human development of the land is intense, human activities are often the main agent of landform change.

Some human activities alter stream discharge and the potential of rivers to modify the landscape. Rivers that are made into canals by lining them with concrete often become inactive as agents of gradation. In many cases, however, waterflow exploits cracks in the concrete and undermines the lining. Where dams and reservoirs are built to store water and reduce flooding, as along the Colorado River, streamflow evens out and becomes less active in landform change. Such reservoirs, however, are liable to silting and their outlet areas to erosion. The natural flow of water through the Florida Everglades was altered by canal building and the extraction of water for metropolitan Miami (Figure 9-27), changes that influenced much of the distinctive environment of southern Florida.

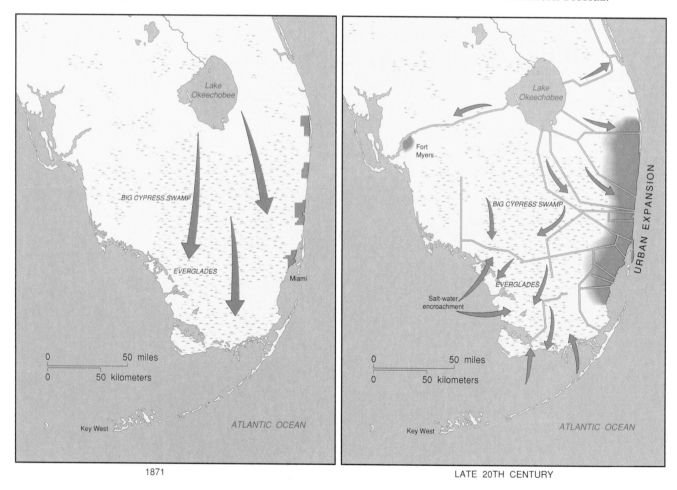

Figure 9-27   Human impact on watersheds: the modification of natural drainage in Florida. Until 1871 the natural drainage was a seasonal flow southward from the overflowing Lake Okeechobee through the Everglades. When canals were dug in the early twentieth century, much of this water was diverted eastward to the coast.

Other human activities alter the hydrologic cycle in a wider sense as the result of changes in land use. *Farmers* remove forest and plow grassland. In Europe this began with the Neolithic and Bronze Age clearance of upland areas nearly 5000 years ago and extended to denser lowland forest over 1000 years ago. Clearing forest reduces interception and increases overland flow, leading to soil erosion and a smaller time lag between rainfall and peak streamflow. Streams deposit the sediment eroded from upstream areas on valley floors downstream. By medieval times much of the Italian peninsula was reduced to bare rocky slopes upstream, while swamps formed downstream on the deposits of eroded soil. The southeastern United States suffered intense soil erosion in the nineteenth century following the cultivation of tobacco and cotton. Upstream erosion and downstream deposition both destroyed good land. In Wisconsin some valley floors contain up to 4 m (12 ft) of sediment eroded from fields in their watersheds since farming began in the 1830s. The effect of forest clearing for farming is particularly disastrous in tropical regions because of the intense and erosive rainfall (Figure 9-28).

Logging also removes forest cover and exposes hillsides to erosion. In Oregon experiments carried out to determine which method of timber cutting results in least erosion showed that clear cutting is best. It produces smaller increases in sediment in streams than patch cutting, which necessitates the building of longer logging roads between the patches. Road building across steep, bare slopes enhances soil erosion.

Mining also has drastic impacts on natural streams. The medieval tin miners of Dartmoor, England (Figure 9-29), altered the shapes of the upper valleys by their surface extraction methods. The lower parts of valleys filled with finer debris washed down in the streams. The California gold miners of the mid-nineteenth century played pressure hoses on the gold-bearing gravels of the Sierra Nevada and washed so much rock debris downstream that it began to fill San Francisco Bay (Figure 9-30).

The *building of large cities* is a major feature of human development and affects the functioning of river systems in several ways. Construction requires the removal of vegetation and topsoil, exposing land to erosion by storms. Local streams become clogged by the sediment carried and the increase in overland flow causes flooding downstream. After development of the urban area, the impermeable cover of roads, houses, commercial buildings, and parking lots drastically increases stormflow. The storm sewers

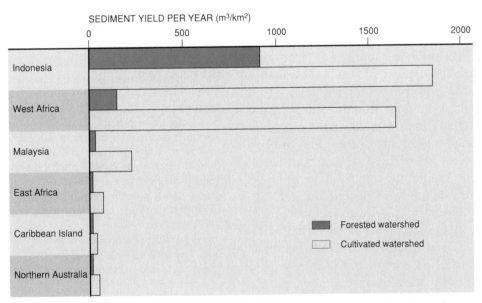

Figure 9-28 Human impact on watersheds: stream erosion of forested and cultivated watersheds in tropical areas. What difference does forest clearance and cultivation make?

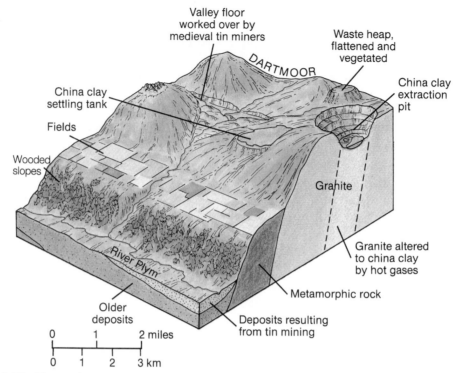

Figure 9-29 Human impact on watersheds: the impacts of surface mining, ancient and modern, on Dartmoor, England. Medieval tin miners dug up the valley floors in their search for tin ore. Modern china clay extraction involves digging huge pits and building extensive waste mounds.

constructed to carry rainwater away from city centers increase the stormflow component of river flow. Figure 9-31 compares aspects of water movements for Moscow, Russia, and its rural surroundings.

The piedmont area of the eastern United States provides an example of human modification of stream processes in a humid midlatitude climate. The area has a history of changing stream response to human activities since the early nineteenth century (Figure 9-32). In the early 1800s, the spread of farming through this area resulted in soil erosion, gullying, and increased sediment loads. In the later 1800s and in the early 1900s, farmland was abandoned and forest reestablished itself, reducing soil erosion and river sediment loads. In the mid-twentieth century, the growing suburbs of Washington, D.C., spread into the area. At first construction caused massive soil erosion and clogged the lower valleys with sediment. After building ceased, the stream load fell to almost nothing.

The shapes and sizes of stream channels are controlled by flows of water and load through them (**surface-relief environment**). The shape of a stream channel also affects how water flows through it. Flood plains and alluvial fans are the major depositional landforms in watersheds.

Stream action varies in different climatic environments. Different rock types (**solid-earth environment**) provide varying degrees of resistance to stream action. Water acting on limestone produces distinctive karst landscapes. Plants (**living-organism environment**) stabilize riverbanks in humid climates.

Humans have the greatest impact on river processes in humid midlatitude watersheds. Farming, logging, mining, and urban construction have major influences on river systems.

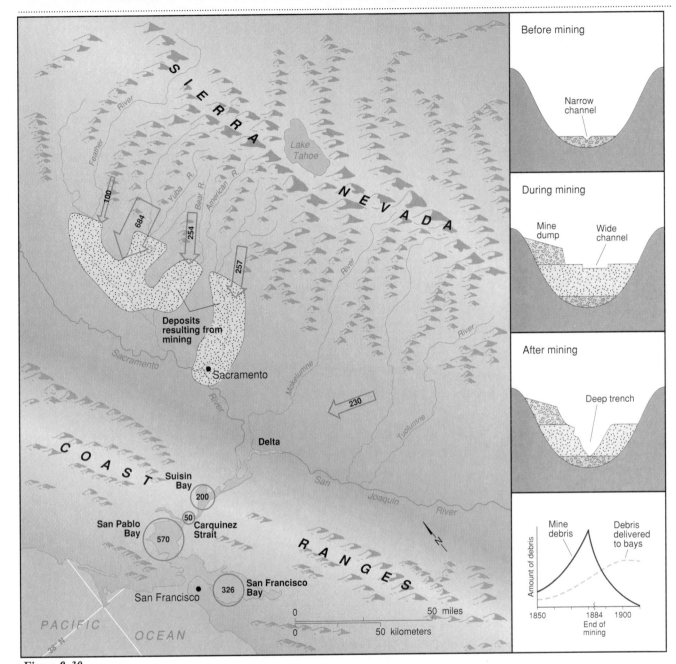

Figure 9-30
Human impact on watersheds: the impact of the California Gold Rush, 1849–84, on the Sacramento River and San Francisco Bay area. The map shows the areas affected by hydraulic mining (using jets of water that washed rock debris downstream); the red arrows show how much sediment (in million cubic meters) was produced between 1849 and 1909. The total was 1700 million m³ (1 million cubic meters equals 35 million cubic feet approx.). The red circles in the bays show how much debris accumulated between 1849 and 1914. The amount totaled 1146 million m³. The inset valley cross sections show the impact on the mined creeks and canyons, and the graph gives an idea of the production of mine waste and its impact on the bays.

Map: G. K. Gilbert, U.S. Geological Survey.

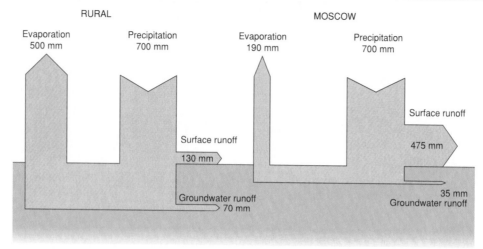

Figure 9-31   Human impact on watersheds: the influence of urbanization on watershed hydrology. Compare the streamflow in rural and urban areas.

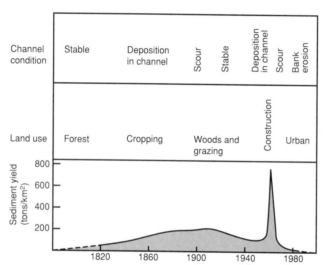

Figure 9-32   Human impact on watersheds: changing land use on the piedmont west of Washington, D.C. Describe the sequence of events and their impacts on waterflow.

ENVIRONMENTAL ISSUE:

Flood Hazards

Some of the issues involved include:
    What are the causes of flooding?
    Why do flood protection schemes fail
    to contain all floods?
    On what do engineers base their
    decisions for designing flood control
    works?

Flooding by rivers and the sea causes more damage and loss of life than any other form of natural hazard. Flooding by rivers may be due to a seasonal maximum of rainfall, as in the monsoon countries, to runoff from melting snow, or to a single massive storm. Seasonal flooding occurs along the Ganges and Brahmaputra Rivers in the Indian subcontinent. It used to occur along the lower Colorado River following spring snowmelt. Sudden storms in arid regions may produce flash floods that rapidly fill formerly dry stream channels and inundate surrounding areas. Flooding often occurs when several factors coincide, such as snowmelt, heavy rains, and high tide at the river mouth. The exceptional flooding in the U.S. Midwest in the summer of 1993 was caused by an unusual period of almost continuous rain.

One problem with efforts to prepare for river floods is that it is difficult or impossible to predict what will be the highest level reached by a flood and when it will occur. The higher flood levels occur less frequently, and engineers classify flood levels by how often a flood of a certain size is expected to occur, on average (see illustration). They refer to 30-year floods, which are smaller than 100-year floods; this concept is used in the design of flood protection works. Such flood categories are based on averages, however, and the actual intervals between large floods may vary. Thus, an area might have three 30-year floods in 10 years or none for 100 years. It is important to realize that flood protection works will not protect from all floods, only those up to the design level. Many flood protection works along the middle Mississippi were designed to contain a

30-year flood, but the 1993 floods that flowed over these barriers were of 100-year proportions.

There are two main approaches to controlling flood waters. Engineers can try to reduce a river's ability to flood, or planners and politicians can attempt to change human patterns of land use so that fewer people and less valuable property are affected. They can try to protect established uses of flood-prone areas or persuade those living and working in such areas to move.

Many attempts to combat flooding center on the control of stormflow entering the river, as the second illustration shows. Such attempts include changes of land use in the upper parts of a watershed where woodland may be planted or arable land grassed to increase interception and reduce overland flow. In urban areas flood control measures include the building of water-retention dams.

The main effects of flooding occur downstream where a common response is to construct engineering works that modify the channel so that it will carry more water. Walls or artificial levees raise the bank levels, wider, deeper, or straighter channels increase their capacity, and diversion channels carry flood flows away from centers of population. The massive concrete channels built to protect the lower elevations of Los Angeles from flooding remain dry for most of the year and are featured in filmed car races and chases.

The only reactions to flooding before the nineteenth century were to bear the losses and costs or avoid living on the flood plain. For many people in developing countries these are still the only options. In European and North American urban-industrial countries, factories and linked worker housing were built in flood-prone zones in the nineteenth century because the land was cheap. Subsequent floods produced disastrous losses of life and property. Emergency actions rescued and rehabilitated victims. Some attempts were made to flood-proof homes and workplaces, but such approaches were often ineffective.

During the twentieth century, effort and money have been devoted to the alteration of stream channels to prevent flooding and improve navigation. In the United States,

government money was poured into such projects. After major flooding along the lower Mississippi River in the 1920s, for example, the U.S. Congress gave the Army Corps of Engineers powers and funds to build dams and artificial levees along the main stream and its tributaries to reduce the risk of further flooding.

During the 1950s, the limited design capabilities of flood protection works led to attempts to control stormflow runoff by upstream land use changes. These efforts, however, met problems in trying to persuade landowners who do not suffer from flooding to alter their practices at cost to themselves. In the 1970s and 1980s, attempts to control flooding shifted again from trying to alter rivers to trying to alter human activities in flood-prone areas.

During the 1970s, U.S. federal programs to cope with flood disasters began to require that local land use zoning agencies should prohibit many uses of the more hazardous zones and that landowners in flood-prone areas must have flood insurance. People do not like leaving familiar areas, however, and resist the additional expense of insurance policies. Progress in flood prevention and protection remains slow, but the United States is still more advanced in this than most countries.

Linkages

Flood hazards illustrate the linkages between the **atmosphere-ocean environment,** which supplies the inputs of precipitation to watersheds, the nature of the ground surface **(surface-relief environment)** and its underlying rocks **(solid-earth environment),** the presence or absence of a vegetation cover **(living-organism environment),** and the behaviour of people.

Questions for Debate

1. Should people be prevented from living or carrying out economic activities in flood-prone zones?

2. Is it best to control flood waters by upstream or downstream measures?

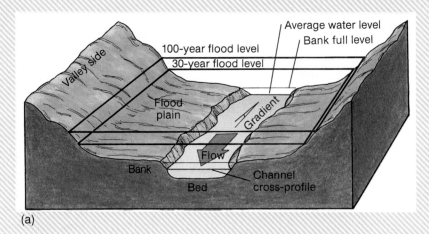

(a) The features of a stream channel and flood-prone zones on either side. (b) Responses to flood hazards in watersheds. The upstream measures attempt to modify the flow and lower the flood peak. Downstream engineering works attempt to cope with a high flood peak once it is generated.

(a)

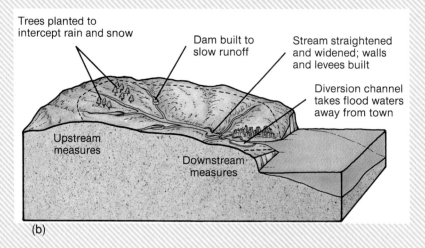

(b)

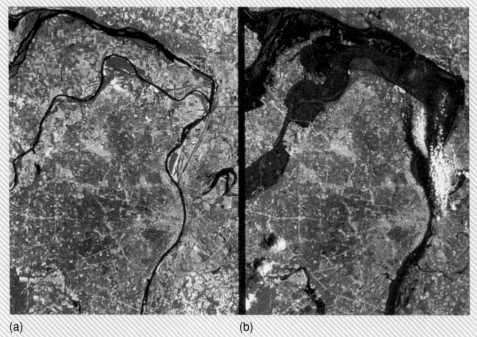

(a)   (b)

Before and After (a) Satellite view of Mississippi and Missouri Rivers at St. Louis before the 1993 flood. (b) The extent of flooded land.

CHAPTER SUMMARY

1. Running water flows mainly in stream channels. Water enters watersheds in precipitation (*atmosphere-ocean environment*) and glacial meltwater and passes into stream channels. Its passage through the watershed is slowed if it encounters vegetation (*living-organism environment*), permeable soil or rock (*surface-relief environment*), or if it is stored in ice masses or lakes.

2. The inputs of precipitation and the pathways of water movement through the watershed control the flow in stream channels.

3. Stream load is the rock material carried in streamflow and includes dissolved, suspended, and bed load fractions (*solid-earth environment*). The movement of water and its load in rivers acts as an agent of gradation through the erosion, transport, and deposition of rock particles.

4. The shapes and sizes of stream channels are controlled by flows of water and load through them (*surface-relief environment*). The shape of a stream channel also affects how water flows through it. Floodplains and alluvial fans are the major depositional landforms in watersheds.

5. Stream action varies in different climatic environments. Different rock types (*solid-earth environment*) provide varying degrees of resistance to stream action. Water acting on limestone produces distinctive karst landscapes. Plants (*living-organism environment*) stabilize riverbanks in humid climates.

6. Humans have the greatest impact on river processes in humid midlatitude watersheds. Farming, logging, mining, and urban construction have major influences on river systems.

KEY TERMS

river   *250*

watershed   *250*

drainage divide   *250*

interception   *252*

infiltration   *252*

overland flow   *252*

permeability   *252*

percolation   *252*

water table   *253*

groundwater   *253*

spring   *253*

baseflow   *253*

aquifer   *253*

discharge   *255*

load   *256*

dissolved load   *257*

suspended load   *257*

capacity   *257*

bed load   *257*

competence   *257*

corrasion   *258*

corrosion   *258*

stream channel   *259*

gully   *261*

knick point   *262*

meander   *262*

braided stream   *263*

floodplain   *265*

alluvial fan   *265*

karst   *272*

doline   *273*

cavern   *273*

CHAPTER REVIEW QUESTIONS

1. Summarize how the features of a watershed modify the passage of water from inputs of rainfall to outputs of streamflow. Refer to interception, infiltration, percolation, stormflow, and baseflow.

2. Compare the annual patterns of water inputs to a river from rainfall, snowmelt, and glacier ice melt.

3. What are the differences between dissolved, suspended, and bed load of streams?

4. Describe and account for the features of meandering and braided stream channels.

5. What are the main landforms of floodplains with meandering streams?

6. What factors affect the sizes of alluvial fans?

7. What are the distinctive landforms in karst landscapes?

8. How do climatic and rock-type variations influence a river's effects on landforms? Refer to examples.

SUGGESTED ACTIVITIES

1. Find out how groundwater is used in your area.

2. Devise a sediment budget diagram for a river in your area. You will need to demonstrate the sources of sediment, show what the river carries at a point or points along its course, and describe where the sediment is deposited. This information can be summarized on a flow diagram.

3. Use a map to examine a local river channel and divide its course into straight, meandering, braided, and anastomosing sections. What are the proportions of each? What other differences do you observe between these sections?

4. Gather further examples of human impacts on watershed processes.

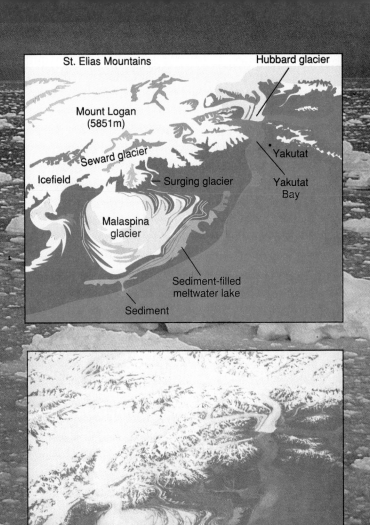

St. Elias Mountains

Hubbard glacier

Mount Logan
(5851m)

Yakutat

Seward glacier

Yakutat
Bay

Icefield

Surging glacier

Malaspina
glacier

Sediment-filled
meltwater lake

Sediment

ICE, WIND, AND LANDFORMS

THIS CHAPTER IS ABOUT:

◆ Ice and wind as agents of gradation

◆ Surface ice and ground ice

◆ Landforms of ice erosion and deposition

◆ Meltwater and landforms

◆ Ground ice and landforms

◆ Environments of wind action

◆ Wind erosion, transport, and deposition

◆ Landforms of wind action

◆ Human activities and wind landforms

◆ Environmental Issue: Living on Permafrost—Do We Understand the Implications?

Glaciers form in cold or high areas where temperatures are too low for snow to melt. The glacier ice, as in the ice from the alpine glaciers of southern Alaska (inset), flows slowly downhill, carving deep valleys and spreading out on lowlands or when it meets the ocean. Antarctic glacial ice breaks into icebergs on reaching the ocean, as seen in the background photo.

ICE AND WIND AS AGENTS OF GRADATION

ce and wind are both important agents of gradation, but they are most significant in the very cold
and very dry parts of the world respectively. In
this geographic distribution, they form a contrast
with river action, which affects almost every part of
the world's land surfaces. Ice and wind act more
slowly than rivers but are effective in producing landforms over time.

This chapter first presents ice in its various
forms, including surface ice masses such as glaciers
and ice that forms in soil and rock beneath the surface (ground ice). It shows how glacial ice moves and
carries out erosion, transport, and deposition of rock
materials, and how these processes create distinctive
landforms. Meltwater and ground ice are also important agents in cold regions and work in conjunction
with ice masses to produce landscapes. The section
concerning wind action describes the parts of the

world where it is an effective agent of gradation and
then discusses the processes involved in wind erosion
and deposition and the landforms produced.

SURFACE ICE AND GROUND ICE

Ice cover presently occurs in high latitudes and at
high altitudes where precipitation is mainly snow,
and temperatures seldom rise above freezing point
(Figure 10-1). **Periglacial environments,** immediately surrounding those covered by ice, often have
permanently frozen ground that contains ground ice.

Surface ice masses covered more land and sea during the glacial phases of the Pleistocene Ice Age than
they do today. Only 15,000 years ago—a short time in
terms of forming relief features—much of northern
Europe and North America was covered by ice, as
shown in Figure 10-1, where Arctic Sea ice extended
south of Iceland. That ice retreated in the present interglacial phase but left behind many landforms that
were created under the ice or at its margins.

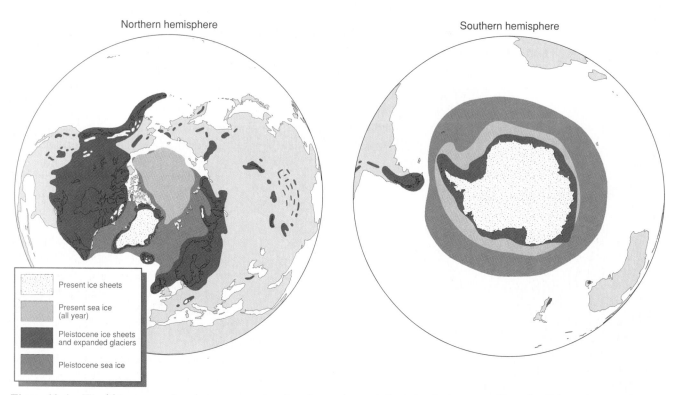

Figure 10-1   World ice cover. Ice sheets are restricted to Antarctica and Greenland at present. How far did ice sheets and sea
ice extend during the Pleistocene Ice Age?

FROM SNOW TO ICE

Snow is precipitated from the atmosphere as tiny, crystalline flakes (see Chapter 4). At temperatures well below freezing, freshly fallen snow often accumulates as a downy powder. At temperatures close to freezing, wetter snow falls. Wet snow is particularly common where high mountains in midlatitudes intercept mild airflows from the oceans.

When snow piles up faster than it melts, it gradually changes to ice. This process can take several years. The increasing weight of accumulating snow compacts the lower layers, and the water from surface snowmelt percolates downward, refreezing to fill pore spaces between the snow crystals. Annual layers of snow often have a thin band of fine rock particles at the top as a result of water flowing during summer melting and dust settling out of the atmosphere (Figure 10-2).

Successive accumulations add to the weight of overlying snow, lead to an increase in the density of buried snow, and change the snow to ice. Packed snow has a density of 0.1 g/cm³. *Firn* is an intermediate stage of dull, white frozen material that still contains trapped air and has a density of 0.5 g/cm³. True ice has a density of 0.8 to 0.9 g/cm³, is crystalline in structure, and contains very little air. Ice is less dense than liquid water (1.0 g/cm³), which is why icebergs float.

The change from snow to ice usually takes three to five years in places like the Yukon of central Alaska, where temperatures fluctuate just above and below freezing for much of the year. Thawing allows water to percolate into the lower snow where it refreezes. By contrast, the change from snow to ice can take up to 200 years in central Greenland, where air temperatures seldom permit surface melting. Ice masses thus accumulate more rapidly in slightly warmer conditions where there are wet snowfall and frequent freeze-thaw cycles.

PERMAFROST AND GROUND ICE

Permafrost is permanently frozen ground. As a minimum, it is ground in which the temperature remains below 0° C (32° F) for more than two years. In many parts of northern North America and Asia, permafrost layers are thousands of years old, having existed since the last glacial phase. Permafrost occurs mainly frozen as **ground ice** that is either dispersed through soil and rock in pore spaces or concentrated in wedges or mounds that influence surface relief. Not all permafrost contains ground ice.

Permafrost affects soil and rock (Figure 10-3). A surface *active layer,* 50 to 80 cm (20 to 30 in.) deep, lies above the permafrost. It is subject to alternate freezing in winter and melting in summer. The *permafrost table* marks the upper limit of permafrost, which extends downward until the effect of Earth's interior heat is greater than that of surface cooling. The lower limit of permafrost is up to 400 m (1200 ft) deep in northern Canada and 1000 m (3000 ft) deep in Siberia.

Continuous permafrost occurs in the northernmost parts of North America and Siberia, in total covering some 21 million km² (8.2 million mi²). *Discontin-uous permafrost* is shallower, and areas with it are separated by unfrozen ground. It covers a total of 17 million km² (6.6 million mi²). Discontinuous permafrost is decreasing in extent as a result of

Figure 10-2 The annual layers of ice shown in a crevasse in an Alaskan glacier. The layers are divided by dirt bands of material washed over the surface or deposited from the atmosphere during the summer.

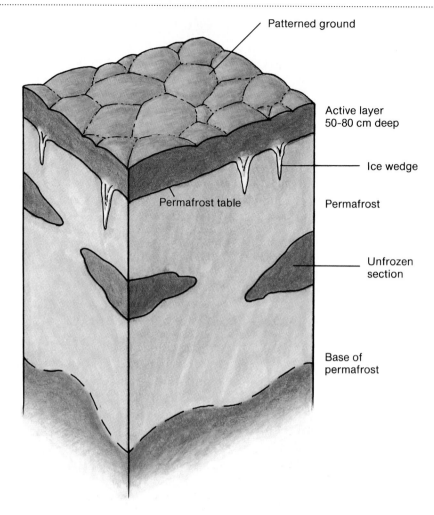

Figure 10-3 Permafrost. The major features include the active layer, in which water melts in summer, and the permanently frozen layer.

atmospheric warming since the mid-nineteenth century, but continuous permafrost areas show few signs of change. Permafrost provides a distinctive environment that people need to treat with care (see Environmental Issue: Living on Permafrost—Do We Understand the Implications? p. 316).

FORMS OF SURFACE ICE

Surface ice occurs in masses with varied shapes and sizes. The largest masses are **ice sheets;** they are often called *continental glaciers* since they cover such huge areas. Earth's two ice sheets occur on Antarctica and Greenland, where they submerge almost the entire landscape in broad ice domes that thin toward the margins (Figure 10-4). These two ice sheets make up 96 percent of the continental area covered by ice. The Antarctic ice sheet covers over 12.5 million sq km (4.9 million sq mi), and the Greenland ice sheet covers 1.7 million sq km (0.7 million sq mi). The weight of the ice in these ice sheets is so great that the crust beneath has subsided by isostasy

(see Figure 7-43). The rocks of central Antarctica and Greenland are buried several kilometers deep by ice and lie below sea level.

Areas of ice covering less than 50,000 sq km (20,000 sq mi) are called **ice caps.** They resemble ice sheets in most respects apart from size. The Vatnajokull ice cap of Iceland is an example.

Alpine glaciers are smaller moving ice masses that are confined by the surface features of high mountains (Figure 10-5). Several types of ice mass are involved. *Cirque glaciers* are bodies of ice occupying amphitheater-shaped hollows just below mountain crests. *Valley glaciers* are moving ice that fills the main part of a valley trough. Such glaciers that may join in a network of tributary glaciers from cirques or high-altitude ice fields. Many valley glaciers do not extend beyond the valley in which they flow. Some, however, flow out on a lowland area often joining other extended glaciers to form a wide area of shallow, stagnant ice at the mountain front. Such lowland glaciers are known as *piedmont glaciers*. A space shuttle view of

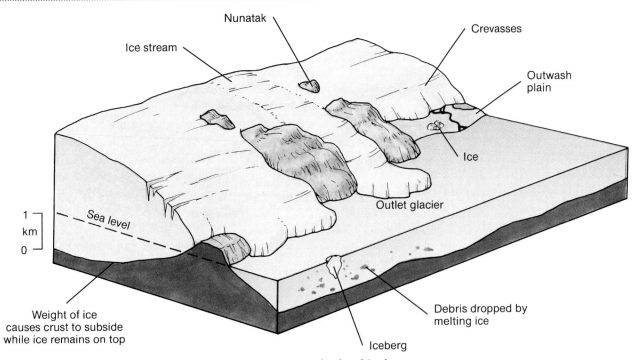

Figure 10-4   Features of a typical ice sheet and its margins on land and in the ocean.

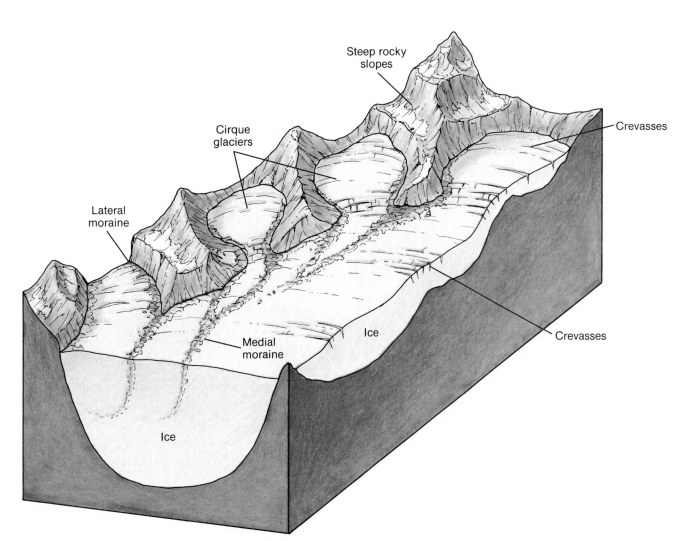

Figure 10-5   Features of a typical alpine glacier system, with valley glaciers fed from cirques.

southern Alaska (Figure 10-6) shows the piedmont Malaspina Glacier and several linked cirque and valley glaciers. The map and photograph on page 293 show the glaciers radiating from Mount Rainier.

Some glaciers occur on the margins of ice sheets. Mountains may protrude through the ice cover and divide the flow into *outlet glaciers,* which often extend back into ice streams. *Ice streams* are zones of concentrated, more rapid flow within ice sheets. They have some of the characteristics of valley glaciers, but they are not bounded by rocky valley sides.

Ice also forms on the oceans in polar regions. Around Antarctica, the floating ice is partly ice that has flowed off the land and partly sea ice. **Sea ice** forms when the ocean surface freezes. The largest areas of sea ice cover the Arctic Ocean and surround Antarctica. At the edges of floating ice, masses break off to form icebergs. *Icebergs* float into warmer areas and gradually melt, but are a hazard for ships while doing so.

ICE FLOW

The ice in ice sheets and glaciers flows from source areas where snow accumulates to zones where ice melts. Snow accumulates on the upper part of an ice mass, where it converts to ice. In these areas, more ice is formed than lost by melting. In the lower part of an ice mass, particularly at its lower end, or *snout,*

ice melts and evaporates, sublimates directly from ice to water vapor, or breaks into icebergs on reaching the ocean. These processes of ice loss are collectively called **ablation.**

When stresses imposed by the increasing weight of accumulating snow and ice reach a critical level ice begins to flow. The pressure imposed distorts individual crystals within the ice. Such **internal deformation** is the basic reason why solid ice can flow like a plastic substance; it moves slowly and changes its shape without fracturing. **Basal sliding** resulting from the presence of water is the second main process that causes ice movement. Ice flow is often assisted by meltwater, which lubricates the base of the glacier and saturates loose rock debris beneath the ice.

Ice sheets and glaciers flow very slowly compared with rivers. Rates vary but are generally between 3 and 300 m per year. In valley glaciers movement rates vary internally. Friction at the base and sides slows ice movement and flow is fastest in the center near the top. Flow rates vary along the length of a glacier, being fastest in the middle where the cumulative volume of ice reaches a maximum. Increasing ablation downstream reduces the volume of ice and its velocity.

Flow in ice sheets is generally outward from the center in all directions. Ice sheets are so large that they often override the influences of relief. The

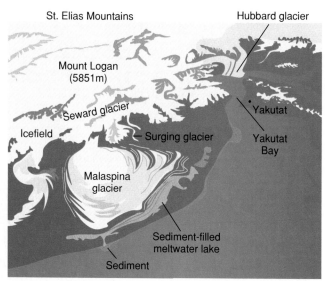

Figure 10-6   Alpine glacier. A space shuttle view of southern Alaska including the Malaspina piedmont glacier, valley glaciers, and mountain ice fields. Details are shown on the annotated sketch.

Mount Rainier, Washington. Mount Rainier is a quiescent volcano that is high enough to have glaciers on its surface. The Nisqually Glacier seen in the photo flows southward and is the one closest to the Paradise Lodge hotel visited by thousands of people each year. Scale of map is 1:100,000.

weight of ice is greatest where the ice is thickest and movement is most rapid a little distance from this point. Ice streams and outlet glaciers are zones of faster movement within ice sheets.

The temperature of ice has an important influence on the movement of ice masses. In **warm glaciers** the ice is just below 0° C (32° F); **cold ice sheets and glaciers** have much lower temperatures. Warm alpine glaciers occur in the mountains of humid midlatitude and tropical regions, such as southern Alaska, the Swiss Alps, the Andes, and the Himalayas. The summer rise in temperature causes surface melting, rapid firn formation, and percolation of water to the base of the ice. Both internal deformation and basal sliding cause the glaciers to flow, but the latter often accounts for 90 percent of the movement. Cold ice sheets and glaciers occur in polar regions and include the ice sheets of Antarctica and Greenland. Flow in cold ice sheets and glaciers is caused totally by internal deformation.

As ice moves, a surface layer up to 60 m (180 ft) thick remains rigid. The weight of this surface layer forces the ice beneath to flow. The rigid surface ice is carried along by the flowing ice, and it fractures when forced to bend over small hills or steps in the valley floor. The cracks are called **crevasses** (see Figure 10-2) and may extend from the surface to the base of the rigid surface layer. Since some are tension cracks that open because of the faster movement in the center of a valley glacier, crevasses provide surface clues to the patterns of glacier ice movement.

Crevasses also play an important role in allowing meltwater and rock debris from the surface to penetrate the glacier ice. When meltwater reaches the base of the glacier, it contributes to basal sliding. Water moving freely downward through ice cuts deep passages by vertical erosion. Where water fills all the spaces in a glacier, flow occurs under pressure and may erode circular passages like those formed below the water table in limestone rocks. The ice carries the rock debris as it moves down the valley, and it may drop the debris from its base.

At times, the presence of large quantities of meltwater within and beneath warm glaciers (Figure 10-7) may cause the ice to *surge* forward—a rapid movement of part of the glacier at speeds 10

Figure 10-7   Meltwater and its deposits at the snout of the Glacier Blanc, French Alps.

to 100 times the normal rate of flow. The Tweedsmuir Glacier, on the border between British Columbia and the Yukon, for example, surged several kilometers in a few days in 1973 and others in Alaska do so frequently. There are over 200 surging glaciers in North America. Some glaciers surge every few decades, but others are unpredictable. Earthquake shocks may act as a trigger in some cases, but in general the reasons for glaciers surging are poorly understood.

LANDFORMS OF ICE EROSION AND DEPOSITION

Moving ice in glaciers and ice sheets is an important agent of gradation. The landforms of high mountains show the effects of alpine glaciation, and much of northern Canada and the United States demonstrates the results of continental glaciation that occurred in the glacial phases of the Pleistocene Ice Age (Figure 10-8). Both types of glaciation create distinctive erosional and depositional landforms.

(a)

(b)

Figure 10-8 Alpine and continental glaciation. (a) The deeply crevassed Mendenhall Glacier spills down from the Juneau Icefield, Tongass National Forest near Juneau, Alaska. (b) Continental glaciation. Bare rock and deranged drainage patterns can be seen in this photo of the ice-scoured shield rocks of Labrador, Canada.

EROSION BY ICE

Since scientists can penetrate only the outer margins of ice sheets and glaciers to observe what happens, they have found it difficult to identify the mechanism by which ice erodes rock. Moreover, ice masses move very slowly and observations must be carried out over long periods of time to obtain worthwhile results.

Glacial erosion affects extensive regions beneath continental ice sheets and forms deep valleys in alpine glaciation. In both, the ice removes regolith formed under different climatic conditions and then scours the underlying rock. Bare rock surfaces are common.

One mechanism of ice erosion is **abrasion,** in which the debris held in the base of the ice scratches and scrapes underlying rocks. The process is rather like the effect of sandpaper on wood. In an experiment in Iceland, blocks of rock were bolted down beneath moving ice. In 3 months, during which the ice moved forward 9.5 m (30 ft), the surface of a marble block was lowered by 3 mm (0.2 in) and that of a basalt block by 1 mm (0.04 in).

Moving ice also erodes rock by the process of **plucking.** As ice moves over a rocky protrusion, it scratches and smoothes the upstream side. On the downstream side, joints and other cracks in the rock fill with ice that is part of the surrounding glacier. When the ice moves forward, it carries with it the ice and attached blocks of rock (Figure 10-9). Plucking can also excavate basins in a well-jointed rock.

Glaciologists debate whether abrasion or plucking is more important in glacial erosion. Ice in the bottom of a glacier carries large blocks of rock rather than fine debris, and this suggests that plucking is more significant. Plucking is most effective, however, in well-jointed rocks. It is likely that abrasion is more important on some rocks and plucking on others.

Beneath ice sheets the smoothing effect of ice abrasion produces **whaleback hills and basins** (Figure 10-10). The hills may be 200 m (600 ft) high and several kilometers long. Smaller forms a few tens of meters

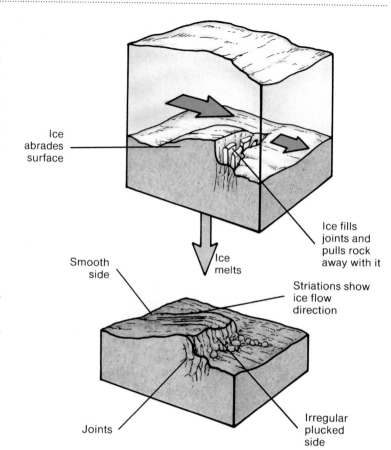

Figure 10-9   Glacial plucking. The upper diagram shows how ice freezes around the blocks of a jointed rock. When the ice moves forward, the blocks are pulled out. After melting (lower diagram), the protruding rock has a smoothed upstream slope and an irregular downstream slope.

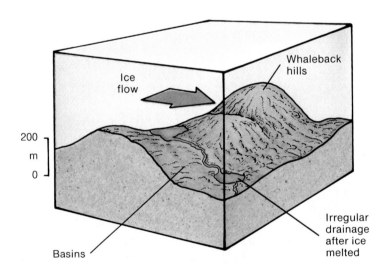

Figure 10-10   Whaleback hill and basin. These features are formed by ice abrasion, often picking out weaknesses in the underlying rocks.

high are known as *roches moutonnées* and often have a smooth abraded upstream slope and an irregular plucked face on the downstream side. Many smoothed rock surfaces have grooves, or *striations,* cut by harder rock fragments in the ice base. Striations on the surface of exposed rocks show the ice flow directions of former ice sheets.

Valley glaciers deepen and straighten their valleys through a combination of ice movement, frost-wedge weathering, and meltwater activity. The resultant troughlike valleys often have an open *U* shape in cross section, and are known as **U-shaped valleys** (Figure 10-11). Along its length, a U-shaped valley often has an uneven floor. Flatter sections alternate with steps formed where more resistant rocks come to the surface. Some valley floors have deep hollows eroded in them and these form lake basins when the ice melts. In Antarctica, U-shaped valleys are up to 50 km (30 mi) across and 3.5 km (2.5 mi) deep. Those in the Rockies and Alps, however, are seldom more than 1 km (.6 mi) across or deep. This difference in size reflects the greater length of ice coverage on Antarctica and the greater thickness of ice there. The drowned mouths of U-shaped valleys form fjords like those along the west coast of Norway (see Chapter 11).

Tributary glaciers are smaller than main glaciers and deepen their valleys less. After the ice melts, the shallower tributary valleys perch above the main valley and are called *hanging valleys* (Figure 10-12). Waterfalls occur where streams flow from a hanging valley into a main valley.

At the heads of valley glaciers, small glaciers create amphitheater-shaped hollows called **cirques.** Some cirque ice flows into valley glaciers, but in many cirques the ice remains in the hollow. Cirques are usually 0.5 to 1 km (1500 to 3000 ft) across and may be up to 1 km deep. Cirque hollows form by a combination of weathering, mass movement, and ice erosion. Hollows where snow gathers and changes to ice are enlarged by frost wedging and mass movement. These processes continue on the slopes above the ice and maintain steep back and side walls. When there is sufficient ice in the hollow for flow to occur, a cirque glacier forms. Rock debris, incorporated in the ice as meltwater washes it into crevasses and down to the glacier base, abrades the bottom of the cirque. This produces the rounded shape of the hollow, and the deepening of the base leaves a lip at the exit of the rounded cirque basin that often traps water in a lake after the ice melts. Many cirques have an orientation so that their outlets face away from the sun's rays at midday. The hollow's shaded position encourages the accumulation of snow and its transformation to ice.

A series of cirques often forms back-to-back on either side of a ridge, which, in its most extreme form, may be a narrow, knife-edge feature between the hollows. Such ridges are called *arêtes.* Sharp peaks rising from a point where three or more cirques are back-to-back have a pyramid-like shape, as in the Matterhorn in Switzerland, and are called *horn* peaks. Isolated peaks or sections of arête that rise above ice are called *nunataks.*

TRANSPORT AND DEPOSITION BY ICE

Ice sheets and glaciers transport large quantities of rock debris. They produce impressive deposits with distinctive shapes beneath the ice and at their margins where ice flow is slower and melting increases.

The size and rigidity of ice sheets and glaciers enable them to carry and move blocks of rock as heavy as several tons. Boulders of this size cannot be moved by streams or the wind. Ice masses act as giant conveyor belts with the uppermost ice and rock fragments borne on the plastic flowing ice beneath. The rock material drops when the conveyor belt slows or stops at the glacier snout.

Glaciers and ice sheets transport an assortment of rock fragments. Some come from rock fragments falling on the glacier edges after frost wedging on the exposed slopes above. The fragments pile up on the glacier surface and some are washed into crevasses. Other rock debris enters the base of the ice through the erosion of rocks, soil, and regolith over which the ice passes (Figure 10-13). The components of the basal load may be rounded by abrasion and the load often contains many fine particles that were once soil. Most rock fragments are, however, altered little by ice transport, although frost action may break down blocks on the surface.

Deposition from ice occurs during both ice movement and melting. The resultant deposits are generally a jumble of poorly sorted particles of varied sizes and compositions, known as **till.** Tills form beneath the ice during basal sliding when rock debris is "plastered" on the ground beneath or where ablation at the snout causes ice to drop rock fragments. Ablation is fast when an ice sheet disintegrates at the end of a glacial phase, resulting in the rapid deposition of the load (Figure 10-14). Tills are commonly mixed with the deposits of meltwater that are layered and better sorted.

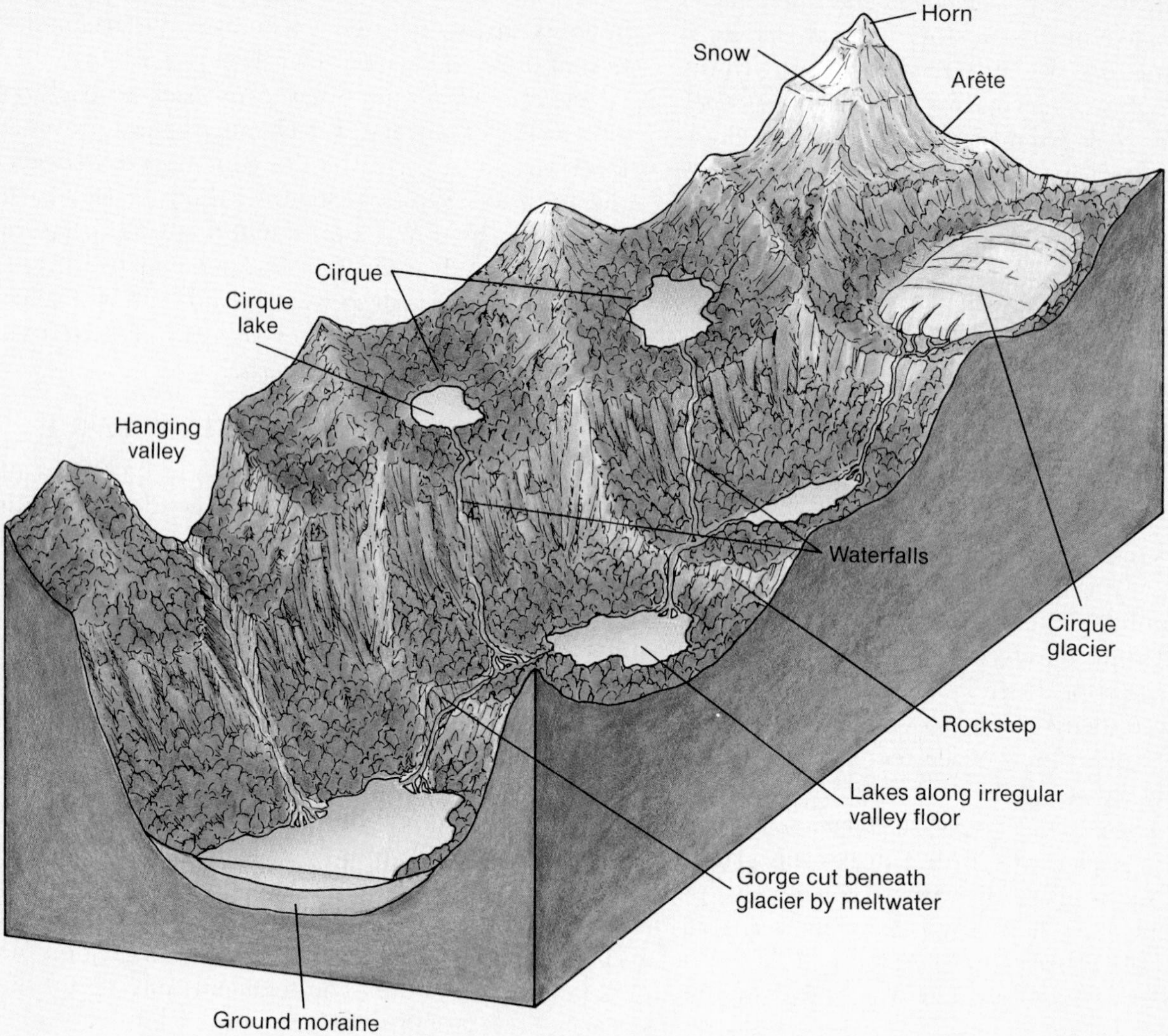

Figure 10-11 Landforms of glacial erosion: an alpine glacier system after most of the ice has melted. Note the U-shaped valleys, hanging valleys, cirques, arêtes, and horn.

Figure 10-12 Hanging valley. The waterfall plunges from a higher valley into a valley carved by a major glacier in Yosemite National Park, California.

© J. P. Lenney

Horn

Snow

Arête

Cirque

Cirque lake

Hanging valley

Waterfalls

Cirque glacier

Rockstep

Lakes along irregular valley floor

Gorge cut beneath glacier by meltwater

Ground moraine

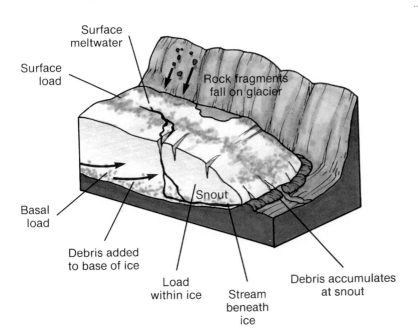

Figure 10-13   Ice transport. The glacier load is supplied from surrounding slopes and the valley floor; it is carried on and in the glacier ice and by meltwater, and then deposited.

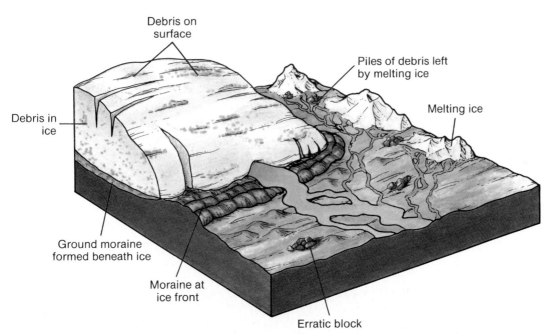

Figure 10-14   Ice deposition. Rock debris is deposited from the base of active, advancing warm ice and from melting ice.

Glacial till deposits create landforms called **moraines.** Piles of rock fragments on the surface of valley glaciers are also termed moraines and may end up as recognizable ridges in the valley when the ice melts. *Lateral moraines* form along the sides of valley glaciers where rock accumulates after falling from higher slopes. *Medial moraines* form where two glaciers meet and their lateral moraines join (see Figure 10-5).

Terminal moraines form following ablation at the snouts of glaciers where the conveyor belt stops. The term refers to moraines that mark the farthest advance of a glacier or ice sheet, but it has been used for any morainic ridge that crosses a valley or runs parallel to the front of a former ice mass. Terminal moraines form when an ice margin remains in the same place for some time so that ablation can deposit sufficient debris to form a ridge. Since glacier and ice sheet margins advance and retreat with climate changes, they may bulldoze the material in a moraine as they advance or leave a series of recessional moraines as they retreat. Most terminal moraines are recessional in origin since the ice would soon override and remove the earlier moraines if it were advancing.

Ground moraine forms beneath an ice sheet and may be tens of meters thick. It comprises a mixture of different sizes and types of rock debris that was partly plastered on the ground beneath moving ice and partly dumped as the ice disintegrated. The surface of ground moraine is often irregular, with hollows occupied by lakes and a deranged river drainage pattern after the ice melts (see Figure 9-22). Such drainage patterns are typical of northern Wisconsin and Minnesota. Ground moraine composed mainly of clay forms a featureless *till plain* such as that of central Indiana and Illinois.

The irregularity or monotony of ground moraine is sometimes broken by masses of smoothed, oval hills known as **drumlins.** These are 1 to 2 km (3000 to 6000 ft) long and up to 50 m (150 ft) high and are usually formed of the same material as the ground moraine (Figure 10-15). Their smooth outline shows

they formed beneath moving ice rather than by ice disintegration. Many drumlins probably formed as ice advanced over older ground moraine and reshaped its surface features.

Areas covered by ice in the past often contain large *erratics,* blocks of rock quite different from those of the local area and often several hundred kilometers from their origins. Other agents, such as rivers or the wind could not have moved such blocks, so glaciers are the only possible explanation of their presence. Erratics demonstrate that the area in which they occur was once covered by ice, and they provide evidence as to the direction of ice flow.

MELTWATER AND LANDFORMS

Meltwater plays an integral role with ice in glacial erosion and deposition. It is important in producing landforms, particularly at the margins of ice sheets and glaciers.

Meltwater under pressure beneath the ice can form deep gorges in glaciated valleys. These gorges may be hundreds of meters deep but only ten meters or so wide—too narrow to be filled by a glacier. A notable example occurs near Grindelwald, Switzerland, where

Figure 10-15   Ice deposition: drumlins in western Yorkshire, England.

such a gorge has been exposed by the retreat of a glacier. Meltwater beneath the extended glacier carved a deep gorge through the rocks at the junction between the mountain and a wide valley where otherwise there might have been a hanging valley waterfall.

Immediately in front of an ice sheet, large quantities of meltwater carve deep and wide valleys and often fill them with debris washed from the melting ice. The valleys leading away from the ice sheet that covered the Great Lakes, for example, including those of the Wabash and Illinois rivers, are a kilometer wide and much larger than the present stream would be able to carve. They once accommodated huge flows of meltwater. Even greater flows of meltwater formed the Channelled Scablands area of Washington State (Figure 10-16), a unique region in the Columbia River basin just south of the U.S.-Canada border. A meltwater lake built up behind an ice dam close to the present international border and eventually burst through,

carving deep, rectangular channels that are used partially, or not at all, by the modern river network.

Meltwater also carries out transport within ice masses. Streams form on the ice surface in summer and their water percolates through the ice or pours down crevasses. This running water carries its own load of fine sand and clay, while large flows may move pebbles. The water flowing out from a glacier front usually has a milky appearance because it contains finely ground rock particles called *rock flour*. Glacial meltwater contains few solutes, since chemical weathering is slow in low-temperature conditions.

Meltwater deposits stand out from other glacial deposits because their layering contrasts with the disorganized mass of debris dumped by melting ice. Meltwater deposits occur both beneath the margins of ice masses and in front of them (Figure 10-17). Deposits from streams beneath ice masses form winding gravel ridges called **eskers** when they are exposed after the ice melts.

Figure 10-16 Meltwater erosion: a dry valley in the Channelled Scablands, Washington. A huge river carved this valley, nearly a kilometer across, when meltwater broke through an ice dam.

Figure 10-17 Meltwater deposits. Meltwater is the main process at work along the margins of ice as the ice disintegrates. The deposits produced are minor landforms that become stabilized by vegetation after the ice disappears.

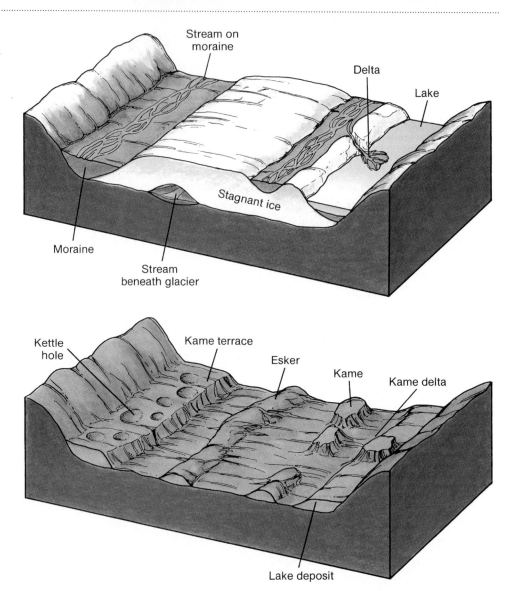

Meltwater streams and lakes on top or along the margins of stagnating and disintegrating ice masses may produce deposits and landforms called **kames.** Deltas formed by meltwater streams entering lakes are called *kame deltas.* Rock debris also settles to the floor of the lakes. After the ice melts, kame deposits often occur in association with hollows, known as *kettles,* formed by the melting of the last blocks of ice.

Streams issuing from the confined tunnels beneath glaciers and ice sheets commonly have their velocity checked, and this affects their capacity to carry load. They spread rock debris over the ground in front of the ice. Such deposits are mainly coarse sands, reflecting the ability of fast-flowing water to transport particles of this size in suspension. The deposits may also contain erratics and be pitted by

kettles. Meltwater deposits that partly fill a valley downstream from a glacier are called a *valley train.* In front of ice caps or ice sheets, the deposits may be thick and extensive with a relatively flat surface and are called **outwash plains** (Figure 10-18).

GROUND ICE AND LANDFORMS

Ground ice occurs most commonly in areas of "wet" permafrost, where there is a considerable amount of water in the soil and rock and it remains frozen on a long-term basis. The expansion of ice when it first forms and as temperatures fall further below 0° C creates a number of small landforms up to tens of meters across in size.

Some landforms are produced in the active layer above the permafrost. Seasonal freezing and melting

Figure 10-18   Meltwater deposits: an outwash plain in front of an ice cap in Iceland.

causes this layer to heave, and the alternate rising and falling of the soil brings rock fragments to the surface. On flat ground, the fragments form polygonal shapes called **patterned ground,** with individual polygons measuring from a few centimeters to 100 m (300 ft) across. On slopes of 6 to 25 degrees, the rock fragments form *stone stripes.*

Ground ice in permafrost produces **ice wedges,** which are bodies of ice up to a meter (3 ft) wide at the top, tapering downward to depths of about 10 m (30 ft). Their most impressive development occurs in the fine sediments of river floodplains in arctic regions, where they combine to form polygonal patterns (Figure 10-19a). Ice wedges form when ice-rich permafrost contracts on freezing, opening up cracks. Meltwater flows into these cracks in early summer and freezes. The process continues and the ice wedge grows.

Another type of ground ice forms **pingos** (Eskimo for "hill"), which are large ice-cored mounds 3 to 70 m (10 to 220 ft) high and 30 to 600 m (100 to 2000 ft) across (Figure 10-19b). One explanation

Surface ice dominates Earth's landscapes in high latitudes and the highest parts of mountains. Ice accumulates where it is too cold for snow to melt and forms ice sheets, ice caps, and glaciers. Compaction and refreezing of melted snow change the snow to ice. Temperature affects the flow in ice sheets and glaciers. Ice moves by internal deformation and basal sliding. The **atmosphere-ocean environment** determines the low temperatures and the amount of precipitation.

The **solid-earth environment** determines the configuration of the continents and oceans and the positions of high mountain ranges. Ice sheets and glaciers may erode or protect the underlying land. They transport large quantities of rock debris. Typical landforms of glacial erosion are roches moutonnées, U-shaped valleys, and cirques. Typical landforms produced by glacial and meltwater deposition include moraines, drumlins, eskers, kames, kettles, and outwash plains.

Ground ice occurs in areas of permafrost. Ground ice produces landforms such as ice-wedge polygons and pingos.

(a)

Figure 10-19 Permafrost landforms. (a) Frost wedges dominate the Mackenzie Delta, northern Canada. These form polygonal features in fine-grained sediment; during summer lakes fill the shrunken areas between the ice wedges. (b) A pingo in the Mackenzie Delta—a long hill formed when ice concentrates and domes up the ground.

(b)

for the origin of pingos is based on the effect of advancing ground ice on trapped water in a permafrost zone. The water is concentrated in a smaller and smaller space, freezes, and expands, doming the surface above. Eventually the soil and vegetation cover of the domed area are broken by the continuing ice expansion and the ice core is exposed to summer melting.

The thawing of ground ice also produces landforms of subsidence known as thermokarst (see Chapter 8). This is particularly effective in areas of discontinuous permafrost.

ENVIRONMENTS OF WIND ACTION

Wind processes and the landforms they produce are often termed **aeolian,** after the Greek god of wind, Aeolus. Winds are most effective as agents of erosion, transport, and deposition in areas where there is little vegetation, such as deserts, glacial ice-front areas, sandy beaches, and plowed fields or other areas where humans have bared the soil (Figure 10-20). Deserts have bare rocky surfaces and loose rock particles with little or no vegetation to bind them together. The areas in front of ice sheets are too cold to support vegetation and much fine rock debris that wind can move is dumped there by melting ice and meltwater streams. Coastal beaches present areas of bare dry sand at low

tide that is available to wind. In many plowed fields wind has access to the surface when it is broken by farming practice into fine particles.

Compared to rivers, glaciers, and ice sheets, wind is less powerful and more intermittent in its effects on landforms. Areas where wind is important in creating landforms generally experience very slow changes in surface features except where loose sand formations are reshaped every time the wind blows.

WIND EROSION, TRANSPORT, AND DEPOSITION

WIND AS AN AGENT OF EROSION

Wind is less powerful than water in its action on the landscape. While water can hold sand in suspension and move boulders, wind seldom moves anything larger than sand and holds only fine dust in suspension. Wind has to blow faster than water flows to move even the finest particles.

There are several reasons for the limited and slower action of wind in erosion. First, wind is merely a movement of air and air has a much lower density than water. Second, wind fluctuates in strength from moment to moment. Strong gusts get particles airborne, and wind turbulence provides vertical lift. A lull in the wind, however, returns the particles to the ground. Third, wind erosion is strongly

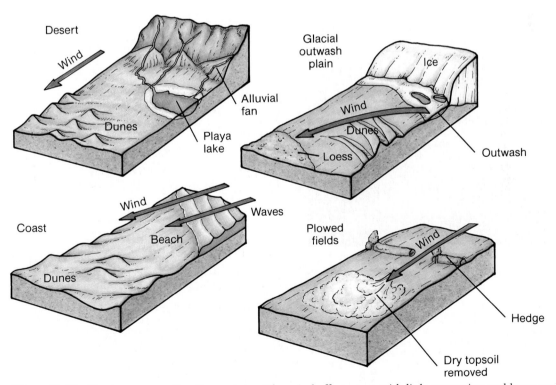

Figure 10-20   Environments of aeolian activity. The wind affects areas with little vegetation and loose rock material such as arid areas, cold areas close to ice sheets, coastal beaches, and plowed fields.

reduced by the presence of vegetation cover and water in surface soil. Both hold the soil surface together and plants also lower wind speed near the ground. The removal of plant cover and lowering of soil moisture makes it possible for wind to become active in erosion (Figure 10-21).

Two main processes are significant in wind erosion of surfaces. Wind uses the particles it carries to **abrade** rock and sediments, wearing them away by bombardment—like sandblasting that is used to clean building stones. Abrasion affects only the 1 to 2 m (up to 6 ft) closest to the ground. Coarse sand is most effective in this process but only rises a few centimeters. Finer silt particles may produce some smoothing of rocks. Abrasion is most effective where the wind picks up sand and blasts it against a weak rock.

Deflation is the second process of wind erosion. It is the removal of loose particles from the ground to produce a depression or lowered surface level. It has most effect on clay and silt particles that can be lifted in large numbers by wind turbulence and can be carried away in suspension. Approximately 130 to 900 million tons of such particles are deflated from continents each year, with 60 to 200 million tons coming from the Sahara alone. Deflation events require loose fine particles and high wind speeds and often produce dust storms.

WIND TRANSPORT

Wind transports and deposits loose clay, silt, and fine sand during surface denudation. In dry and unvegetated areas, the main factors affecting wind transport are wind speed and particle size.

Wind moves particles by a combination of lift and drag. *Lift* causes particles to rise from the surface. Lift forces depend on pressure differences within the air like those over an airplane wing. *Drag* is a force that resists horizontal movement along the surface and depends on differences in pressure between the upwind and downwind sides of a particle. Both upward and forward movement occurs after a threshold wind speed is reached. The turbulence of gusty winds is important in starting movement.

The main load of wind is very fine particles (less than 0.1 mm, 0.004 in, diameter), including dry clay, silt, and salt crystals. This load resists movement at first, since the fine particles present little surface roughness to encourage turbulence and often stick to other particles of similar size if moisture is present. Although the finest particles are more difficult to lift off the ground than larger particles (up to 1 mm, 0.04 in, diameter), once they are in the air they remain there. Very fine particles can be blown long distances. For instance, large quantities are blown from the Sahara over the Atlantic Ocean where rain brings them down to the surface and they sink to the ocean floor.

Figure 10-21   Wind erosion. Wind has destroyed this field by removing fine clay particles and blowing the coarser sand into ripples.

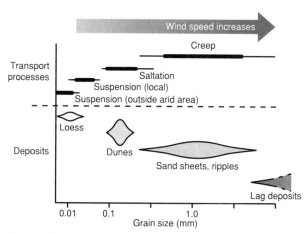

Figure 10-22   Wind transport and deposition. How does increasing wind speed affect the size of particles moved, the type of movement, and the resulting deposits?

Wind speeds, mechanisms of particle movement, and particle size affect the types of deposit produced (Figure 10-22). The finest particles are lifted into suspension and removed from their source area. Slightly larger particles are also lifted into the air but do not escape the source area since the wind lifts fine sand grains (up to 0.4 mm in diameter) for a short time, and they then fall to the ground. Wind moves sand in a bouncing motion called **saltation** (Figure 10-23). Most saltation takes place within a few centimeters of the ground, but individual grains may rise to a height of 1.5 m (5 ft). Once turbulence raises a few grains off the ground, the higher wind speeds a few centimeters above the surface carry them downwind. The forward motion causes particles to disturb any grains they land on, setting other particles in motion.

Surface creep is a rolling and sliding mechanism that affects larger grains. It is caused by wind drag and the bombardment of saltating grains. Even in coarse sand, surface creep is seldom responsible for moving more than 25 percent of the total material in motion.

The volume of sand transported is proportional to the third power of wind speed. If wind speed doubles, the number of particles transported can increase by up to eight times. Most sand, therefore, moves when wind velocities are high. Even the strongest winds, however, cannot move particles larger than 2 mm (0.08 in) in diameter.

WIND DEPOSITION

Deposition occurs when wind speeds fall below the threshold necessary to transport particles. Particles larger than 2 mm (0.08 in) in diameter stay on the ground after the finer matter has been removed, forming *lag deposits* that may be influenced by other agents such as running water. Sand deposits move but stay close to their source area. They have a remarkably even-sized composition after the clay and silt are removed and larger fragments are left behind. Clay and silt particles often move outside the source area and remain in suspension until rain brings them down to the ground. In the Sahara, for example, there are extensive areas of lag deposits, sand accumulation, and bare rock; but there are few deposits of clay or silt particles.

LANDFORMS OF WIND ACTION

LANDFORMS OF WIND EROSION AND TRANSPORT

Bare bedrock areas subject to wind erosion are called **hammadas.** Wind erosion creates polished surfaces and small grooves or pits up to a few centimeters deep on these surfaces. Where a rocky cliff faces into a wind that has access to a sand supply, sand abrasion produces a sharp break of slope at the base.

Abrasion can polish or groove pebbles and boulders in lag deposits. Such rock fragments are known as *ventifacts* and have facets similar to those of a cut gem. In the Mojave Desert of southern California, ventifacts are being worn down by sand, but elsewhere finer particles may smooth the rock surfaces.

Somewhat larger landforms include deep grooves, depressions, and tapered hills. A *yardang* is a landform that is typically up to 10 m (30 ft) high and 100 m (300 ft) long and is aligned in the direction of the prevailing wind. It often has a steep slope facing the wind and tapers downwind (Figure 10-24). Yardanglike landforms over a kilometer high, a kilometer wide, and hundreds of kilometers long occur around the Tibesti Mountains in the

Figure 10-23   Wind transport. Saltation occurs when particles rise into the air, are carried along by the wind, and fall back to the ground. They may bounce up again or cause others to move. The rougher the surface, the higher the sand grains bounce.

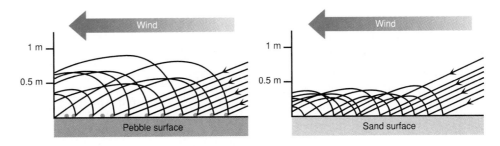

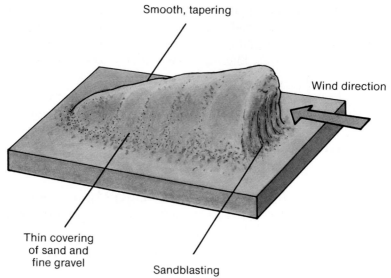

Smooth, tapering

Wind direction

Thin covering
of sand and
fine gravel

Sandblasting

Figure 10-24 Wind erosion. A yardang carved in mud deposits on the
floor of a depression in Egypt.

central Sahara. Yardangs commonly form in groups
in soft sediments of former lakes although some in
Egypt occur in solid rocks including granite. The
features of yardangs suggest an origin by wind erosion, including deflation from intervening troughs
and abrasion of the wind-facing slopes. In the Sahara, yardangs occur where sand is being transported
and so provides the abrasive materials. Weathering
and surface runoff may also modify yardangs.

Deflation hollows occur in a range of sizes. Small
hollows a meter deep and a few meters across may
form in plowed fields. Blowouts are short-lived deflation hollows in sand dunes. The 1930s Dust Bowl
disaster in the southwestern Great Plains left behind
a number of medium-sized deflation hollows.

Deflation basins can be several hundred kilometers long. Depressions in South Africa 100 m (300
ft) deep and up to 100 km (65 mi) long were
formed at least partly by deflation since they are
elongated in the direction of the prevailing wind.
The most remarkable concentration of basins
caused partly by deflation occurs in Egypt (Figure
10-25). An area of 70,000 sq km (27,000 sq mi)
contains basins with depths averaging 250 m (800
ft). The Qattara Depression reaches 134 m (420 ft)
below sea level and has been formed partly by faulting and river erosion. It contains soft sediments,

Figure 10-25 Wind erosion. Hollows in Egypt that have
been formed at least partly by deflation.

however, that are liable to deflation. A lowering
rate of 90 mm (3.5 in) per 1000 years would be sufficient to create most of the depth.

LANDFORMS OF WIND DEPOSITION

Deposition after wind transport produces landforms
such as sand dunes, loess plains, and loess hills. The
selective removal of fine particles during transport

leaves behind gravel and boulders in lag deposits. Such lag deposits cover a large proportion of many desert areas. Where lag deposits form a continuous cover, the feature is known as **desert pavement,** or *reg* (Figure 10-26). Individual particles are faceted or polished. These coarse deposits may also protect finer particles from removal by the wind.

Sand Dunes   Although sand dunes are popularly seen as the main landforms of deserts, they generally make up a small proportion of a desert's landscape. These landforms cover less than 1 percent of the arid southwest of the United States and reach a maximum of only 25 percent on the vast plains of the Sahara (Figure 10-27). The extensive "sand seas" there are known as *ergs* (Figure 10-28) and occur where sand is concentrated by the convergence of prevailing winds. The sand in these deposits is almost entirely quartz particles, suggesting a long period of movement during which the more soluble and less resistant minerals were destroyed by weathering.

Sand dunes also occur along coasts, where sand deposited on beaches dries, and onshore winds blow it above the high-tide level. The winds blow dry sand into piles at the back of a beach. Like the dunes in deserts, coastal dunes form by wind action, but coastal dunes are often stabilized by vegetation.

A **dune** is a mound or ridge of windblown sand. Dune shape and size depend on the supply of sand, the constancy of wind strength and direction, and in some cases the presence of an obstacle to trap the sand

Figure 10-26   Wind deposition. A gravel-covered desert surface (reg) in Algeria.

Figure 10-27   Wind deposition: the sand seas (ergs) of the Sahara, showing their uneven distribution and the associated wind directions. A few have been named.

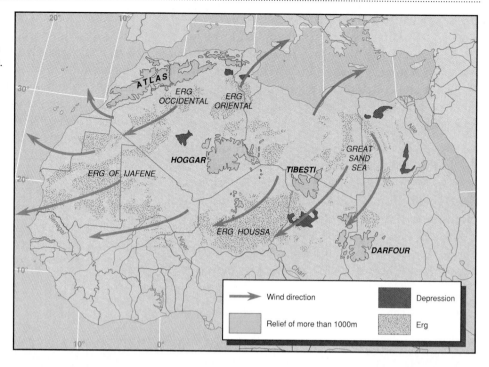

Figure 10-28   Wind deposition: a sand sea (erg) in eastern Algeria.

(Figure 10-29). Dunes are mobile features. The wind blows sand by saltation and creep up the windward side of the dune and the sand falls down the slip face on the leeward side (Figure 10-30). Obstacles such as trees, bushes, or buildings may prevent such movement.

Dune forms vary in size from ripples a few centimeters high up to 50 cm (20 in) high and 5 m (15 ft) across to dunes that are 5 to 30 m (15 to 100 ft) high and 50 to 300 m (150 to 1000 ft) across. *Ripples* develop when saltation and surface creep act on slight irregularities in the sand. Small asymmetric

ridges form at distances determined by the patterns of saltation. Coarser sand forms larger ripples.

Larger dunes are single or composite features. Crescent-shaped *barchans* form in areas where there is a limited sand supply and a constant wind direction. The random thickening of a sand patch (Figure 10-31) leads to saltation and creep acting on the windward surface increasing the height of the dune. The crescent shape develops as easier movement along the thinner edges of the main mass of sand causes them to move forward faster and form the arms. Once formed, turbulence along the dune margins maintains the shape. The barchan moves downwind, the speed of movement depending on the size of the dune and the wind strength. Smaller forms move more rapidly than larger. Barchans occur on rock pavement or fuse in belts with crescent-shaped units if the sand supply increases.

In areas of more sand, barchans grade into winding *transverse dunes* (Figure 10-32a, b). Wind turbulence and eddies within the wind flow determine the waviness of these ridgelike dunes. Transverse dunes may grade into *star-shaped dunes* when wind blows from different directions at different seasons of the year (Figure 10-32c). As wind turbulence increases, transverse dunes are reoriented into *linear dunes* that run parallel to the wind direction. Eddying air movements maintain hollows between the ridges (Figure 10-32d). The *seif dune* is a linear dune that lies at an angle to the prevailing wind in places where a secondary wind direction becomes significant (Figure 10-33). Seif dunes have greater relief than other linear dunes, reaching over 100 m (300 ft) high in Egypt and over 200 m (600 ft) in southern Iran.

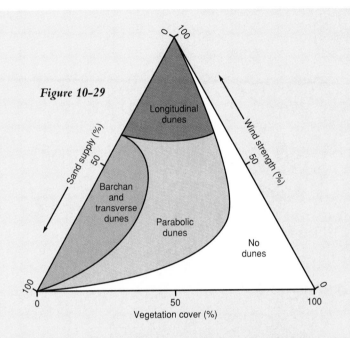

Figure 10-29

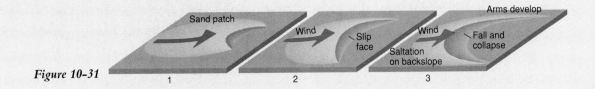

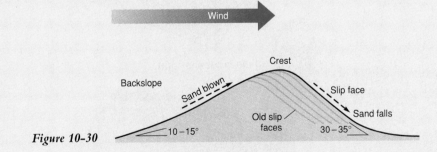

Figure 10-30

Figure 10-31

Figure 10-29   Wind deposition. A classification of dune types in the Navajo country of northeastern Arizona based on factors that determine their shape—sand supply, wind strength, and vegetation cover.

Figure 10-30   Wind deposition. The basic features of a dune. This cross section shows how the gentler backslope is formed by sand transported in the wind, while the steeper slip face forms when sand falls over the crest.

Figure 10-31   Wind deposition: the formation of a barchan dune.

The largest dunes are called *draa* in the Sahara and form by combinations of other dunes. They often have smaller dunes on top.

Parabolic dunes (see Figure 10-32e) occur where vegetation or moisture partly anchors the sand while the center is blown out. Such dune forms occur at White Sands, New Mexico, and cover nearly 30 percent of the Thar Desert in northwestern India. They are also common among coastal dunes subject to blowouts (see Figure 10-32f). Other fixed dunes form behind obstacles such as trees, bushes, rocks, or small lakes that supply moisture to anchor the sand.

Loess Deposits   **Loess** is a structureless, uncemented, light-colored, wind-borne silt deposit. Up to 90 percent of the particles in loess are between 0.05 and 0.5 mm across. The high silt content makes loess deposits compact in humid regions. Along the lower Mississippi valley, loess forms vertical cliffs.

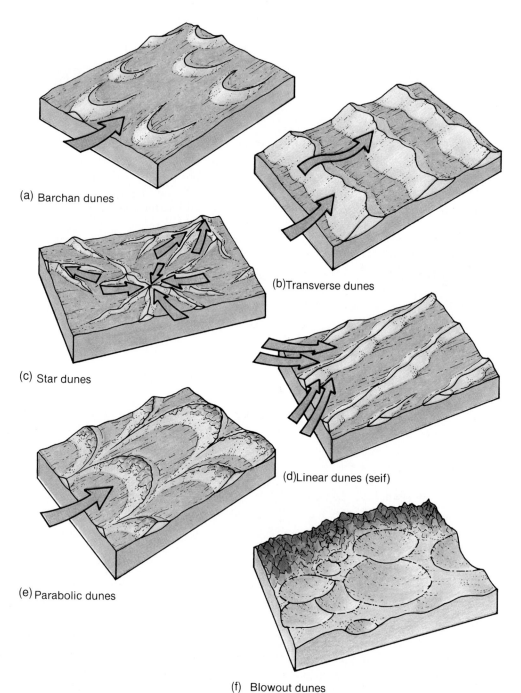

(a) Barchan dunes

(b) Transverse dunes

(c) Star dunes

(d) Linear dunes (seif)

(e) Parabolic dunes

(f) Blowout dunes

Figure 10-32   Wind deposition. (a–f) The major types of dune forms and the associated wind directions.
Source: U.S. Geological Survey.

Figure 10-33 Wind deposition: a space shuttle view of linear dunes in Saudi Arabia.

Loess is subject to rapid erosion by overland flow, streams, or wind since it is formed of small particles that are not cemented together. Once deposited, however, vegetation soon covers loess as it provides an almost ideal growing medium. A vegetation cover makes loess less susceptible to erosion.

The largest deposits of loess occur in central China, central North America, and from central Europe into Russia (Figure 10-34). Loess covers some 10 percent of Earth's land surface with depths ranging up to 100 m (up to 300 ft). The greater depths of loess blanket and bury previous landforms producing a flat relief. Loess is the basis of many of the best soils in the U.S. Midwest.

Most loess formed when strong winds deflated the unvegetated outwash plains surrounding the Pleistocene ice sheets (Figure 10-35). The deposits thin rapidly with distance from their sources, indicating that they were not blown far from the outwash plains.

In the United States, deposits that are farther from the former ice sheets were first carried the extra distance by meltwater streams. Winds then deflated the braided stream channels. Thick loess deposits also occur around the margins of deserts.

HUMAN ACTIVITIES AND WIND LANDFORMS

Human activities increase the significance of wind erosion by plowing soil in semiarid regions and by placing stress on the grasses that bind coastal dunes. The surface of bare, plowed land dries out and breaks down the cohesion of surface soil particles, allowing wind action to carry away the finer particles. This occurred on a huge scale in the Great Plains "Dust Bowl" of the 1920s. Millions of tons of topsoil blew away over a period of dry years. To correct this problem, farmers began to maintain a grass or stubble cover between plantings, and to

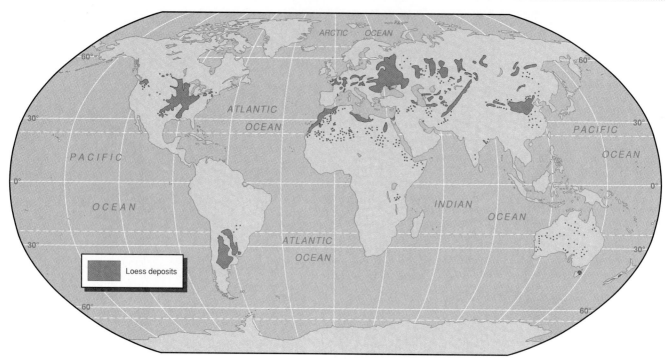

Figure 10-34   Wind deposition: world map of loess deposits. Some can be linked to former ice sheet margins and others to the margins of arid regions.

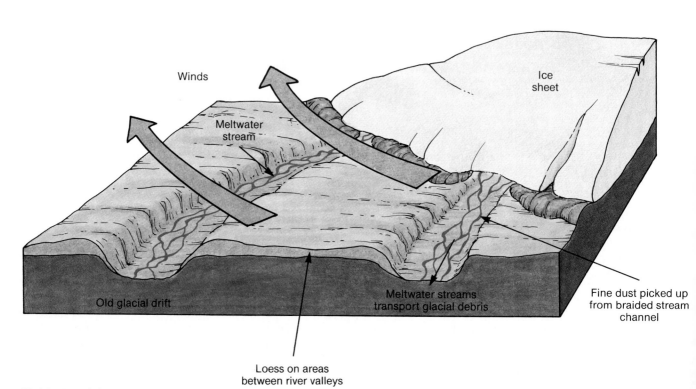

Figure 10-35   Wind deposition: the formation of loess around an ice sheet margin. Strong winds picked up dust particles from wide stream channels and bare soils. Some streams transported the fine particles for hundreds of kilometers before the wind affected it.

plant in strips with alternate strips left unplowed. Wind also erodes plowed fields as it blows through gaps in walls or fences.

A major increase in the impact of wind activity has occurred during the twentieth century along the southern edge of the Sahara where plowing of the land for commercial crops such as groundnuts resulted in disasters during periods of drier years. Old sand dunes have been exposed and reactivated. Attempts to stem the advance of mobile dunes by planting trees have not had the desired effect.

In coastal areas the tramping of many people across vegetated dunes leads to a loss of surface resistance and liability to blowouts of the unvegetated sand. Local authorities try to prevent blowouts by fencing off dune areas while the plants grow again and by laying boardwalks across the affected area.

Wind is a less effective and rapid agent of gradation than running water or flowing ice. Its greatest effects are in deserts, along coasts, in cold landscapes with little vegetation cover, and on plowed fields when the surface dries. Plants (*living-organism environment*) provide resistance to wind erosion.

Wind action depends on air movement in the *atmosphere-ocean environment* and acts on rocks and minerals provided by the *solid-earth environment.* Wind transports fine dust easily and moves it long distances. Wind moves fine or medium sand only a few kilometers. Wind does not move particles larger than 2 mm (0.8 in) across.

The landforms of wind erosion vary in scale from small to very large. They include ventifacts, yardangs, and deflation hollows. Wind deposits include dunes and loess.

ENVIRONMENTAL ISSUE:

Living on Permafrost—Do We Understand the Implications?

Some of the issues in this debate include:
What are the special problems that affect building and routeway construction in permafrost areas?
Can people apply their understanding of other areas to the development of permafrost regions?

Permafrost areas in northern Alaska, Canada, Scandinavia, Russia, and Siberia are drawing increasing numbers of people and human activities, especially for the extraction of valuable mineral products. The surface of permafrost areas, however, is extremely unstable and once disturbed is difficult to return to its natural state. This was discovered by trial and error as houses, roads, bridges, and airports were built in arctic lands.

Houses built directly on permafrost soils without any insulation underneath collapsed into a pool of mud because their fires and cookers melted and deepened the active layer. Modern buildings on permafrost are built with a gap beneath to allow cold air to circulate, or with heavy insulation to prevent such effects on the active layer. The photos show buildings at the Prudhoe Bay oil center in Alaska.

Construction for transportation also brings problems. A road cut into a hillside will cause water to seep from a new active layer on the sides of the cutting, freezing on the surface of the road. Attempts to drive straight across the tundra result in vehicles' wheels or tracks cutting through the thin vegetation cover and opening the active layer to further melting. Such cuts do not heal, but open further. Frost heaving causes airstrip surfaces to become uneven and dangerous; railroad tracks move up, down, and sideways.

Building the Trans-Alaska pipeline raised a great public outcry, in part because of the potential danger from building across permafrost. Oil is transported at relatively high temperatures to ease its flow through the pipe. The pipe had to be designed so that this heat would not increase the active layer. Most of the pipeline was built on supports above the surface that were flexible to allow for frost heaving. The rest was buried, but had to be heavily insulated at great cost. Gas pipelines are less of a problem in permafrost, since gas is transported at low temperatures to concentrate the flow.

Other problems of permafrost include clearing the land for farming and waste disposal. Clearing the land for farming often causes the formation of thermokarst as the active layer deepens, thaws, and the water flows out at a lower point, taking soil with it. Waste disposal cannot use holes dug into the permafrost, and the waste decomposes very slowly in the very low temperatures.

Linkages

Permafrost is a condition in the **surface-relief environment** that is induced by intense cooling in high-latitude climates (**atmosphere-ocean environment**).

Questions for Debate

1. Are the extra costs incurred in devising engineering solutions to living on permafrost worthwhile?

2. Should the many problems of living on permafrost result in limits being placed on the numbers of people allowed to live in such conditions?

Buildings at Inuvik, North West Territories, Canada, raised above the ground to prevent melting of the permafrost.

CHAPTER SUMMARY

1. Surface ice dominates Earth's landscapes in high latitudes and the highest parts of mountains. Ice accumulates where it is too cold for snow to melt and forms ice sheets, ice caps, and glaciers. Compaction and refreezing of melted snow change the snow to ice. Temperature affects the flow in ice sheets and glaciers. Ice moves by internal deformation and basal sliding. The *atmosphere-ocean environment* determines the low temperatures and amount of precipitation.

2. The *solid-earth environment* determines the configuration of the continents and oceans and the positions of high mountain ranges. Ice sheets and glaciers may erode or protect the underlying land. They transport large quantities of rock debris. Typical landforms of glacial erosion are roches moutonnées, U-shaped valleys, and cirques. Typical landforms produced by glacial and meltwater deposition include moraines, drumlins, eskers, kames, kettles, and outwash plains.

3. Ground ice occurs in areas of permafrost. Ground ice produces landforms such as ice-wedge polygons and pingos.

4. Wind is a less effective and rapid agent of gradation than running water or flowing ice. Its greatest effects are in deserts, along coasts, in cold landscapes with little vegetation cover, and on plowed fields when the surface dries. Plants (*living-organism environment*) provide resistance to wind erosion.

5. Wind action depends on air movement in the *atmosphere-ocean environment* and acts on rocks and minerals provided by the *solid-earth environment.* Wind transports fine dust easily and moves it long distances. Wind moves fine or medium sand only a few kilometers. Wind does not move particles larger than 2 mm across.

6. The landforms of wind erosion vary in scale from small to very large. They include ventifacts, yardangs, and deflation hollows. Wind deposits include dunes and loess.

KEY TERMS

periglacial environment   *288*

permafrost   *289*

ground ice   *289*

ice sheet   *290*

ice cap   *290*

alpine glacier   *290*

sea ice   *292*

ablation   *292*

internal deformation   *292*

basal sliding   *292*

warm glacier   *294*

cold ice sheet, glacier   *294*

crevasse   *294*

ice abrasion   *296*

ice plucking   *296*

whaleback hill and basin   *296*

U-shaped valley   *297*

cirque   *297*

till   *297*

moraine   *299*

drumlin   *300*

esker   *301*

kame   *302*

outwash plain   *302*

patterned ground   *303*

ice wedge   *303*

pingo   *303*

aeolian process   *305*

wind abrasion   *306*

deflation   *306*

saltation   *307*

surface creep   *307*

hammada   *307*

desert pavement   *309*

dune   *309*

loess   *312*

CHAPTER REVIEW QUESTIONS

1. What are the similarities and differences between ice sheets, ice caps, valley glaciers, piedmont glaciers, and cirque glaciers?

2. List the evidence that shows glacier ice can erode the land surface.

3. Compare warm and cold glaciers in terms of flow mechanisms and landforms produced.

4. Comment on the role of the active layer above permafrost as a key to the understanding of periglacial landscapes.

5. Why is human influence less obvious in the landforms of glaciated and wind-dominated areas than in humid midlatitudes.

6. Show how dune forms are determined by sand supply, wind strength, and wind direction.

7. Describe the materials, conditions of formation, and locations of erg and loess deposits.

SUGGESTED ACTIVITIES

1. Compare flowing water, moving ice, and wind action in their gradational roles of erosion, transport, and deposition.

2. Investigate examples of wind action in an area near you.

CHAPTER

COASTAL LANDFORMS

THIS CHAPTER IS ABOUT:

◆ Coastal landform environments
◆ Waves and tides
◆ Cliffs
◆ Beaches
◆ Deltas and estuaries
◆ Salt marshes and mangroves
◆ Reefs
◆ Climate and coastal landforms
◆ Human impacts on coastal landforms
◆ Environmental Issue: The Aral Sea—Is It Dead?

Land-based and ocean processes meet at coasts. Sand brought to the coast by rivers builds up to form beaches that protect wetland areas of marsh grasses, like those on the shores of southern Mozambique in Africa (inset). Ocean currents smooth the outer margins and give the beaches a straight line.

ZIMBABWE

Beira

Vilancoulos

Ilha do Bazoruto

Púnta Sâo Sebastiâo

MOZAMBIQUE

SOUTH AFRICA

Maputo

COASTAL LANDFORM ENVIRONMENTS

he meeting of land and sea produces a set of landforms influenced by waves, tides, and currents as well as the surface-relief processes described already—weathering, mass movements, running water, moving ice, and wind. In combination these processes attack the rocks forming the continental coasts. The result is a landform environment of almost continuous change; as one section wears away, another is built up. Figure 11-1 shows a section of coast in south Devon, England. The action of the sea produced the pebble beaches and vertical rocky cliffs, and slope processes formed the grassed areas above the vertical cliffs.

This chapter presents the processes at work along coasts and then examines how the landforms of distinctive coastal environments—cliffs, beaches, deltas, estuaries, salt marshes, mangroves, and reefs—form.

The variations imposed by climate patterns and changes over time and the impacts of human activities on coastal landforms complete the chapter.

Marine processes affect landforms in the **coastal zone** (Figure 11-2). This zone extends from a little above the water's edge at high tide to a little below the low tide level. It is linked to larger, adjacent onshore and offshore areas. Many large lakes and inland seas, as well as the oceans, have coastal zones. Large lakes, such as the Great Lakes of North America, take coastal landforms into the heart of the continent.

Within the coastal zone, a number of distinct landform environments exist. Coastal landforms are the outcome of interactions among waves, tides, currents, fresh water, salt water, rocks, sediment, and structures built by living organisms. Some of these interactions are shown in Figure 11-3. Interactions between the sea and land produce the coastal landform environments of cliffs and beaches.

Figure 11-1 Coastal landforms: cliffs and beaches along the coast of south Devon, England. This section of coast faces east.

Wetland area trapped behind beach

Cliff

Beach

Beach

Hallsands: old village destroyed by storm

Waves approach parallel to shore

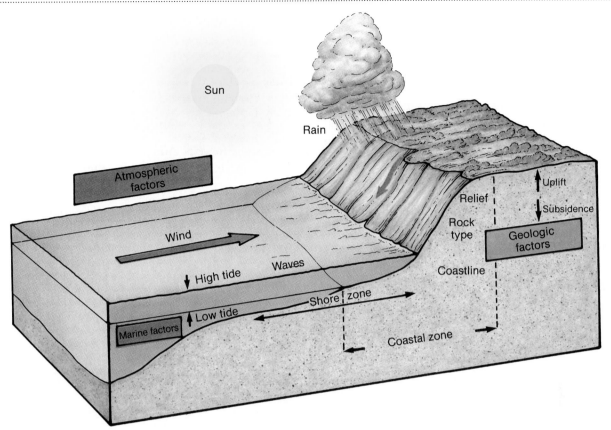

Figure 11-2   The definition of a coastal zone and some of the forces acting on it. A combination of marine and atmospheric processes act on geologic characteristics.

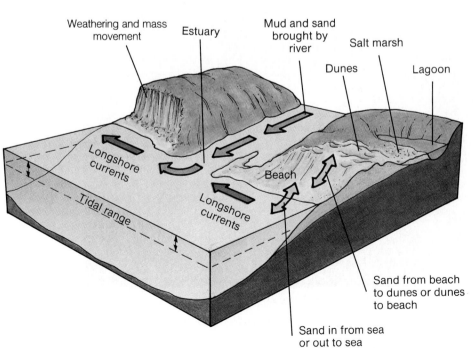

Figure 11-3   Coastal environments: links between the movement of sediment and marine processes.

Coastal environments where fresh water and salt water meet include deltas and estuaries. Coral reefs and mangrove swamps are distinctive tropical and subtropical coastal landform environments produced by organic structures.

Human activities have increasing effects on coastal areas. The 1990 census of the United States showed that over half the U.S. population lives within 75 km (45 mi) of an ocean or Great Lakes coast. This proportion is expected to rise to 75 percent by A.D. 2010. At the same time as people crowd into coastal locations, erosion is affecting the coasts in all 30 states with coastal boundaries. This makes the study of coastal landform environments especially urgent.

WAVES AND TIDES

The sea is a mobile medium that continuously changes its surface form. Waves are sea surface features that indicate how these movements cause water to be thrown against a coast. Tides are daily fluctuations in ocean level that cause the level of wave attack to rise and fall. Both waves and tides cause strong currents of water to flow into the coastal zone.

WAVE PROCESSES

A **wave** is a ridge or swell on the sea surface that moves through the water (Figure 11-4). Waves are described by characteristics such as height from crest to trough, wavelength (distance between crests), velocity, steepness, and period (frequency of crests) (Figure 11-5).

Waves in Deep Water Sea waves are caused by winds blowing over the sea surface, or, more rarely, by earthquake shocks. They transfer the energy imparted to them efficiently compared to water flow in rivers, since there is little loss by friction. Wave forms travel through water, but at sea the water itself does

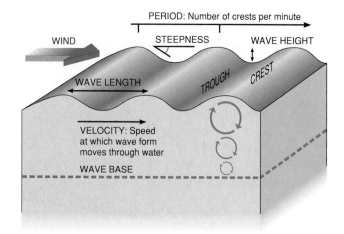

Figure 11-5 The features of sea waves. The wave form moves through deep water causing water to circulate. This movement dies out with depth and ceases at wave base, approximately half a wavelength deep.

Figure 11-4 A wave breaking as it approaches the Oregon coast.

not move forward in the wave but has a circular path as the wave passes. Wave movement decreases below the surface and dies out at *wave base,* which is approximately half a wavelength deep.

Wave size and shape depend on the wind speed, the wind *fetch*—or length of passage over open water—and the duration of wind blowing at that speed. The strongest winds produce the tallest waves. Waves over 5 m (15 ft) high form in gale-force winds over 54 kph (35 mph). Such conditions are common in the midlatitudes where frequent cyclonic storms generate large waves (Figure 11-6). **Storm waves** are tall, closely spaced waves with high levels of energy. The northern Pacific and Atlantic oceans mainly experience intense storm wave activity in winter, but the Roaring Forties zone of the southern hemisphere is stormy all year.

Swell waves affect more of the world's coasts than storm waves, but they have only moderate levels of energy. They have longer wavelengths but are not as tall as storm waves, and they are more consistent in shape and timing between crests. Swell waves begin as storm waves that move out of the stormy environments and gradually become longer and lower as they move through the oceans. Southern hemisphere storm waves in the Pacific Ocean have a passage of 11,000 km (7000 mi) before they arrive as swell waves on the coasts of southern California. During

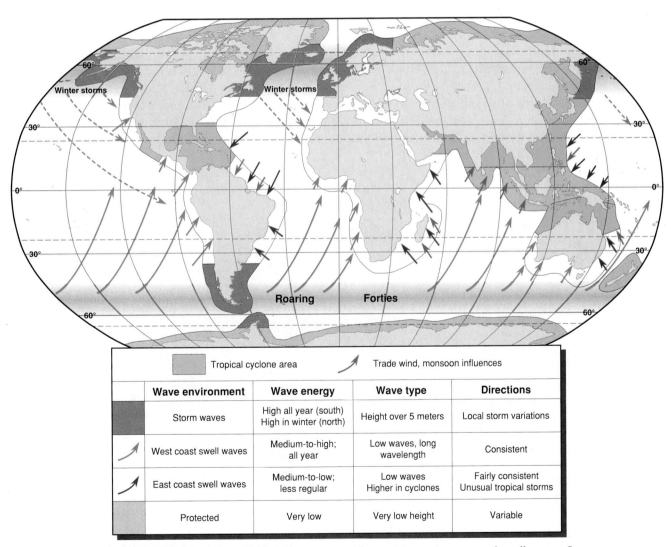

Wave environment	Wave energy	Wave type	Directions
Storm waves	High all year (south) High in winter (north)	Height over 5 meters	Local storm variations
West coast swell waves	Medium-to-high; all year	Low waves, long wavelength	Consistent
East coast swell waves	Medium-to-low; less regular	Low waves Higher in cyclones	Fairly consistent Unusual tropical storms
Protected	Very low	Very low height	Variable

Figure 11-6 World wave environments in which the main types of wave are storm waves and swell waves. Some coasts are protected from both and experience little strong wave action.

this journey the wave height decreases from between 4 and 6 m (12 and 18 ft) to between 2 and 3 m (6 and 9 ft).

Coastlines in some parts of the world are protected by islands from storm and swell waves. Wave energy is generally low in such regions, reducing the ability of the sea to create distinctive landforms. Protected coasts in the tropics, however, may be affected by sudden and potent, though short-lived, wave attack during the passage of a tropical storm or hurricane.

Winds require a fetch of a few hundred kilometers before they generate maximum wave height—10 to 15 m (30 to 45 ft). The English Channel boat-crossing of 34 km (21 mi) from Dover to Calais is much rougher when winds blow from the west and bring tall waves from the long Atlantic fetch, than when winds of equal force blow from the east out of Europe and over a few kilometers of water.

The duration of high wind speeds is also important in the generation of tall waves. Wave heights increase for up to forty-eight hours if high wind speeds continue. Once waves reach the maximum wave height for a particular wind speed, they do not grow further. When the wind dies down, waves travel for some distance before the wave height decreases to any great extent.

The largest waves of all result from earthquake shocks or occasionally from volcanic eruptions. *Seismic sea waves* occur almost exclusively in the Pacific and Indian oceans but are not very common, being recorded on average once every five to ten years. Major seismic sea wave events occur once in twenty-five to fifty years in Hawaii. An earthquake shock produces a large wave form that moves at 640 to 960 kph (400 to 600 mph) in water over 3000 m (10,000 ft) deep. This wave crosses the Pacific in a few hours, slowing around volcanic islands and submerged volcanoes, but increasing in speed across deep ocean trenches. Although seismic sea waves are seldom more than one meter high in the open oceans, they may rise to 15 m (50 ft) as they reach a shore or higher if funnelled into estuaries. The seismic sea wave that followed the Alaska earthquake of 1964 washed away trees up to 50 m (150 ft) above sea level along coastal inlets.

Waves Approaching the Coast   When waves approach the coastal zone, the shallowing bottom relief affects their shape and direction. Their paths are diverted, or *refracted,* concentrating wave energy on headlands and spreading it out in bays (Figure 11-7).

As waves enter shallow water, the water moves forward as well as the wave form. This produces

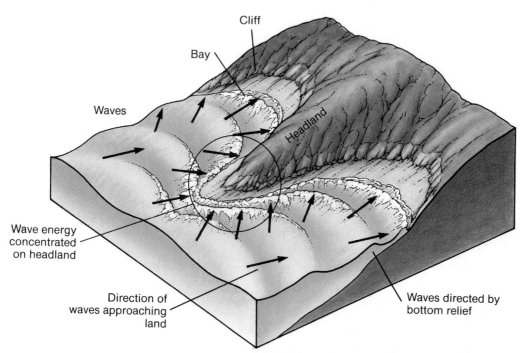

Figure 11-7   Waves are refracted by coastal relief. Water slows on entering shallow water, and the waves turn toward a headland but spread out in bays. This focuses wave energy on headlands and dissipates it in bays.

wave-induced currents, which are the main influences on coastal landforms. The increasing friction with the underlying shore slows the water, so the waves grow taller than the depth of water beneath them until the wave crests topple over in the *breaker zone.* The water rushes shoreward through the *surf zone* and becomes a shallow sheet known as *swash* that flows up the beach until it slows and stops by friction. Some of the water sinks into the beach materials and the rest returns down the slope of the beach as *backwash,* gathering momentum as it goes.

Waves that are tall and plunge steeply on breaking bring more energy to the coast and tend to move sand and pebbles from the beach out to sea in a strong backwash. Waves that are low direct water up the beach. They deposit sand on the upper part of the beach as water drains into the underlying material and their backwash is not so strong.

For much of the time, wave crests arrive at the shore at a slight angle after refraction. This causes a flow of water along the shore in the breaker and surf zones. This flow is in a *longshore current.*

The nearshore zone of many beaches has a circulating system of currents. Water moves through narrow corridors toward, along, and away from the shore (Figure 11-8). Offshore movements called *rip currents* move water seaward across and beneath breaking waves and then spread out, losing impetus. The strongest rip currents occur when high energy waves approach the shore.

TIDAL PROCESSES

Tides are regular daily rises and falls of the ocean surface caused by the gravitational pull of the sun and Earth's moon. This effect varies in a monthly cycle that depends on whether the sun and moon pull together, producing a high tidal range, or *spring tide,* or oppose each other, producing a low tidal range, or *neap tide.*

Tidal Environments Tides vary in their height range according to the arrangement of continents and coastal relief. The **tidal range** is the vertical difference between the highest and lowest water levels and defines the tidal environment. Coasts with tidal ranges over 4 m (13 ft) are said to be *macrotidal;* those with tidal ranges under 2 m (6.5 ft) are *microtidal;* those with ranges between 2 and 4 m (6 and 13 ft) are *mesotidal.* Tidal range is as much as 17 m (55 ft) in the Bay of Fundy in eastern Canada (Figure 11-9) but scarcely 1 m (3 ft) in enclosed seas such as the Mediterranean.

The tidal range affects the width as well as the height of tidal environments. The zone affected by wave action and alternate wetting and drying is generally wider along macrotidal coasts than microtidal coasts. It may be up to 1 km (3000 ft) wide in macrotidal wetlands or beaches. Local relief and the slope of the shore also affect the width of tidal environments. Steeply sloping shores with cliffs have a narrow width of tidal environment.

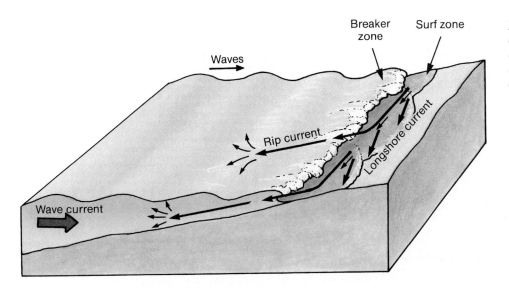

Breaker zone Surf zone

Waves

Rip current

Longshore current

Wave current

Figure 11-8 Wave-induced currents. An onshore wave current produces longshore currents, and these turn offshore in a rip current.

Figure 11-9   Tidal range. High and low tide in a Bay of Fundy cove in eastern Canada.

Tidal Currents   Areas with high tidal ranges also have strong **tidal currents,** which are horizontal flows of seawater in response to the rise and fall of tides. Tidal currents are more limited in geographic extent than wave-induced currents but locally may be almost as effective in modifying coasts. One type of tidal current is that which flows between Manhattan and Long Island, New York. The high tide arrives from the east along the north side of Long Island while the low tide is still ebbing to the west. Flow is rapid and turbulent as the high tidal flow tumbles into the low-tide area. Such currents prevent sediment from accumulating in narrow channels and may enhance erosion of the channel margins.

Another type of tidal current forms as water flows in and out of river mouths and bays. During a falling tide, water and sediment flow outward; on a rising tide, they flow upstream. Where an estuary narrows inland, the tidal flow upstream at spring tide may form a wave a meter or so high, called a *tidal bore.*

The sea plays an important part in producing coastal landforms in conjunction with other surface-relief and living-organism environment processes. Waves, tides, and currents are marine processes that act on coasts. The intensity of wave action varies around the world according to the strength and continuity of storm-force winds (**atmosphere-ocean environment**). Variations in the heights of tides cause marine processes to affect different vertical ranges on coasts.

CLIFFS

Cliffs are steep rock faces and are common coastal landforms. The cliff landform environment has several distinctive features. First, marine erosion is the main agent at work in the formation of cliffs, and their vertical or sloping rock faces are most prominent in storm wave environments (see Figure 7-1). On more protected coasts, cliffs have only short vertical sections and in humid regions the upper slopes are covered by vegetation that stabilizes the slopes (see Figure 11-1). Second, cliff forms are common along convergent plate margins (see Chapter 7) where hills and mountain ranges form areas of high coastal relief and deep offshore water.

Third, cliffs also occur along some coasts where the opening oceans cut across former mountain ranges. The severed ranges often run at right angles to the coast, ending in spectacular cliffs, as in southwestern Ireland.

Coastal cliffs form by a combination of processes that begin with erosion of the rock at the foot of a slope when it is attacked by high-energy waves. The impact of waves breaking against rock is a powerful force and it compresses air between the wave front and the shore and in the rock joints. The combination loosens blocks and particles of rock. More rock is then worn away by *abrasion* as waves fling pebbles at the cliff base. Large boulders are moved only by the largest waves and are less important in cliff erosion than pebbles, which are moved more often and by waves of moderate energy.

One result of wave impact and abrasion by pebbles is a *notch* at the cliff base below high-tide level. As rock is excavated from this notch, the cliff above becomes unstable and rockfalls occur. The falls maintain the steepness of the cliff and cause it to retreat. The proportions of erosion by wave impact and abrasion depend on the type of rock being attacked and on the presence of joints and other fractures.

Since cliffs are usually steep rocky slopes, the processes at work in cliff environments include weathering and mass movements. Some cliffs form in unconsolidated materials such as glacial drift. Such cliffs are subject to rapid removal of material by the force of the waves and they may also be eroded by running water (Figure 11-10).

Shore platforms are horizontal, or nearly horizontal, surfaces formed in rock at the base of a cliff. There are several types. Shore platforms that result from cliff erosion are called *wave-cut platforms,* although other processes such as weathering are often at work helping to mold them. They occur below high tide and slope seaward. As a wave-cut platform is eroded and gets wider, marine erosion of the cliffs slows. This is because the waves enter shallow water across the platform, slowing their momentum. A second type of shore platform occurs along coasts that are naturally protected from storm waves. Such coasts are common in the tropics and often have horizontal shore platforms, where wetting and drying of rock surfaces between tides causes finely layered rocks to flake.

Rock type influences the detailed shapes of cliffs. In areas of resistant or recently uplifted rocks, cliffs

Figure 11-10   Cliffs. Waves undercut the base of these glacial deposits on the New England coast. Pebbles are dislodged and the angle of the cliff is determined by the fragments that compose it. Streams carve gullies across the weak material.

tend to be higher. Where weaker and more resistant rocks alternate and where strong rocks are highly fractured, features such as caves, stacks, and arches form (Figure 11-11). Waves erode *sea caves* along lines of weakness at the base of a cliff and extend the process of undermining farther than a notch. *Stacks* are rock pillars isolated as waves remove the rock between them and the backing cliff. *Arches* are stacks still joined at the top to the backing cliff; when the upper section of an arch collapses, a stack remains.

BEACHES

Beaches are coastal depositional landforms composed mainly of loose sand or pebbles. The main beach landform environments occur along coasts of low relief, but beaches also form along coasts of higher relief where sediment accumulates in bays.

Sand produces wide, flat beaches, and pebbles form narrower, steeper features. Most pebble beaches occur in middle and higher latitudes where storm-wave environments erode larger rock fragments from cliffs and where melting ice sheets at the end of the last glacial phase deposited larger rock fragments. The Grand Banks of Newfoundland and the floor of the North Sea have pebble deposits resulting from glacial deposition and now supply material that forms pebbly beaches along surrounding coasts. The central and southern Atlantic states of the United States, however, were not affected by glacial processes and their offshore sand deposits were supplied from the erosion of inland areas by rivers. These states, therefore, have sandy beaches. Tropical beaches are commonly formed of fine sand brought to the coast by rivers combined with much shelly material that reflects the climatic warmth and the low energy waves. The shells dissolve and their calcium carbonate cements the sand to form *beach rock*.

The landforms of beach environments are often small, measuring up to a few meters in size (Figure 11-12). At the highest point of the beach there may be a pebbly and steeply sloping *storm beach* several meters high, where the pebbles are moved only by storm waves. The vertical relief of features on the sandy part of the beach is often only a few centimeters, reflecting the tidal range and the modest impact of breaking waves. *Runnels* are small, flat-floored channels formed by low-tide runoff. They alternate with *rippled* surfaces and low sandy *bars* formed by

Figure 11-11 Cliffs. Cave, arch, and stack on the north French coast.

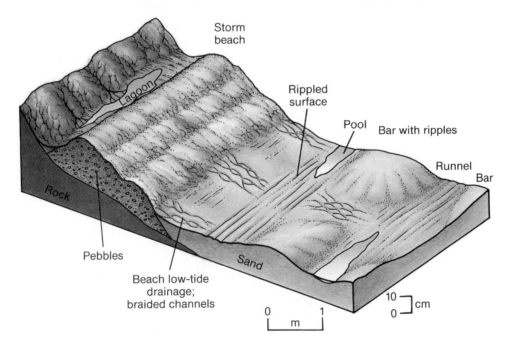

Figure 11-12 Beaches. Typical beach landforms shown in a composite diagram. The features of a particular beach are determined by the wave and tidal environments and the supply of sand and pebbles.

breaking waves. The ripples and bars generally cover most of the beach and the runnels are more localized. *Beach cusps* are crescent-shaped sand or pebble accumulations around semicircular depressions that may be caused by powerful swash and backwash movements. Most minor beach features exposed at low tide are removed during the following high tide (Figure 11-13).

Beaches change form more frequently than cliffs, often over a yearly or seasonal cycle. For example, winter storm waves may pull sand out to sea, while gentler summer waves return the sand to the beach. In macrotidal environments where there is an abundant supply of sand, strong and consistent onshore winds drive high waves that produce wide, sandy beaches with several sequences of bar-runnel

Figure 11-13   Beaches: low-amplitude bars and runnels at Cape Hatteras, North Carolina.

alternations. The high-energy waves take sand far up the beach, while the strong backwash currents also take sand back down the beach slope, spreading it out to form a wide feature. Weaker winds, microtidal environments, and a small sand supply produce narrower beaches with steeper gradients and a single bar-runnel sequence.

Longshore currents move sediment parallel to the coast in a process known as **longshore drift.** It occurs in the breaker and surf zones where waves reach the shore at an angle to the water's edge and is most active with steep, storm waves. In longshore drift, sediment is pushed up the beach at the angle of the incoming waves (Figure 11-14); backwash returns the sediment down the steepest beach slope. The overall result is that the sediment moves along the beach and produces several distinctive beach forms. If the coast consists of alternate bays and rocky headlands, the sandy sediment accumulates in the heads of the bays. On low coasts, on the other hand, the sediment moves along the shore and produces different beach landforms that dominate the coastal landscape.

Along the coasts of the eastern United States and around the Gulf of Mexico, barrier beaches extend for over 4300 km (2700 mi) (Figure 11-15). **Barrier beaches** are elongated sand or pebble banks lying parallel to, but usually separate from, the coastline. Several factors may be involved in their formation. First, a good supply of sand moved by longshore drift is deposited when carried into deeper water. Second, dune ridges may be drowned by a rising sea level, as along the eastern coast of the United States following

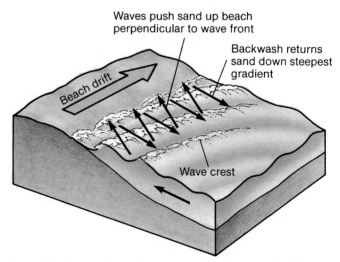

Figure 11-14   Beaches: the process of longshore drift. Waves push sand up the beach in swash; backwash takes it down the beach's steepest slope. The result is that sand moves along the beach.

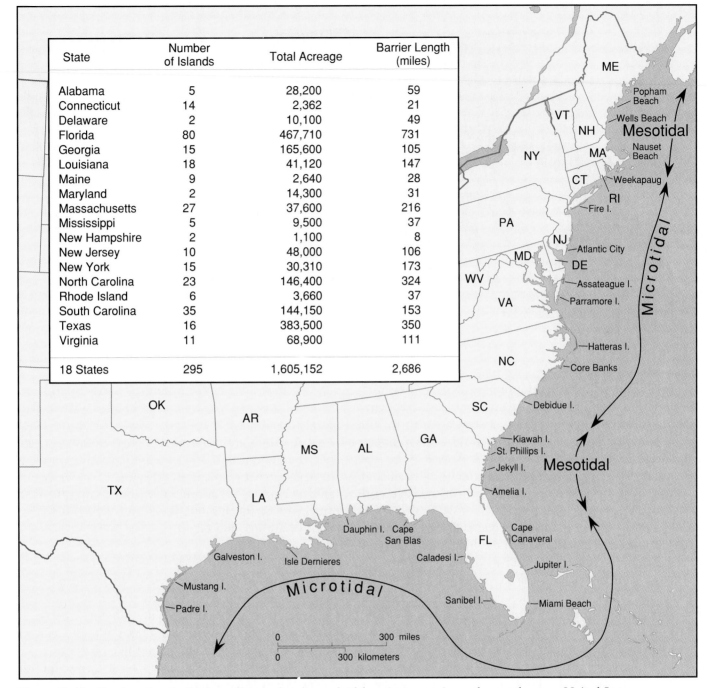

State	Number of Islands	Total Acreage	Barrier Length (miles)
Alabama	5	28,200	59
Connecticut	14	2,362	21
Delaware	2	10,100	49
Florida	80	467,710	731
Georgia	15	165,600	105
Louisiana	18	41,120	147
Maine	9	2,640	28
Maryland	2	14,300	31
Massachusetts	27	37,600	216
Mississippi	5	9,500	37
New Hampshire	2	1,100	8
New Jersey	10	48,000	106
New York	15	30,310	173
North Carolina	23	146,400	324
Rhode Island	6	3,660	37
South Carolina	35	144,150	153
Texas	16	383,500	350
Virginia	11	68,900	111
18 States	295	1,605,152	2,686

Figure 11-15   Beaches: the distribution of barrier beaches and tidal environments in southern and eastern United States.

melting at the end of the last glacial phase. Third, deposition from longshore drift may combine with sediment brought by currents toward the shore from offshore deposits. This last factor is important in the formation of *offshore bars,* which are semi-submerged sand deposits just outside the breaker zone. Some offshore bars are built above sea level, as along the Texan Gulf coast of the United States.

Barrier beaches enclose and protect areas of water called *lagoons.* Lagoons often have freshwater rivers draining into them and develop into wetland marshes as vegetation grows around their margins, gradually filling them.

Barrier beaches vary in their degree of attachment to the land (Figure 11-16). *Bay barriers* are connected at both ends to headlands. They are usually

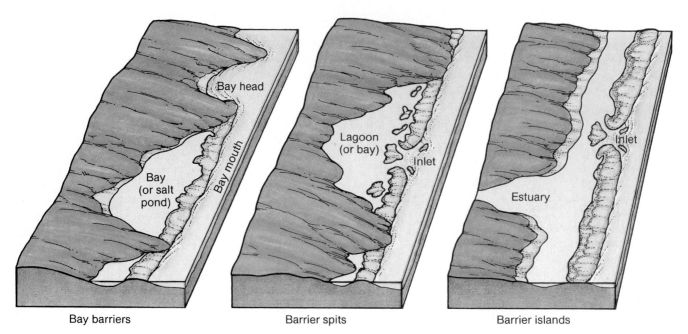

Figure 11-16   Beaches: the main forms of barrier beaches.

small and may have temporary inlets across them. Bayhead barriers occur at the back of a bay. Bay-mouth barriers connect the headlands with a lagoon behind and are common in microtidal environments where inlet formation is inhibited. A *tombolo* is a type of bay barrier that connects an island to the mainland. The map and photo on page 337 show the baymouth bar at Morro Bay, California.

Barrier spits are attached at one end and extend into open water. They form from the longshore drifting of sediment into the mouths of bays or estuaries where the coast changes direction and the water deepens. For instance, Sandy Hook on the northern New Jersey coast occurs where the coast turns into the Hudson estuary. Longshore drifting from the south supplies sand to a macrotidal environment. The tip of Sandy Hook curves in response to waves approaching from several directions. Water ponded behind the spit formed a lagoon in which a marsh developed, while the exposure of sand above high-tide level led to the formation of dunes.

Barrier islands are not attached to the mainland. They form as dune ridges and offshore bars emerge. Their features depend on the tidal environment and sand supply. A small sand supply in a microtidal environment leads to long, narrow, low-lying islands that are vulnerable to storm action. With a greater supply of sand, several parallel ridges form on the island, giving greater stability. The Outer Banks of North Carolina and Galveston Island, Texas, are typical examples. Another example is shown in the space shuttle photo of the Mozambique coast (Figure 11-17).

In mesotidal environments, barrier islands are short because the greater rise and fall of water creates tidal inlets subject to daily low-tide scour. Such smaller islands are more stable because water is channelled around them. The bays and lagoons behind are filled by marshy wetlands. Mesotidal barrier islands are less common than microtidal ones along the eastern coast of the United States and occur in Maine, South Carolina, and Georgia.

DELTAS AND ESTUARIES

Distinctive coastal landform environments occur where river and marine processes interact. This happens commonly where river mouths form breaks in cliff and beach environments. Deltas and estuaries are the main landforms produced in these conditions.

A **delta** is a fan-shaped deposit at a river mouth. Deltas occur in both upland and lowland coastal environments and in lakes. The essential condition for delta formation is that the rate of deposition on the delta surface exceeds the ability of marine processes to remove the sediment to deeper water. The resulting accumulation of sediment produces new land.

Deltas vary in size and shape. The differences depend on, first, whether most sediment is dropped on the delta surface or at the seaward margin, and, second, how marine processes modify the features at the seaward edge. Delta landforms are determined by the relative importance of river, wave, and tide (Figure 11-18). Stream action dominates

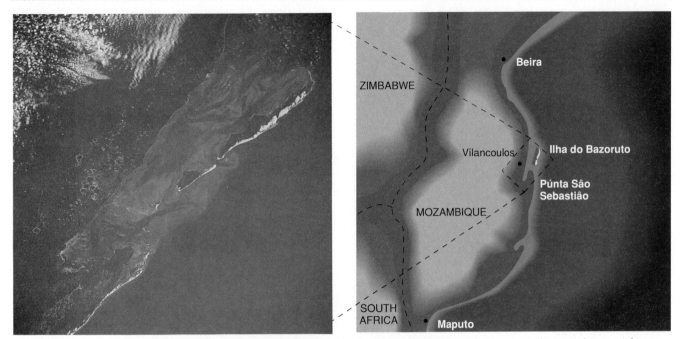

Figure 11-17   Beaches: a space shuttle photo of a barrier island backed by tidal marshes on the coast of Mozambique, Africa, near Vilancoulos.

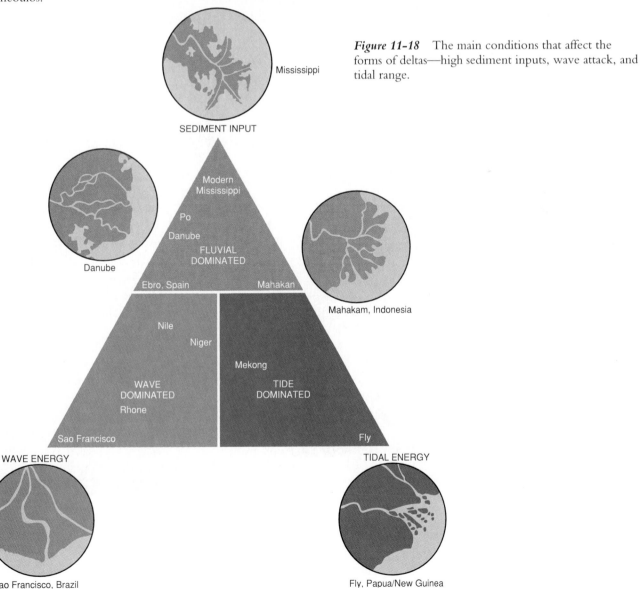

Figure 11-18   The main conditions that affect the forms of deltas—high sediment inputs, wave attack, and tidal range.

(a)

(b)

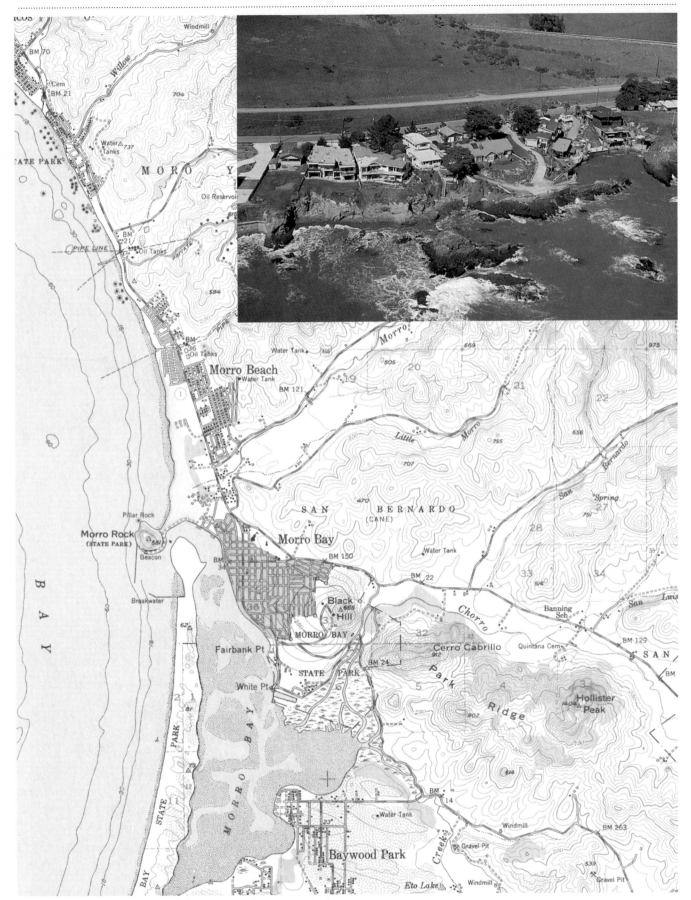

Coastal features at Morro and Cayucos, California. The map shows Morro Bay with its baymouth bar, silting bay, and wetlands where the Chorro Creek enters. Farther north the coast has low cliffs with hills rising inland behind Cayucos (photo).

where the marine currents are relatively weak and the tidal range is small. Deposition of sediment occurs along the margins of the channel as the stream slows on meeting the sea. The distributary channels extend out to sea. The Mississippi River delta is of this type, and its shape is often described as "bird-foot." Wave action smoothes the shape of the delta front by beach deposition, as in the Nile delta of Egypt and the São Francisco delta in Brazil. In macrotidal environments tidal scour forms deep channels across the delta, as in the Ganges delta in Bangladesh or the Fly River delta in Papua, New Guinea. Examples of deltas photographed from space shuttles are shown in Figure 11-19.

The weight of sediment deposited on the delta surface leads to compaction of the underlying layers and subsidence of the surface. If deposition does not keep pace with subsidence, some or all the delta is drowned by the sea. Deltas are thus unstable features

(a)

Figure 11-19   Deltas. Types of deltas photographed from a space shuttle. (a) The Mississippi River delta, Louisiana. (b) The River Po delta, northern Italy. (c) The River Nile delta, Egypt. (d) The River Ganges delta, Bangladesh, often called the Sundarbans.

(b)

where there is a complex pattern of balances between sediment accumulation and subsidence, inputs and losses of sediment, and marine erosion and deposition.

An **estuary** is the mouth of a river that broadens into the sea and is affected by tidal processes. Estuaries are common in coastal environments where the rivers carry small quantities of sediment. Estuarine sections occur in most river channel mouths, and similar processes affect delta distributaries.

Estuaries in macrotidal environments are alternately filled by seawater at high tide and then flushed by fresh river water on the falling tide. At high tide, fresh water is pounded back by incoming seawater and clay particles in the river load react with salt water. This causes the clay particles to stick together and fall to the bottom of the channel forming a thick layer of mud. At low tide the river flow carves a channel in the estuary mud (Figure 11-20).

(c)

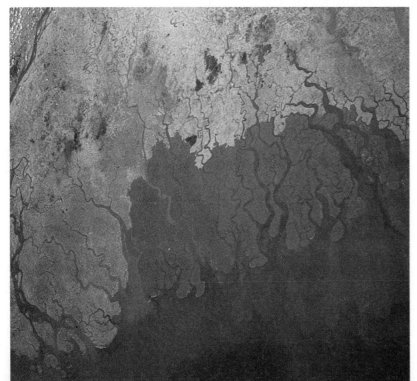
(d)

Figure 11-20 The main features of an estuarine tidal environment. Mudflats form by deposition at high tide and are colonized by salt marsh plants when they rise above the level of average high-water neap tides.

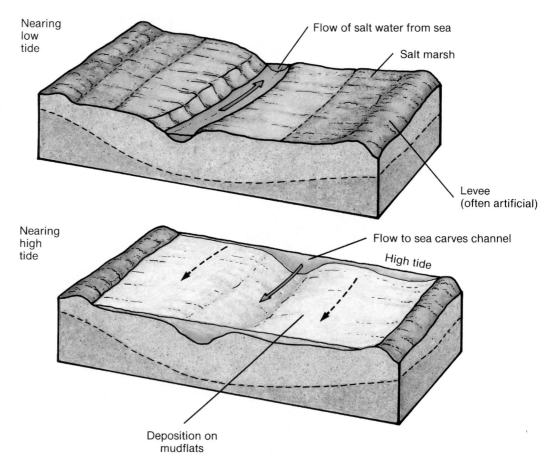

Estuaries in microtidal environments experience little back-and-forth movement of this type and remain filled with brackish water for much of the time. *Brackish water* is a mixture of seawater and fresh water with a salinity of 15 to 30 parts per thousand.

Muddy sediment, composed of clay and silt particles with a high water content, is characteristic of both estuaries and lagoons on delta surfaces and behind barrier beaches. *Mudflats* are sections of a shoreline covered by silt and clay that are submerged at high tide.

SALT MARSHES AND MANGROVES

Coastal environments that are protected from wave activity encourage the development of plants that have important local influences on coastal landforms. Salt marshes and mangrove colonies form in such environments. Climate is the main factor that decides whether a coast has salt marshes or mangroves.

Salt marsh plants colonize midlatitude mudflats in estuaries and lagoons behind barrier beaches when the mudflat surface reaches above the average neap tide high-water level. Salt marsh plants tolerate alternate exposure and submergence by changing tides and water of changing salinity. Grasses trap further muddy sediments, raising the level of sediment. This allows other grasses and shrubs to get a roothold, and the area may be built above tidal level so that it is only occasionally flooded. Trees that tolerate wet conditions occasionally grow on higher island areas within the wetland.

Mangroves colonize low-lying coasts and estuaries in the tropics and subtropics. Mangrove roots extend through water and into bottom sediment, while the leaves and branches rise several meters above high-tide level. The roots trap sediment in a brackish water environment and often have a filtering effect by removing solutes from the water. The sediment beneath mangroves builds toward high-tide level. In many cases, mangroves protect a coast from erosion and longshore drifting of sediment, but in southern Florida they are sometimes ripped out by hurricane-induced surges.

REEFS

Coral reef structures form in clear, shallow water in subtropical and tropical coastal environments. They are built by animals that live in colonies and secrete skeletons of calcium carbonate. The skeletons of many coral colonies accumulate, building on dead former colonies, and form a reef. Coral reefs are up to several thousand meters deep and several kilometers wide.

The growing part of a coral reef is just below sea level. The world distribution of coral reefs (Figure 11-21) shows that the greatest coral growth occurs where the sea-surface temperatures are between 22° C and 29° C (72 to 84° F). In such temperatures, the tiny plants associated with corals flourish. Coral reefs grow most rapidly on the windward sides of land areas where breaking waves increase levels of dissolved oxygen. Reefs also grow in quieter conditions, but more slowly. Reef debris broken in storms often piles up to form an island. Some islands, such as Barbados, were formed by the uplift of coral reefs.

Many coral reefs in the Indian and Pacific oceans formed around volcanic islands. During the world voyage in the 1830s that led to his theory of evolution, Charles Darwin noted a sequence of three types of reef that form around a volcanic island before it becomes extinct and subsides (Figure 11-22). *Fringing reefs* are attached to land without an intervening lagoon. *Barrier reefs* are separated from land by a lagoon. *Atolls* form where there is no land after the central volcanic island subsides beneath sea level. Figure 11-23 shows some typical coral reefs as seen from the space shuttle.

Not all coral reefs form around volcanic islands. Those around Florida and in northeastern Australia, for example, formed along continental coasts that provided a source of calcium carbonate in the water and a stable anchorage for the reef colonies.

Coral reefs demonstrate the effects of both internal and external Earth processes on surface landforms. The formation and subsequent history of the volcanic Hawaiian Islands (see Figure 7-9) affected the coral reefs around each island. The youngest islands, where lava erupts today, have a few coral reefs that are all close inshore—the beginnings of fringing reefs. The other Hawaiian Islands have extinct volcanoes but well-developed fringing reefs. Northwest of the Hawaiian Islands, barrier reefs surround smaller islands. Beyond these, there are atolls, such as Bikini Atoll. Cores drilled through Bikini Atoll showed that the reefs there built up over 50 million years, rising in level as the central volcano subsided to maintain coral growth at the ocean surface. Even the sea level fluctuations of the Pleistocene Ice Age did not destroy the reefs completely, although karst-like caverns formed when the coral limestone was exposed to the atmosphere during periods of low sea level.

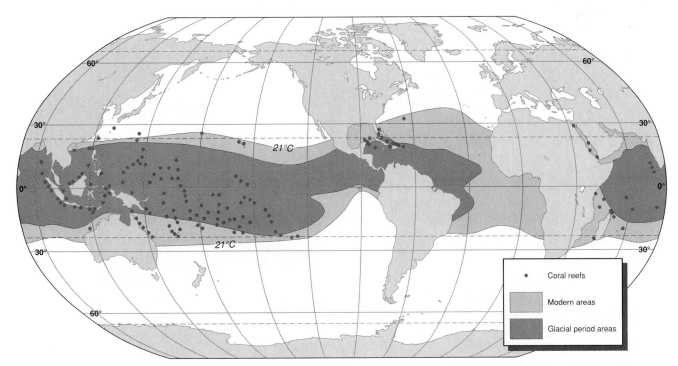

Figure 11-21   Coral reefs: World map showing how active reefs occur in water at relatively high temperatures and the reduced area of reef formation during the Pleistocene Ice Age. A sample of the 425 active reefs is shown.

Source: Data from M. J. Selby, *Earth's Changing Surface,* Oxford University Press, Oxford, 1985.

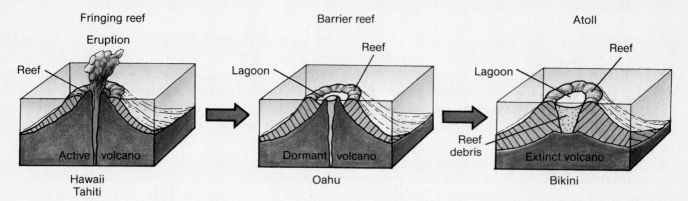

Fringing reef · **Barrier reef** · **Atoll**

Figure 11-22

Figure 11-22 A developmental series linking the three main types of coral reefs. Initially, reefs attach to an island provided by volcanic eruptions. After the volcano becomes extinct, it subsides and the barrier reefs grow away from the land separated by a lagoon. Atolls are left when the central island disappears from view.

Figure 11-23 Different types of reef in the Society Islands in the Pacific Ocean, as photographed from a space shuttle. Barrier reefs surround Tahaa-Raitea in the southeast; the atoll in the northwest is Motu Iti.

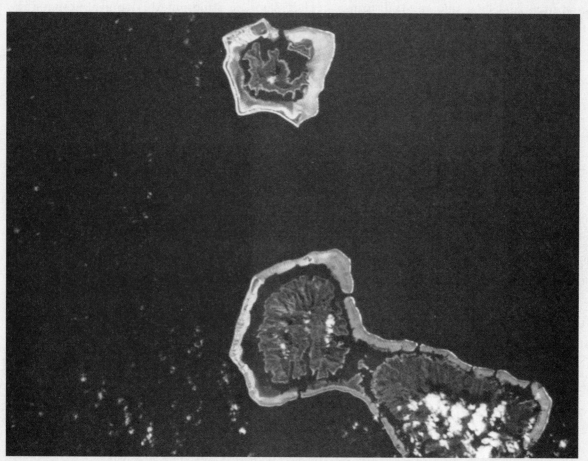

Figure 11-23

CLIMATE AND COASTAL LANDFORMS

Global patterns in climate have a smaller impact on coastal landforms than on other surface-relief processes. Steep cliffs and pebbly beaches are most common in midlatitude storm-wave environments. Coral reefs occur only in the tropics. Salt marshes occur in midlatitudes and mangroves in the tropics. Sandy beaches and estuarine mudflats, however, occur where wave and tidal influences and sediment supply let them develop in all climatic environments.

Climate changes, however, have major impacts on coastal landforms when they influence changes in sea level. Wave action is concentrated in a narrow zone between high and low tide, and changes in sea level shift this zone up or down. Old cliffs, wave-cut platforms, beaches, and reefs are left high and dry when sea level falls. Rises in sea level drown older coastal landforms and the mouths of river valleys.

Climate changes during the Pleistocene Ice Age have left significant marks on coastal landforms. The formation of ice sheets withdrew water from the oceans and lowered sea level. When the ice sheets melted, sea level rose. Such worldwide changes of sea level caused by changing volumes of water are known as *eustatic* changes. At the same time, the formation of ice sheets added a load to the continents, causing them to sag by *isostasy* (see Chapter 7). Since the melting of ice, continents that once had an ice cover have begun to return to their former level. Both eustatic and isostatic sea level changes occurred during the Pleistocene Ice Age, and it is often difficult to unravel the distinctions between them.

The main result of sea level changes is to produce coastal landforms of emergence or submergence. **Emergent coastal landforms** occur where sea level falls relative to the land. They include raised beaches and marine terraces. A *raised beach* has landforms typical of beach and cliff environments but at a height well above the present sea level (Figure 11-24). A raised beach may have sandy or pebbly deposits, shells, or a wave-cut platform, with a cliff behind. A *marine terrace* is a former wave-cut platform and may be part of a staircase of terraces, as along the coast and offshore islands of southern California.

Submerged coastal landforms are the result of sea level rising relative to the land. The lower parts of valleys fill with water, and water depth offshore increases. Such landforms have been common since the rise of sea level when the ice sheets melted between 15,000 and 10,000 years ago. San Francisco Bay and Puget Sound are examples of drowned valleys on the west coast of the United States; the Hudson estuary and Delaware and Chesapeake Bays are examples on the east coast. The drowned river valley mouths in the hilly region of northwest Spain are called *rias,* and this name has been applied widely to such features.

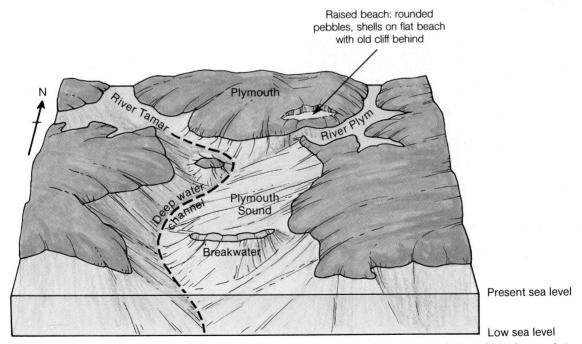

Figure 11-24   Changing sea levels: coastal landforms at Plymouth, England. Plymouth Sound is a drowned river mouth, or ria. A raised beach above the present sea level shows that the level was once higher.

In areas of formerly glaciated high relief, **fjords** form deep ocean inlets. They occur where glaciers excavated U-shaped valleys and the rising sea level drowned the lower parts (Figure 11-25). Fjords are most common on west-facing midlatitude coasts where heavy precipitation produced very active glaciers in glacial phases of the Pleistocene. Such areas include southern Alaska, British Columbia, Norway, South Island of New Zealand, and the west coast of Scotland.

Drowned valleys and raised beaches may occur together since the ups and downs of sea level during the Pleistocene often left their marks side by side. Plymouth Sound in southern England, for instance, is a deep ria, but raised beaches occur tens of meters above the present waterline.

The threat of possible global warming and a rise in sea level is a current issue (see Environmental Issue: The future—Global Warming or Cooling?,

Figure 11-25   Changing sea levels. Misty Fiords National Monument in the Coast Range of southeastern Alaska is a fjord.

p. 165). A rise in sea level of only a meter or so would flood low-lying coastal areas such as deltas, estuaries, reefs, barrier beaches, salt marshes, and mangrove colonies. It would also affect cities in coastal locations. Sea level changes may result from other factors than climate changes (see Environmental Issue: The Aral Sea—Is It Dead? p. 348).

HUMAN IMPACTS ON COASTAL LANDFORMS

Human beings make intensive use of the coastal zone and have major impacts on coastal landforms. Harbors link land and ocean transport systems and many forms of recreation focus on the coastal zone. Increasing numbers of people wish to live in coastal locations, and their activities increasingly affect coastal landform environments. In particular, people try to prevent coastal erosion that affects their own property or livelihood, and some try to create new land.

Attempts to reduce the loss of land by erosion in one area of concentrated human activity often result in losses of land elsewhere. The attempts to protect coastal land include the building of *breakwaters,* which are substantial walls built to shelter harbor entrances. *Groins* are low fences or walls built at right angles across a beach to trap sediment and to prevent the area behind from being attacked by waves. The result of trapping sediment at one place, however, is often to remove it from other places that then become subject to wave attack. *Sea walls* are built to isolate a section of vulnerable coastline from marine action. They may affect wave action by increasing backwash, thus lowering the beach level and undercutting the sea wall.

Changing coastal environments within the United States illustrate some of the impacts of human activities. The coast of San Mateo County, California, which is south of San Francisco, is an example of attempts to protect a coast (Figure 11-26). The resistant granite headland north of Pillar Point contrasts with the softer sedimentary rocks, landslide debris, and unconsolidated river and beach deposits that form the coast around much of Half Moon Bay. The bay coastline is retreating naturally at rates of 3 m (9 ft) per ten years. Building a breakwater in 1959 at the northern end of the bay caused the most intense marine erosion to shift southward and maximum erosion rates then rose to 13 m 346(40 ft) per ten years.

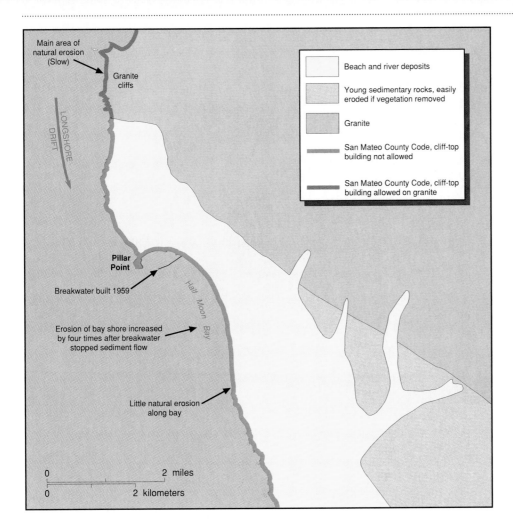

Figure 11-26 Human activities and coasts. The building of a breakwater at the northern end of Half Moon Bay, San Mateo County, California, led to increased erosion in the weak rocks to the south, which caused increased restrictions on building construction.

The Mississippi delta is an area where human actions have intervened in the constant battle between deposition, erosion, and subsidence. The river crosses its flat delta surface in a series of distributaries that have a history of shifting channels (Figure 11-27). In the natural state, high flows along the Mississippi River caused flooding. The overbank flows formed natural levees that grew higher than the backswamp areas between distributaries. At times the river broke through the natural levees, depositing sediment in the backswamp areas and even forming new outlets to the sea.

The first French settlers at New Orleans built small artificial levees to give greater protection to their homes. From the beginning of the twentieth century, the U.S. Army Corps of Engineers constructed a series of massive artificial levees and floodwater outlets to protect the large metropolis of New Orleans and to maintain the shipping channel through its outlet in the most recent of the deltas. The waters of the engineered channel now take most of the sediment into the Gulf of Mexico. Overbank floods are not allowed to occur for most of the time; sediment is no longer deposited on wetlands. The building of dams in the upper tributaries of the Mississippi-Missouri watershed further reduced the supply of sediment to the delta. During the 1980s, the subsidence of coastal lands along the south of the delta became a cause for concern. The reduction of sediment supply upset the balance between deposition and subsidence, and the rate of delta subsidence now exceeds the rate of deposition on the surface.

Humans have also been active in draining tidal lands to provide more land around high-value urbanized areas. Most large cities around the Great Lakes

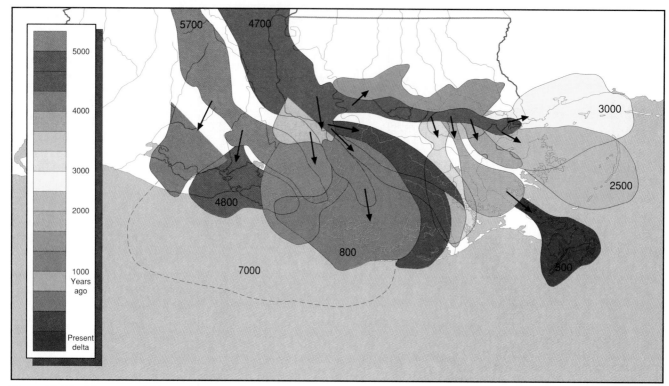

Figure 11-27 Human activities and coasts. The Mississippi deltas have changed much over the last 7000 years. Engineering works preserve the most recent delta and cause it to extend farther south while other sections subside for lack of deposition.

extended their area by building out over the lake. The City of Boston filled in much of its bay, creating the Back Bay residential area in the late nineteenth century and the Logan International Airport in the mid-twentieth century (Figure 11-28).

The Hackensack Meadows in New Jersey were once tidal wetlands covering 72 sq km (28 sq mi) and stretching for 25 km (15 mi) along the Hackensack River. The area was used for sewage works and as a dump for solid waste. In the 1960s, several local jurisdictions worked together to plan its reclamation in a grandiose urban project that involved building the Meadowlands Stadium where the New York Giants now play.

To build Miami and Miami Beach, large sections of mangrove swamp between the outer beach and the mainland were filled. Artificial islands were constructed in the wider southernmost section. More recently, extensive tracts of single-family homes with waterfront access were built along the coast south of Miami. Since the mangroves no longer fulfill their filtering role, there are problems with stagnant water and sewage disposal.

Marine cliffs typically have steep rocky faces and a wave-cut platform at the base (*solid-earth environment*). Beach environments are mostly sandy but may be muddy or pebbly. Beach landforms change rapidly because they are formed in unconsolidated sediment (*surface-relief environment*).

Estuarine and deltaic landforms occur where river water and load meet the ocean. Salt marshes form in midlatitude estuaries and wetlands. Mangroves and reef landforms occur in tropical areas and involve structures built by living organisms (*living-organism environment*).

Climate change causes sea level to rise or fall, leaving coastal landforms at levels not now affected by marine action. Human impacts on coastal landforms locally increase rates of erosion, deposition, subsidence, and the destruction of mangroves, wetlands, or reefs.

(a)

Figure 11-28 Human activities and coasts. (a) The reclamation of land in Boston harbor has reduced the harbor area by half since A.D. 1800. (b) Photo showing the area reclaimed for Logan International Airport.

(b)

ENVIRONMENTAL ISSUE:

The Aral Sea—Is It Dead?

Some of the issues involved in this controversy include:

How could the Aral Sea exist in a desert?

What actions led to the sea shrinking?

What are the outcomes for the people living in the area?

The Aral Sea is an inland sea in the Kazakhstan desert. Its water comes from melting snows in mountains along the Afghanistan border. The sea affects the local climate as it is the source of massive evaporation, and the column of moisture in the air above the sea helps to keep out drying winds. When the moisture is transferred through the atmosphere it produces precipitation on the mountains to the east, and this feeds streams that flow back into the Aral Sea, as the map shows. The sea is thus part of a self-contained and balanced circulation of moisture.

Between 1850 and 1965, the level of the sea was in natural equilibrium and fluctuated only between 50 and 53 m (160 and 170 ft) above world sea level. Moreover, despite rapid evaporation, the waters were only moderately saline (10–14 parts per thousand of salt). The sea was navigable along its 250 km (150 mi) length. Fishing was an important occupation, while the fringing reeds and the vegetation of the Amu Darya delta provided raw materials and pasture.

Changes in the area's hydrology began in the 1960s following an increased intensity of human intervention. The U.S.S.R. began to demand higher productivity from its lands and irrigation was increased to grow cotton. Canals were constructed to take water hundreds of kilometers from the rivers that led into the sea. The irrigated sector of Kazakhstan tripled in area between 1950 and the late 1980s. Flows of water into the Aral Sea fell in this period from 50 km^3—average for 1930–1960—to 35.7 km^3 in 1970, 10 km^3 in 1980, and 1–5 km^3 through the 1980s. By 1988, the sea's level had dropped 12 m (40 ft), the volume of water had decreased by 60 percent, and the surface area had shrunk to half of its 1960 coverage. The salinity tripled. And yet, new water withdrawals are still being planned and built, some to take water outside the watershed draining to the Aral Sea. By early 1990 the sea had split in two, and by A.D. 2010 it may no longer exist. Navigation and fishing have already ceased.

Expanding irrigation had further consequences. Reduced evaporation from the shrinking sea lowered the relative humidity of the air and the precipitation on surrounding hills. As a result, summer temperatures rose and less precipitation fell causing glaciers in the mountains to become sluggish and to supply less meltwater to the rivers. The column of humid air became too weak to prevent dry winds from scouring the region, and they brought salt particles in dust storms. Once the salt particles settled on the soils, the soils became alkaline and millions of acres became unsuitable for growing cotton. Lowering the groundwater levels by pumping from aquifers to make up for the reduction in river water killed vegetation, caused homes to collapse as the ground subsided, and lowered water quality, which led to disease.

Many of these outcomes had been predicted and charted by Soviet scientists and farmers for years, but the government administrators would not listen to them. They were intent on producing millions of tons of cotton at any price. Their folly has led to one of the world's major environmental disasters and rendered much of a region that has witnessed many important events in human history unsuitable for habitation.

Linkages

The Aral Sea was maintained by linkages between the **atmosphere-ocean environment** that evaporated water and carried humid air to deposit precipitation on the mountains, the **solid-earth environment** that raised the mountains, and the **surface-relief environment** that returned the water in rivers to the Aral Sea.

Questions for Debate

1. Is the Aral Sea disaster a result of inadequate knowledge or an inadequate administrative system?

2. Can the disaster be halted at this late stage?

(a)

(b)

(c)

(a) The Aral Sea in Kazakhstan and its surroundings. The sea is a basin of inland drainage in an arid region. It is supplied by meltwater from the mountains to the southeast. Space shuttle views of the Aral Sea in 1985 (b), showing some of the changes since 1960, and in 1989 (c), showing a close-up of the division of the sea into two parts as the level fell. Ice covered the surface at this time.

CHAPTER SUMMARY

1. The sea plays an important part in producing coastal landforms in conjunction with other surface-relief and living-organism environment processes. Waves, tides, and currents are marine processes that act on coasts. The intensity of wave action varies around the world according to the strength and continuity of storm-force winds **(atmosphere-ocean environment).** Variations in the heights of tides cause marine processes to affect different vertical ranges on coasts.

2. Marine cliffs typically have steep rocky faces and a wave-cut platform at the base **(solid-earth environment).** Beach environments are mostly sandy but may be muddy or pebbly. Beach landforms change rapidly because they are formed in unconsolidated sediment (surface-relief environment).

3. Estuarine and deltaic landforms occur where river water and load meet the ocean. Salt marshes form in midlatitude estuaries and wetlands. Mangroves and reef landforms occur in tropical areas and involve structures built by living organisms **(living-organism environment).**

4. Climate change causes sea level to rise or fall, leaving coastal landforms at levels not now affected by marine action.

5. Human impacts on coastal landforms locally increase rates of erosion, deposition, subsidence, and the destruction of mangroves, wetlands, or reefs.

KEY TERMS

coastal zone   *322*

wave   *324*

storm wave   *325*

swell wave   *325*

wave-induced current   *327*

tide   *327*

tidal range   *327*

tidal current   *329*

cliff   *329*

shore platform   *329*

beach   *330*

longshore drift   *332*

barrier beach   *332*

delta   *334*

estuary   *339*

salt marsh   *340*

mangrove   *340*

coral reef   *340*

emergent coastal landform   *343*

submerged coastal landform   *343*

fjord   *344*

CHAPTER REVIEW QUESTIONS

1. What are the typical features of sea waves?

2. To what extent do nonmarine processes mold landforms in coastal environments?

3. Compare the features of storm wave coasts, coasts affected by swell waves, and those that are protected; explain their differences and similarities.

4. How are waves and coastal currents linked?

5. How does tidal range affect coastal landforms?

6. Why do some river mouths have estuaries and others deltas?

7. Describe the main features in the development of the Mississippi delta over the last 5000 years.

8. Describe the features of coral reefs, and explain the differences between fringing, barrier, and atoll reefs.

SUGGESTED ACTIVITIES

1. Observe coasts you visit—lake shores as well as along the ocean. Note the landforms present and try to explain their origins. Notice if human activities affect the coastal landforms.

2. Devise a classification of coastal landforms based on the dominance of erosion, deposition, uplift, or subsidence.

CHAPTER 12

ENVIRONMENTS OF LIVING ORGANISMS

THIS CHAPTER IS ABOUT:

- ◆ Ecosystems: interactions, organization, and scale
- ◆ Energy flow in ecosystems
- ◆ Nutrient cycling
- ◆ Patterns in ecosystems
- ◆ World vegetation regions
- ◆ Human impacts on ecosystems
- ◆ Environmental Issue: Deforestation—Does It Destroy the Environment?

Grassland is one of the major types of vegetation cover on Earth's surface. It forms in conditions of moderate or seasonal rainfall. In midlatitudes it often helps to produce fertile soils that can be plowed to grow grain crops. A farm in German Valley, Illinois, is shown in the inset photo.

arth's plants and animals live in the **bio-sphere,** the thin layer of soil, water, and air at the planet's surface that can support life. The biosphere is formed by the interaction of the atmosphere-ocean environment, the solid-earth environment, and the surface-relief environment. Variation in the nature and interaction of these environments around the world creates spatial patterns among living things. The biosphere is a precariously small part of the planet, but it is home to an amazing diversity of plants, animals, and microscopic organisms. Humans are a small but important part of this living environment and depend closely on its other parts for survival. Understanding the processes that operate in the biosphere can help humans monitor and control their own interactions with the whole Earth system.

ECOSYSTEMS: INTERACTIONS, ORGANIZATION, AND SCALE

The study of the biosphere is an important subdivision of geography called biogeography. Biogeography is closely linked to the science of ecology (the study of living things and their environment). The main difference between the two subjects is the geographer's emphasis on spatial pattern. The central concept of both biogeography and ecology is the **ecosystem.** An ecosystem is a group of living things and the environment in which they live. The living things are linked to each other and to their environment by transfers of energy (food) and other vital materials such as water and soil minerals.

Studying an ecosystem shows how living things depend on each other and on the other Earth environments (Figure 12-1). Green plants, for example,

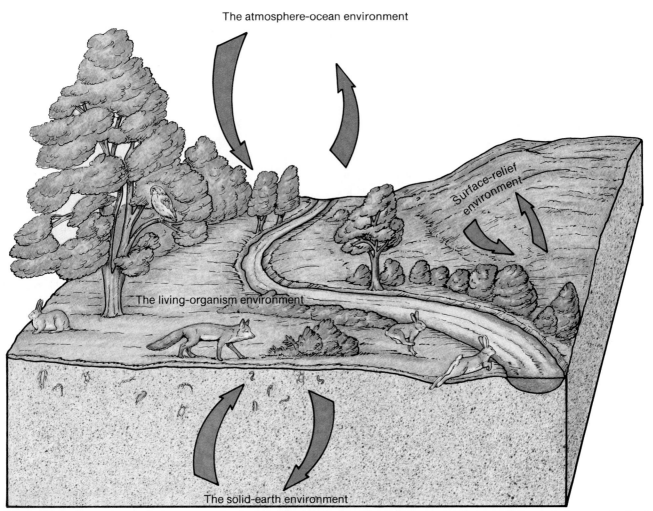

The atmosphere-ocean environment

Surface-relief environment

The living-organism environment

The solid-earth environment

Figure 12-1   Part of a wooded ecosystem in the humid midlatitudes. Plants draw energy from sunlight, and plants and animals exchange gases with the atmosphere-ocean environment. Minerals are weathered from the rocks of the solid-earth environment and drawn up by plant roots. The surface-relief environment delivers water and influences local climate.

make food from sunlight, atmospheric gases, water, and other chemical elements. The amounts of sunlight and water available and their variation over the year depend on a place's climate and therefore on the workings of the atmosphere-ocean environment. The supply of water is also affected by the surface relief; plants on steep slopes, for example, usually have less water around their roots than those in valley bottoms. The amount and type of chemical elements available in the soil depend mainly on the type of rock beneath the surface and how this is weathered (see Chapter 8).

Grazing animals in the ecosystem eat green plants and draw on the energy and chemical elements stored in the plant tissue. Meat-eating animals eat the grazers. Plant and animal waste and dead tissue fall to the ground at all stages in the chain. Other organisms eat the waste and dead tissue and break it down into chemical elements that often enter the soil and can be taken up again by plants.

The amounts of energy and nutrients available at a place determine what plants and animals can live there. In the tropical rainforest, for example, solar energy and rainwater are abundant all year round. These conditions allow many types of green plant to flourish. The plants, in turn, provide food for thousands of organisms. In contrast, only a few plants can survive the dry climates of desert regions, and there is less energy available for other organisms. Energy and nutrient transfer are, therefore, two very important processes in the biosphere; they link together different parts of the ecosystem and help to decide the structure and distribution of its living components.

The idea of the ecosystem can be used at any scale (Figure 12-2). At a global scale the world can be divided into large regions or **biomes** in which the climate, soil type, and vegetation are more or less uniform. Within a biome there will be some variation due to patterns in slope, aspect, or elevation. The plants themselves also affect the environment. Evergreen forests, for example, create acid soils and year-round shade at ground level. Few plants can live beneath them. In contrast, trees that lose their leaves at some time of the year let light penetrate to the forest floor, create less acid soil, and plants can grow beneath them.

Studying ecosystems sounds quite easy. In practice it is difficult to decide what is "in" and what is "out" of an ecosystem. Migrating birds, for exam-

ple, link ecosystems on different continents and the seeds of plants can be carried many thousands of miles. As an example, Figure 12-3 shows some alternative ways of defining a pond ecosystem.

Earth's plants and animals live in the biosphere, the thin layer of soil, water, and air at the planet's surface that can support life. The biosphere is formed by interactions among the **atmosphere-ocean, solid-earth,** and **surface-relief environments.** It is organized by the transfer of energy and chemical materials among these environments and living things. The ecosystem is an important framework for studying the biosphere. It can be applied at any spatial scale but may be difficult to delimit in practice.

ENERGY FLOW IN ECOSYSTEMS

Everything that "works" needs a supply of energy. A car needs a tank of gasoline; a lightbulb needs electricity; our bodies need food. All living organisms in the biosphere need energy to go about their normal, everyday work of eating, breathing, and reproducing.

PHOTOSYNTHESIS, FOOD CHAINS, AND FOOD WEBS

All organisms in the biosphere, including meat-eaters, gather their energy from green plants. Green plants carry out a complex chemical process called **photosynthesis,** in which the raw materials of water, carbon dioxide, and the sun's energy are used to build molecules of energy-rich carbohydrates, usually sugar (Figure 12-4). Oxygen is given off as a by-product. Photosynthesis is fundamental to life on Earth; without photosynthesis there would be less oxygen in the atmosphere and no usable energy for organisms in the ecosystem.

The energy fixed by photosynthesis is stored in the chemical bonds among the carbon, hydrogen, and oxygen atoms in the carbohydrate. It is released in the chemical reaction of respiration. Respiration breaks down the carbohydrate molecule to simpler parts, usually with the help of oxygen. The energy released from the chemical bonds is used for work in the body and eventually becomes heat. It is the chemical reaction of respiration that raises body temperature after exercise. Carbon dioxide is a by-product of respiration.

(a)

(b)

(c)

(d)

Figure 12-2   The nested ecosystems of the biosphere. The northern coniferous forest (a), the tropical rainforest (c)–(d), and the tropical grasslands (b) shown here are three of the regional ecosystems or biomes. Each biome can be subdivided into smaller ecosystems supported by slightly different niches in the physical environment. In the tropical forest, for example, some plants grow on the branches of standing trees; others live in rotting trunks on the forest floor.

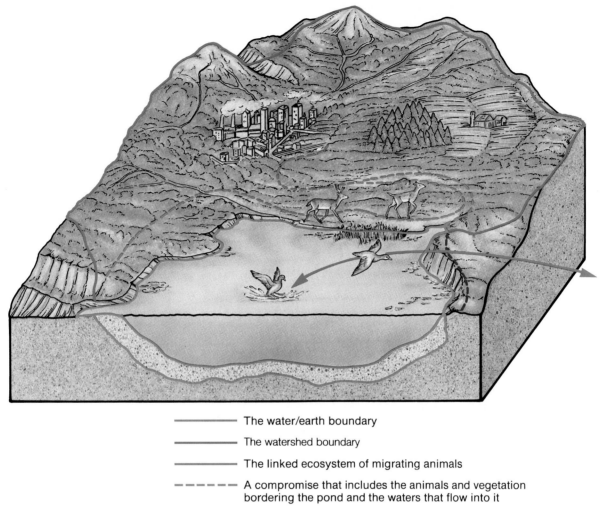

──────── The water/earth boundary

──────── The watershed boundary

──────── The linked ecosystem of migrating animals

- - - - - - A compromise that includes the animals and vegetation
bordering the pond and the waters that flow into it

Figure 12-3   Different ways of defining a pond ecosystem. What other boundaries could be used?

Photosynthesis is carried out by green plants on land and by tiny floating plants and algae in both fresh and salt water. All these organisms contain a green pigment called **chlorophyll** which, when hit by sunlight, triggers the chain of chemical reactions that is photosynthesis. Plants that contain chlorophyll and can photosynthesize are called **autotrophs,** or self-feeders, because they make their own food using the sun's energy. All other organisms are called **heterotrophs** (other-feeding) and take their food in a ready-made form from the autotrophs.

Energy is passed to the heterotrophs along two pathways (Figure 12-5). First, energy can be passed to animals that live at least partly above the ground. In this pathway the plants are eaten by **herbivores,** which are animals such as rabbits, bison, or gazelles that eat only plants. A herbivore may be caught and eaten by a **carnivore** (an animal such as a fox, a lion, or a leopard that eats only meat) or an **omnivore** (an animal such as a human that eats both plants and animals). The plants and animals linked by this transfer of energy form a **food chain.** Each organism in the chain builds the energy stored in plants or its prey into its own body for future use.

The second pathway for energy transfer takes place mainly underground. It starts when a plant or animal excretes waste or dies before being eaten by the next link in the food chain. On land the dead tissue then becomes food for the army of microscopic organisms in the soil, mostly bacteria and

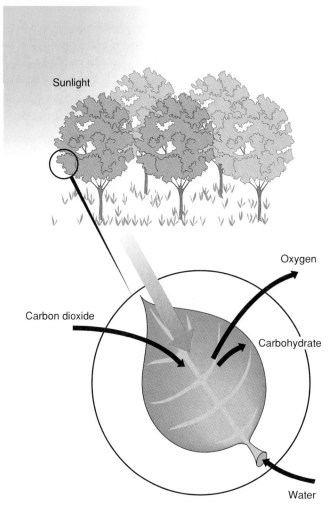

Figure 12-4   The process of photosynthesis in green plants. Carbon dioxide is drawn from the air; water usually comes from the soil. Sunlight starts a chemical reaction that produces carbohydrates and oxygen.

fungi. Organisms that feed on dead and decaying tissue in the ecosystem form their own food chain and are given the general name of **decomposers.** A similar waste-disposal system operates in the oceans.

In most ecosystems there is a variety of autotrophs and therefore a choice of menu for the herbivores. Similarly, there is usually a choice of grazing animals for the carnivores and so on. In practice, then, the food chain is more accurately described as a diffuse **food web** (Figure 12-6). Being able to take food from a number of sources is a less risky strategy for most organisms; if one type of food is wiped out, there is always an alternative.

The giant panda (Figure 12-7) is at risk from not adopting this strategy. The panda feeds almost exclusively on bamboo. At intervals as far apart as sixty years, the entire bamboo population flowers simultaneously. The plant then dies and the pandas are at risk from starvation. This situation last arose in 1984 when only a massive international rescue operation saved these vulnerable creatures.

The transport of energy from the sun to living organisms in the above-ground food chain does not use all the energy available. Therefore, the process is not totally efficient. To start with, much of the energy radiated by the sun is filtered out by the

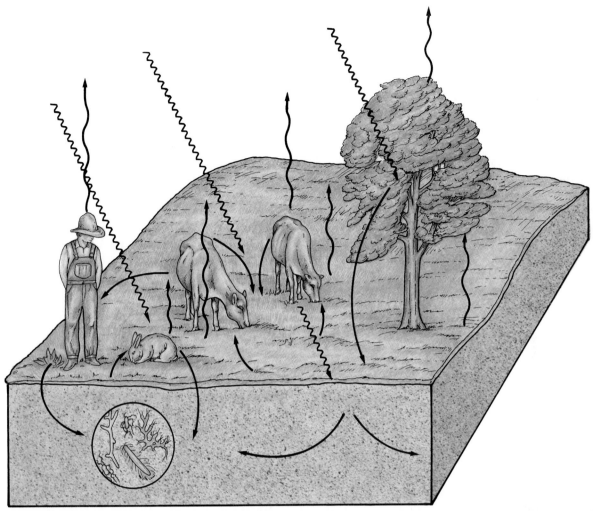

Figure 12-5 Food chains. The green plants, the grazing cows, and the farmer are linked together in a food chain above the ground. The decomposers gain their energy from dead plant and animal tissue, mainly in the soil.

atmosphere or is reflected back to space from clouds and Earth surfaces (see Chapter 3). Second, chlorophyll is sensitive to light in only the blue and the red parts of the visible spectrum. Autotrophs can therefore use only a fraction of the energy that reaches them. Third, not all the energy captured at one level of the food chain is available to the next. Animals use up some of their energy in day-to-day living, and many individuals are food for decomposers rather than the next link in the above-ground chain. Overall only about 10 percent of the energy available at one level in the food chain is passed on to the next. This rapid reduction limits the number of links in a food chain to about three or four and has important implications for human diets. In many countries crops are grown specifically to feed animals such as cows, sheep, and pigs, which humans then eat. It is much more energy-efficient for humans to eat grasses and grain plants directly or to eat animals like earthworms that are more efficient converters of energy.

ECOSYSTEM PRODUCTIVITY

The autotrophs are also called the ecosystem's primary producers. The **net primary production** of an ecosystem is the amount of energy the primary

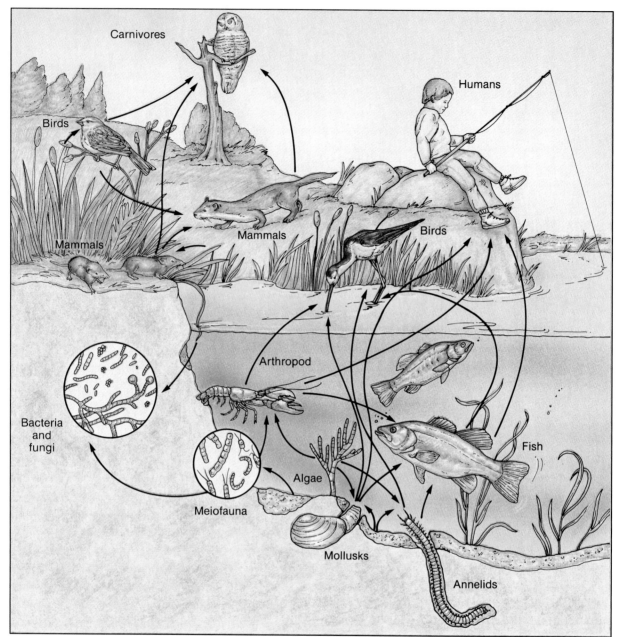

Figure 12-6 The food web in a pond ecosystem. The human and some of the birds are omnivores and feed on both plants and animals.

producers have fixed over and above their own requirements. It can be thought of as a credit balance of energy that is available for other organisms to use. It is "visible" in the ecosystem as plant growth. Net primary production figures are a useful way of comparing the energy-fixing abilities of different ecosystems (Figure 12-8, Table 12-1). Production is highest in the tropical rainforests, where water, sunlight, and carbon dioxide, the raw materials of photosynthesis, are plentiful. Figure 12-8 shows that productivity over land broadly follows the pattern

Energy enters the ecosystem through the process of photosynthesis and is released by respiration. Photosynthesis creates energy-rich carbohydrates from water, carbon dioxide, and solar energy and produces oxygen (**atmosphere-ocean environment**). Photosynthesis is carried out by green plants (autotrophs) in water and on land. The autotrophs' energy is transferred to other organisms (heterotrophs) through the food web. The transfer of energy is not totally efficient.

Figure 12-7   The giant panda, symbol of the Worldwide Fund for Nature. There are now thought to be less than 1200 pandas living in the wild. The panda's natural habitat is the bamboo forests of southwestern China, and the bamboo is virtually its only source of energy. Human use of the forest has restricted the panda's range, and when the bamboo flowers (as described in the text) the pandas face starvation.

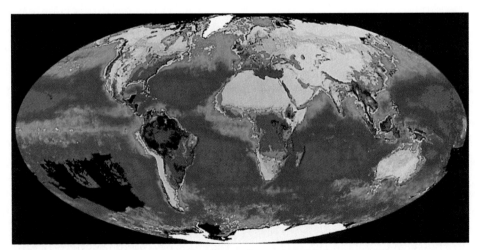

Figure 12-8   This composite satellite image shows the global patterns of primary productivity. On land green indicates high productivity. In the oceans, areas of high productivity are shown in orange and red.

TABLE 12-1   NET PRIMARY PRODUCTION IN BIOME ECOSYSTEMS

Biome	Area (10^6 km^2)	Average NPP g/m^2/yr	% World NPP
Tropical rainforest	17.0	2200	21.6
Tropical seasonal forest	7.5	1600	6.9
Temperate forest: deciduous	5.0	1300	3.7
Temperate forest: evergreen	7.0	1200	4.8
Boreal forest	12.0	800	5.5
(Mediterranean) wood and scrub	8.5	700	3.5
Tropical grassland	15.0	900	7.8
Temperate grassland	9.0	600	3.1
Tundra and alpine tundra	8.0	140	0.6
Desert and semidesert scrub	18.0	90	0.9
Extreme desert: rock, sand, ice	24.0	3	0.04
Cultivated land	14.0	650	5.2
Swamp and marsh	2.0	3000	3.5
Lakes and streams	2.0	400	0.5
Oceans	361.0	155	32.3

TABLE 12-2   COMMON NUTRIENTS AND THEIR AMOUNTS IN THE HUMAN BODY

Element and Symbol	Approximate % of Earth's Crust by Weight	% of Human Body by Weight
Oxygen (O)	46.6	65.0
Silicon (Si)	27.7	trace
Aluminum (Al)	6.5	trace
Iron (Fe)	5.0	trace
Calcium (Ca)	3.6	1.5
Sodium (Na)	2.8	0.2
Potassium (P)	2.6	0.4
Magnesium (Mg)	2.1	0.1
Hydrogen (H)	0.14	9.5
Manganese (Mn)	0.1	trace
Fluorine (Fl)	0.07	trace
Phosphorus (P)	0.07	1.0
Carbon (C)	0.03	18.5
Sulfur (S)	0.03	0.3
Chlorine (Cl)	0.01	0.2
Vanadium (V)	0.01	trace
Chromium (Cr)	0.01	trace
Copper (Cu)	0.01	trace
Nitrogen (N)	trace	3.3

of global climate. Over sea the areas of highest productivity are where upwelling currents bring nutrients to the surface.

NUTRIENT CYCLING

Nutrients are chemical elements needed for life. *Macronutrients,* such as carbon, hydrogen, and oxygen, are those needed in large quantities (Table 12-2). The combination of hydrogen and oxygen as water is particularly important. In the human body, atoms found as water molecules outnumber those in all other molecules, and life stops very quickly if the water supply is cut off. *Micronutrients,* or trace elements, are those needed in only very small quantities. Micronutrients important to humans include cobalt (a part of vitamin B12), zinc, and molybdenum (both parts of enzymes).

The flows of energy and nutrients in the ecosystem are closely linked. This is particularly clear for carbon, hydrogen, and oxygen, the building blocks of carbohydrates. There is, however, one important difference. The movement of energy in the ecosystem is one-way. The supply is constantly replenished by the sun. Energy is fixed into the food web, used by the

organisms in it, and ultimately released into the atmosphere as heat. Once released as heat, the energy cannot be reused by the organisms in the ecosystem. In contrast, the supply of chemical elements in Earth environments is finite, and they must be recycled.

BIOGEOCHEMICAL CYCLES

Nutrients in the ecosystem move in a continuous cycle from the nonliving to the living environment and back again (Figure 12-9). The pathways along which they move are called **biogeochemical cycles.** *Bio* refers to the living phase of the cycle in organisms, and *geo* refers to the nonliving phase in soils, rocks, and water.

Nutrients enter living things by several routes. Green plants exchange gases (carbon dioxide and oxygen) with the atmosphere-ocean environment in photosynthesis. Plants and animals do the same in respiration. Nutrients are released into the soil by rock weathering and may be taken up by plant roots. Nutrients leave living things through respiration, photosynthesis, and decomposition. Nutrients released into the soil can follow one of two routes.

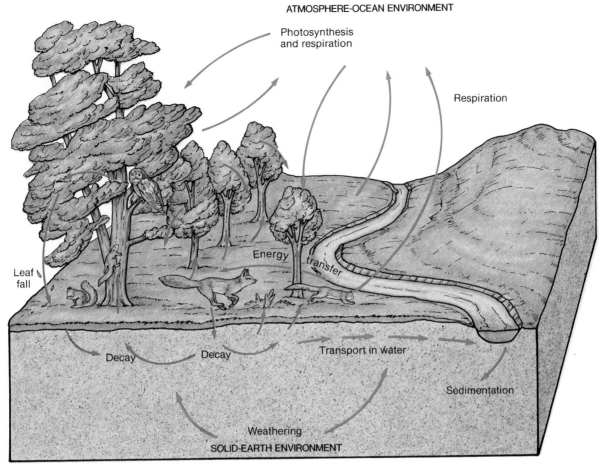

ATMOSPHERE-OCEAN ENVIRONMENT

Photosynthesis
and respiration

Respiration

Energy transfer

Leaf
fall

Decay　　Decay　　Transport in water

Sedimentation

Weathering
SOLID-EARTH ENVIRONMENT

Figure 12-9 Biogeochemical cycles. Nutrients are exchanged among the atmosphere-ocean environment, the biosphere, and the solid-earth environment. In this diagram blue arrows represent transfers to and from the atmosphere-ocean environment, and red arrows represent transfers to and from the solid-earth environment. Green arrows represent transfers within the living-organism environment.

First, the nutrient can stay in the soil and be taken up by a plant root to reenter the *bio* part of the cycle. Second, rainwater can wash the nutrient out of the soil in solute form so that it enters a stream. From there it may enter a lake or sea and eventually become part of a solid rock again. The world's stock of nutrients consists of individual atoms in different positions in the cycle and at different places around the world.

Nutrients are more numerous and spend longer periods of time in some parts of the cycle than others. In general the atmosphere-ocean and the solid-earth environments offer long-term storage for large quantities of nutrients. These reservoirs are not easily depleted. Plants and animals, dead or alive, and soil hold smaller and more variable amounts of nutrients for shorter time periods.

Biogeochemical cycles show how the solid-earth environment is linked to living things. Sediments are laid down and consolidated to form solid rock. The

rocks are uplifted by tectonic activity (see Chapter 7), exposed to the atmosphere, and weathered (see Chapter 8). Some of the molecules and atoms released by rock weathering find their way into the soil and sometimes into the bodies of living things. The atoms that make up your bones may once have been part of a mountain range.

GASEOUS AND SEDIMENTARY CYCLES

The precise pathway of each nutrient depends on its main reservoir in the biosphere. Nutrients in *gaseous* cycles move mainly between living things and the atmosphere-ocean environment. Gaseous cycles operate on a global scale. A local excess or deficit is soon balanced by transport within the atmosphere-ocean system. Nutrients that move in a gaseous cycle are therefore about equally abundant everywhere in the world. They include the macronutrients of carbon, nitrogen, oxygen, and hydrogen (see Chapter 2).

Figure 12-10 illustrates the gaseous cycles of oxygen and nitrogen. Oxygen is usually exchanged directly between the atmosphere or ocean and living things, either as the gas oxygen or as water. In contrast, atmospheric nitrogen cannot be used directly by most plants and its exchange depends on the work of specialized bacteria. Nitrogen-fixing bacteria in the soil convert atmospheric nitrogen to a form usable by plants, nitrates. Some of the bacteria live on the roots of a group of plants, called *legumes,* that includes peas, beans, and clover. The bacteria fix more nitrogen than the legumes will need. Legumes are therefore often planted in rotation with other crops to avoid or decrease the need for nitrogen fertilizer. The nitrogen taken up by plants is eventually returned to the soil as dead tissue. It is broken down to nitrates by a second specialized group of bacteria. The nitrogen can then be reused by plants or released to the atmosphere by a third group of bacteria.

Nutrients in sedimentary cycles are exchanged between living things and the solid-earth environment and are usually micronutrients or trace elements. Sedimentary cycles are controlled by the processes of sedimentation and weathering (see Chapters 7 and 8) and therefore operate over long timescales. Variation in the mineral content of rocks and their rate of weathering may cause local excess or deficit in these nutrients. For example, phosphorus (Figure 12-11) enters the soil very slowly but is easily washed away by rainwater. Phosphorus that enters the ocean can be brought back to the land through seabirds; the birds' deposits are mined directly as a fertilizer.

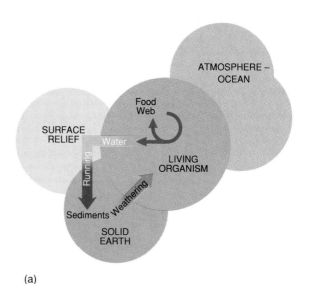

(a)

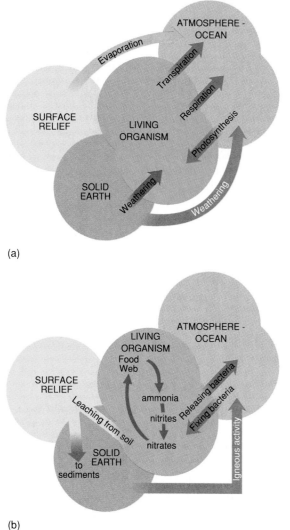

(b)

Figure 12-10 Gaseous cycles. (a) Oxygen cycle. (b) Nitrogen cycle. How do these two cycles differ?

(b)

Figure 12-11 Sedimentary cycles. (a) The phosphorus cycle. (b) Phosphate-rich sediments are formed by the concentration of *guano* (bird droppings) over hundreds of years. These pelicans are on a small island in the Gulf of California, Mexico. The phosphorus in their droppings comes from the fish they eat; the fish in turn gain phosphorus from the ocean waters.

Nutrients are chemical elements necessary for life. The supply of nutrients on Earth is finite and is constantly recycled. Nutrients are exchanged among living things and the surrounding Earth environments. Most nutrients have one main reservoir, either the **solid-earth** (most micronutrients) or the **atmosphere-ocean system** (most macronutrients).

PATTERNS IN ECOSYSTEMS

Conditions in the local physical environment control where plants live and with which other plants they live. Each plant has specific requirements for water, sunlight, physical support, and a whole range of other chemical and physical conditions. If a particular patch of ground can supply these requirements, a plant can grow there. If the conditions are not met, the plant cannot live there. The interaction of plants with their environment leads to patterns of vegetation distribution in space. The distributions change over time as the environment changes.

Green plants are at the start of the food chain. What plants grow where and in what numbers determine the types and numbers of heterotrophs that can be supported in the ecosystem. The whole structure of the ecosystem, therefore, depends on the physical nature of the biosphere at a place and how plants interact with it and with one another.

PATTERNS IN SPACE

Every plant grows most strongly in a narrow ideal range of environmental conditions, for example, temperature or water availability. A plant can live in conditions slightly outside these limits, but will grow less strongly. The set of ideal and tolerable conditions varies for each plant, which is why some plants live only in boggy places and others flourish on well-drained soils, and why plants common in the tropics are not found near the poles. Most plants can also survive short periods outside their limits in, for example, the case of a long drought or a late frost.

Plants do not, however, grow in isolation. They must compete with other plants for the sunlight, water, and chemicals they need and, in spatial terms, for the sites where these resources are in best supply. Plants grow in places where the environmental conditions are within their ideal or tolerable ranges and where they can compete successfully with other plants. This set of conditions is a plant's **ecological niche.** The biosphere is a mosaic of such niches, each formed by a combination of climate, local conditions of relief, the availability of nutrients, and the actions of plants themselves. Certain niches occur again and again in the landscape, and the set of plants best suited to them will also be repeated. A set of plants growing together under similar environmental conditions is called a *plant association*. The distribution of niches and plant associations in the biosphere leads to spatial patterns in natural vegetation from the local to the global scale (Figure 12-12 and see Figure 12-2).

PATTERNS IN TIME: ECOLOGICAL SUCCESSION AND EVOLUTION

Ecosystems grow and change over time; such change is called **ecological succession.** Succession is driven by changes in the autotrophs' ecological niche and can arise in two ways. First, the environment can change through an outside force such as a forest fire or a landslide. Such changes are rare and are usually sudden and catastrophic. Second, the plants themselves can alter their site. Plant-guided changes occur slowly, everywhere, all the time.

A plant adds organic material, such as dead plant tissue, to the soil while it is growing. Extra organic matter allows the soil to hold more water and nutrients and make them available to the plants. The local climate of sunlight, air movement, and humidity also change as the plant grows. In total, these changes often mean that the plant's home is no longer tolerable for its own offspring. A different species will provide the next generation of plants at this site. One species replaces another in this way until further modification of the environment is so limited that the species "in post" can perpetuate itself. At this point succession will only start again if some outside variable, such as a forest fire or a change in climate, triggers it.

The combination of continuous, gradual change in the environment and occasional catastrophes cause ecologic succession. After succession lasting hundreds of generations, very different surfaces can support similar plant associations and ecosystems (Figure 12-13).

"Young" and "mature" ecosystems have rather different characteristics of energy flow and nutrient

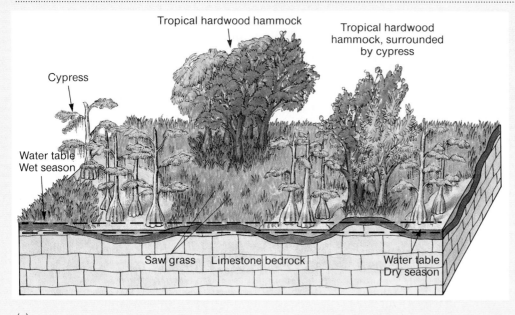

(a)

Figure 12-12   Plant associations in the Everglades, Florida. (a) The presence of water is the single most important control on plant associations in the freshwater part of the Everglades. Tropical palms and hardwood trees grow in hammocks in places where the limestone bedrock is just two feet above that of the surrounding grassland. Cypress heads develop where the bedrock surface is slightly lower than its surroundings, creating waterlogged conditions all year round. Cypress heads can be a dry season refuge for alligators. (b) Hardwood hammocks and cypress heads in the saw-grass prairie of the Everglades.

(b)

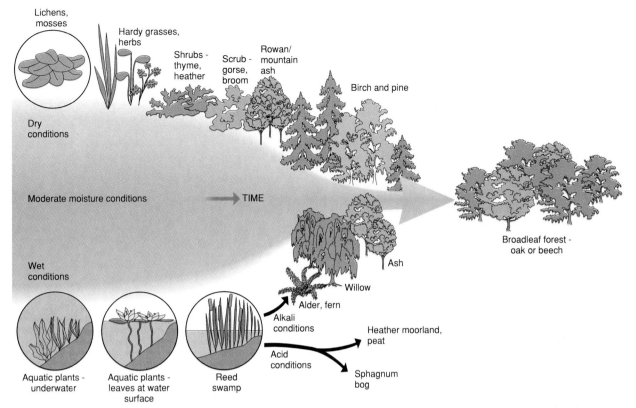

Figure 12-13   The process of ecological succession. A very dry area and a very wet area may, over time, develop similar ecosystems.

cycling. A young system contains only a few species and a short, simple, food web. Many of the species are adapted to life in stressful and unstable environments. They pour much of their energy into producing seed to ensure the survival of the next generation and into growing as quickly as possible. Young ecosystems have a very high energy-fixing ability (a high net primary production).

Mature ecosystems have a much less frenzied existence. Individual organisms tend to be bigger, live longer, and produce proportionately less seed. There are many more ecological niches, more species, and a complex food web. The amount of energy fixed by the system more or less matches its needs and the rate of growth is slow. A mature ecosystem has developed **homeostatic mechanisms** that help maintain its equilibrium. For example, a diverse food web means that if one species is wiped out, most heterotrophs have an alternative food supply (Figure 12-14). The store of nutrients in the organisms and the soil is a buffer against a disrupted supply from rock weathering.

Ecological succession acts over timescales of tens to hundreds or thousands of years, depending on the life spans of the organisms involved. Over longer time periods the distribution of living things around the world has changed more dramatically.

Plants interact with each other and with their local environment. They grow in locations that provide an ideal or tolerable set of resources and in which they can compete successfully with other plants. Recurring ecological niches in the biosphere contain similar plant associations.

Over time, some plants modify their environment so that it is no longer possible for their offspring to live there. External conditions such as a landslide also create new environments for colonization. A new species takes over in the process of ecological succession. Succession stops when the environment is stable.

PATTERNS IN TIME: EVOLUTION AND ISOLATION

Many organisms seem well-suited to their physical environment. Chimpanzees, for example, are at home in the trees of tropical forests, and the shaggy white coat of the polar bear protects it from the severe cold of the arctic winter. There are, however, some puzzling gaps and anomalies in the matching of organisms and niches. Apes are found in Africa but not in the similar forests of South America. Polar bears live only in the Arctic and penguins only around Antarctica. Kangaroos and other pouched animals, called marsupials, are

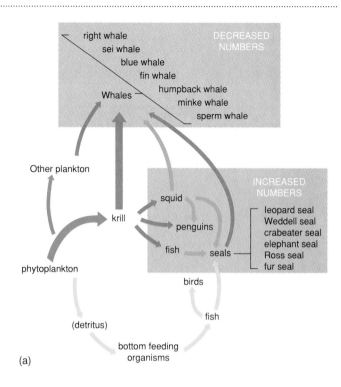

Figure 12-14   Food webs in the Southern Ocean. (a) Overfishing of whales in the Antarctic has led to an oversupply of krill, the tiny shrimplike creatures that whales eat. The numbers of seals and penguins have increased as a result. There are plans to increase the direct harvesting of krill. What effect will this have on the recovery of the whale population? (b) Southern fur seals on the South Georgia Islands in the South Atlantic. This species has increased its population as the number of whales has reduced. The large number of seals coming ashore is, however, damaging the moss and lichen carpet that covers part of the ground surface.

(b)

found mainly in Australia although they could live in similar niches on other continents. In total, some organisms are clearly suited to the places in which they live but are absent from other, very similar places from which they are geographically separate.

This distribution of living things cannot be explained solely by the ideas of ecological niches and ecological succession. The current distribution of Earth's organisms is controlled by two other, interrelated processes. The first is the evolution of different forms of life over time. The second is the isolation of groups of plants and animals by barriers such as land, sea, deserts, and high mountains as the solid-earth environment pushes the continents around over Earth's surface.

Ecological succession and evolution are very different concepts. Ecological succession gives existing organisms the opportunity to move in and become part of an ecosystem. **Evolution** is the development of new species or new forms of life. A *species* is a type of living organism that is able to interbreed with others of the same species and produce fertile offspring. Each known species is described by two Latin names. The full name for the human species, for example, is *Homo* (Latin for man) *sapiens* (Latin for wise).

Each living organism carries the details of its size, shape, metabolism, and all its other characteristics in its genes. In reproduction the genes are copied and passed on to the next generation. A slight error in copying the genes can lead to the birth of a slightly different form of life. If this form of life is successful, it will flourish and will pass on the winning genes to its own offspring. Over time and several generations, a slight tendency to, for example, a long neck, can lead to the development of a new species like the giraffe. Charles Darwin, a scientist and naturalist who lived in the 1800s was the first to publicly recognize this process. He called it *survival of the fittest* because the surviving offspring and species were those most fit to use the environment around them.

The distribution of species in the world today is the result of evolution and the changing patterns of oceans, continents (see Chapter 7), and sea level (see Chapter 6). The marsupials, for example, are thought to have evolved in South America and travelled to Australia when these continents were still linked by a relatively warm Antarctica. The marsupials left behind in other continents were eventually replaced by the more competitive placental mammals, whose young develop within the body, that evolved later. By this time Australia was a separate continent and the placental mammals could not reach it. The isolated Australian marsupials had no competition and have survived and

flourished. In the last glaciation when sea level was low, the formerly isolated Alaska and Siberia were connected across the Bering Straits. Arctic species were able to migrate between the American and Eurasian continents. This land bridge is thought to be the route by which *Homo sapiens* entered the North American continent.

Ecological succession, evolution, and the changing positions of land and ocean cause changes in the distribution of species over time. Ecological succession leads to change in the ecosystem at a place as new species, better adapted to the environment, replace others. Evolution leads to new forms of life (*living-organism environment*). Movements of Earth's crust (*solid-earth environment*) distribute species around the world and, as in the case of the marsupials, may isolate them from competition with species that evolve later.

WORLD VEGETATION REGIONS

The world is divided into a series of large, regional ecosystems, or biomes, that occupy one level in the biosphere's spatial hierarchy (see Figure 12-2). At this scale climate controls the basic nature of the biosphere, and the other Earth environments create local variation. In each biome the climate and soil are more or less uniform and so is the type, or growth-form, of the dominant vegetation. *Growth-form* describes the physical size and shape of plants. The most common division of growth-form is into trees, shrubs, and herbs. Trees are plants with a single woody stem (trunk) that raises most of the green vegetation above the surface. Shrubs are also woody plants but are smaller than trees and have stems that branch near the ground. Herbs are small plants, such as grasses, that grow close to the ground and have no woody tissue.

Because the type and numbers of green plants determine the structure of the whole ecosystem, biomes are usually described by the characteristics of their dominant vegetation. The most important control on the growth-form of vegetation is climate, particularly the annual and seasonal patterns of temperature and precipitation (Figure 12-15). Tropical rainforests exist over a range of rainfall conditions, for example, but need a specific range of temperatures. In contrast, grasslands exist over a range of temperatures but are limited by their water requirements. The biomes are distributed around the world as broad bands roughly parallel to the equator (Figure 12-16). The pattern is distorted by the distribution of land and sea and by mountain ranges.

Figure 12-15 The conditions of temperature and rainfall that delimit each biome.

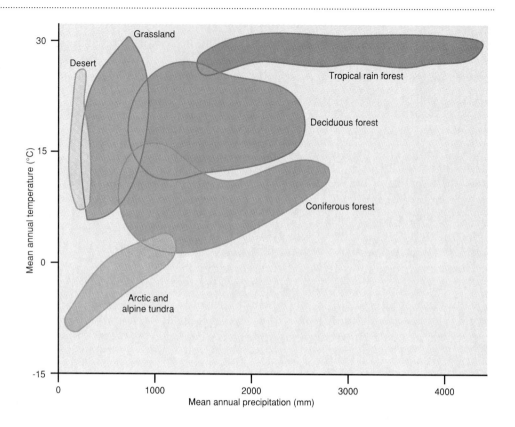

Figure 12-16 The distribution of world biomes.

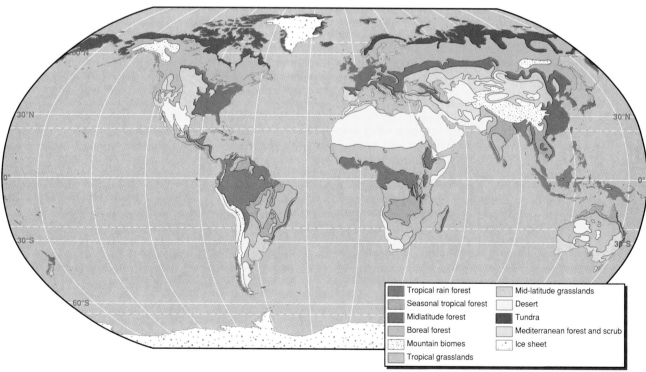

PATTERN AND DIVERSITY IN BIOMES

The characteristic climate, soils, and vegetation of a biome are strongly related to each other (see Chapter 13 for discussion of soils). They are, however, slightly out of time with each other because soils and vegetation react slowly to changing conditions of climate. Many present-day plant communities, for example, are still adjusting to climatic changes in the 10,000 years since the last glacial phase. If Earth's climate remains stable for a few million years the processes will catch up with each other and the soils and vegetation in a biome will become a better mirror of its climate.

Biomes are usually described and mapped as areas of uniform conditions that have clear boundaries on the ground. This is, of course, a considerable simplification. Temperature usually varies from north to south within a biome, and there is often a strong east to west moisture gradient across midlatitude continents. Individual biomes blend gradually into their neighbors. The typical conditions of climate, soils, and vegetation in a biome, therefore, provide a general overview and change gradually in all directions from its center. Exactly what grows where, on a local scale, still depends on how the green plants interact with each other and their environment.

Almost all biomes have gaps in them where humans have altered the natural vegetation. Some biomes, like midlatitude deciduous woodlands, have been almost completely destroyed. The descriptions that follow are of how the world vegetation regions would exist in a completely natural state.

Biomes are regional ecosystems recognized and described by the growth-form characteristics of their dominant vegetation. Biomes are usually thought of as more-or-less uniform areas. Their environmental conditions, however, vary gradually over space and their boundaries are not easily recognizable on the ground.

FOREST BIOMES

Forests are the most complex and productive of the world biomes. They are also the most demanding on the local environment. Trees need to fix large amounts of energy to live, so they grow only in areas where there is a good supply of water and sunlight, the raw materials for photosynthesis, and enough nutrients in the soil.

Tropical Rainforest   **Tropical rainforests** are found at and around the equator, where sunshine, carbon dioxide, and water are in abundant supply all year. The tropical rainforests are the world's most productive natural ecosystem, and they contain a great diversity of species at all levels in the food web. The autotrophs in the rainforest grow and reproduce all year round and are arranged into three or four vertical layers (Figure 12-17). The upper layers of vegetation are inhabited by birds and bats, the middle layers by different birds and by climbing animals like sloths, monkeys, and squirrels (Figure 12-18). Only a small amount of light filters through to the forest floor, so the shrub and herbaceous layers are patchy. Common houseplants, like the African violet, grow wild in these gloomy conditions. Animals such as deer and pigs live on the ground, eating fruits and other materials that drop from the upper layers. Thousands of species of insect are found at all levels. Tropical rainforests are valuable to humans and are under increasing pressure from agriculture and logging operations (see Environmental Issue: Deforestation—Does It Destroy the Environment? p. 385).

Tropical Seasonal Forest   **Tropical seasonal forest** is a transitional type of vegetation between tropical rainforest and tropical grassland; it is usually found poleward of the tropical rainforest. The main characteristic of its vegetation is the change from a continuous cover of "evergreen" tree species in the rainforest to a more patchy cover of deciduous trees, which lose their leaves at some time in the year, in the areas with a low total rainfall or a marked dry season. Total rainfall decreases, and both temperature and rainfall vary

Above ground

(a)

Figure 12-17   Stratification of vegetation in the rainforest.
(a) Species at different levels in the canopy need different
amounts of sunlight. (b) From above, the rainforest canopy is
densely packed.

(b)

(a)

(b)

(c)

Figure 12-18 Animal life in the tropics. (a) The three-toed sloth pictured here uses its claws to hang from a tree, sound asleep for about 18 hours out of 24. Its ungroomed coat is a home to green algae, which are food in turn to moths. (b) In contrast to the sloth, the spider monkey is more active and can move rapidly from tree to tree in search of food. (c) Birds of paradise are among the colorful bird species found in the upper canopy of the rainforest.

more markedly with the seasons with increasing distance from the equator. The structure of the forest changes gradually as the trees adapt to survival through the increasing length or intensity of the dry season. At the wetter end, some features of the rainforest's vertical structure remain although the tree canopy is more open. At the drier end, stunted trees are grouped together in dense thickets. Here the trees may be deciduous or develop small, almost watertight leaves that are kept all year round.

Midlatitude Forests The **midlatitude forests** overlap in moisture requirements with the tropical forests but have considerably lower annual temperatures that also fluctuate dramatically with the seasons. Most of the trees in this biome are deciduous, dropping their leaves in fall with spectacular displays of color (Figure 12-19a). Leaf-fall in these biomes is related to the seasonal drop in temperature. The

temperature change and associated lack of food also affect organisms higher up the food web. Some birds migrate to warmer countries. Other animals, like squirrels, hibernate or feed from supplies stored in the summer months.

Midlatitude forests on different continents have similar growth-forms but contain different species. Hickory and maple, for example, are common in North America but are rare in Europe where beech and oak are dominant. Local associations of species, such as the oak-hickory forests of the piedmont plateau east of the Appalachians, are common in particular combinations of climate, soils, and topography. All midlatitude forests have the same basic structure of an open canopy and little vertical stratification of the tree layer (Figure 12-19 b). The combination of good soils, hardwoods, and opportunities for hunting made the midlatitude forests an attractive option for early settlers and farmers.

Figure 12-19 Midlatitude forest. (a) Fall foliage, Croton, New York. Leaf-fall in midlatitude forests creates a spectacular display of red and yellow shades. What might cause the change in color? (b) The open canopy of these deciduous trees allows sunlight through to the forest floor.

(a)

(b)

Mediterranean Forest and Scrub   As the name suggests, the typical area for this biome fringes the Mediterranean Sea, but it also occurs in California, South America, southern Africa, and southwestern Australia. **Mediterranean forest and scrub** develops where, compared to the midlatitude forests, winters are mild and rainy and summers are hot and dry. Where trees can survive, they have small toughened leaves that help to reduce summer water loss. Elsewhere the dominant vegetation is dense thickets of shrub species (Figure 12-20), called *chaparral* in America. Fire, started accidentally or deliberately, is an important control on vegetation in this biome. The scrub species regenerate rapidly after fire from buried seeds or directly from the burnt roots and stems. The rapid and vigorous regrowth prevents colonization by other species.

Cold Climate (Boreal) Forests   The biome is also called the *taiga* and, in North America, the northern coniferous forest. The typical vegetation of the **boreal forest** is coniferous forest made up of spruce, fir, pine, and larch (Figure 12-21). The climate is extremely harsh; the winters are long and cold and the summers are short and cool. Average monthly temperatures may be below 0° C (32° F) for six months of the year and the ground in many places is permanently frozen. There is a more or less continuous ground layer of heath species in dry places and mosses and boggy species in wet places.

The harsh climate and the nature of the vegetation means that the soils are acidic and contain few nutrients. Compared to tropical and midlatitude forests, the energy-fixing ability and the species diversity of the boreal forests are low. The harsh climate and poor soils have effectively protected the boreal forest from human disturbance. Early settlers hunted animals for their fur; modern settlers are more likely to work for logging or mining companies.

In North America two forms of coniferous forest extend into the midlatitudes. One follows the Pacific coast from Alaska to northern California and includes the giant redwoods and Douglas fir in the south. The other is found at high elevations in the western mountain ranges such as the Sierra Nevada and the southern Rockies. Compared to the true boreal forest to the north, conifers in these two types of forest enjoy a long growing season with higher temperatures and more available water.

Figure 12-20   Chaparral vegetation including mountain mahogany, juniper, and oak in the dry climate of Utah.

Figure 12-21   The moose finds food and shelter in the boreal forest during the winter (a), migrating poleward to the tundra for the summer (b).

(a)

(b)

GRASSLAND BIOMES

Grasslands occur where trees cannot grow because of the temperature or rainfall but where conditions are not severe enough to produce desert. In most cases the limiting factor on tree growth is a seasonal deficit in moisture. Some scientists feel, however, that the current distribution of grassland has been shaped to some extent by deliberate burning, or trampling by domesticated animals in the past. Recent experiments have shown, for example, that when protected from burning and grazing, prairie grassland in eastern North America will develop into woodland. *Homo sapiens* is likely to have had the same effects in tropical grasslands, which have a long history of human occupation.

Tropical Grasslands (Savanna) The name **savanna** is given to any tropical vegetation that is dominated by grasses (Figure 12-22a). The term is used mainly to describe open grassland communities with isolated trees, but also includes places that have a nearly continuous tree canopy above the grass layer and pure grass communities. The reasons why grasses and trees occur in different proportions in different places are not understood fully. In some areas the mixture can be related to climate (Figure 12-22b); in others there

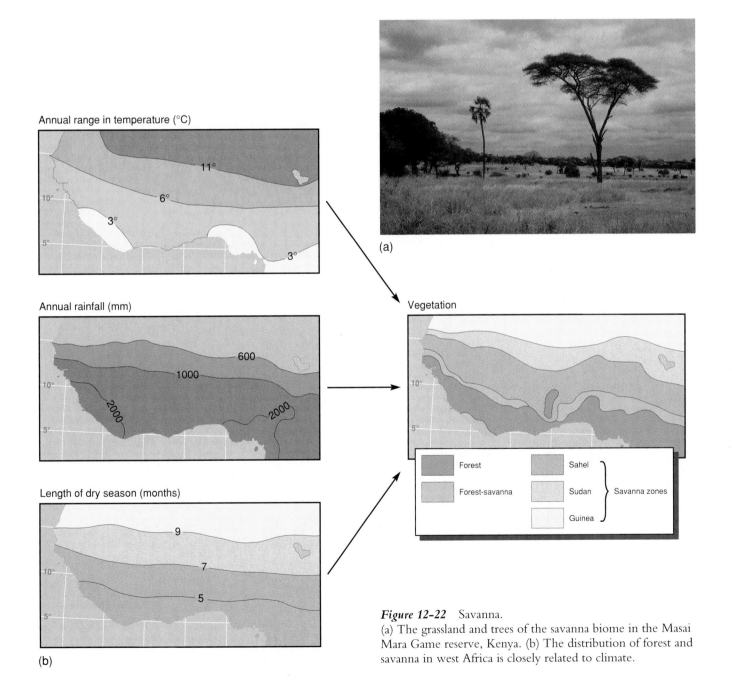

Figure 12-22 Savanna.
(a) The grassland and trees of the savanna biome in the Masai Mara Game reserve, Kenya. (b) The distribution of forest and savanna in west Africa is closely related to climate.

seems to be no logical pattern. Some researchers point to the effects of the surface-relief and soils in storing water; others feel that fire is a more important control. Dead grass accumulates in the dry season and is easily set on fire by lightning or by human intervention. Burning promotes the growth of the young and nutritious grass shoots and, in most cases, kills tree seedlings. Burning is therefore advantageous to the grazing animals and to the carnivores, including humans, that hunt them.

Compared to the tropical forests, the savanna is a simple ecosystem. It is, however, highly productive. In Africa it supports large numbers of herbivores—such as zebra, antelope, giraffe, and wildebeest—and carnivores—such as lions and hyenas. In Australia, kangaroos and wallabies dominate (Figure 12-23).

Human use of the savanna has shifted from a hunter-gatherer culture to domesticated herds and cultivation around permanent settlements. Overgrazing and overcultivation in the drier areas can help

Figure 12-23 Animal life in the savanna. (a) and (c) Zebra, wildebeest, and gazelles are typical herbivores of the African grasslands. They are food for the carnivorous lions. (b) In Australia the dominant herbivores are marsupials, such as the kangaroos shown here.

(a)

(b)

(c)

desertification, the process by which grasslands become deserts (see Environmental Issue: Desertification—Human or Climatic Disaster? p. 408).

Midlatitude Grasslands   The **midlatitude grasslands** develop in areas where the winter is cold but comparatively short and the growing season is long. The rainfall is seasonal and there is often a water deficit in the summer. These grasslands include the prairies of North America, the steppes of eastern Europe, and the pampas of South America. In North America there is a clear east-west sequence of tallgrass prairie, mixed, and short grass prairie that is related to moisture and a north-south species gradient that is related to temperature. The tallgrass prairie in the east contains patches of hickory-oak woodland. The mixed-grass prairie is the most typical prairie vegetation (Figure 12-24). Short-grass prairie is found at the dry western edge of this biome and contains drought resistant plants like the prickly pear.

The soils of the midlatitude grasslands are typically deep and fertile and are now largely cultivated

(a)

Figure 12-24   (a) Midlatitude grassland in its natural condition. (b) Most of this biome has been converted into cropland.

(b)

for grain crops. Before farming became widespread, bison and the pronghorn antelope roamed the prairies in large herds. Both animals have been hunted close to extinction and are now protected.

Tundra **Tundra** is a Finnish word that means "barren land." The tundra biome lies in the constantly cold conditions poleward of the boreal forest. Alpine tundra is a similar community found in the cold temperatures of high mountains. The climate of most tundra areas is inhospitable to humans and most other organisms. Precipitation is low, between 200 and 500 mm (10 and 20 in.) per year, the subsoil is permanently frozen, and there are only about fifty days in the year when the surface temperature is above freezing. The tundra is also frequently swept by strong winds. Not surprisingly, few plants survive here. The vegetation is mainly mosses and lichens, with some very hardy grasses, herbs, sedges (grasslike plants), and occasional dwarf trees. All these plants grow close to the ground (Figure 12-25). In the short summer, however, their productivity is high enough to support low densities

Figure 12-25   The low-growing vegetation of the tundra biome.

of grazing animals like caribou. The caribou go back to the boreal forest in winter. Burrowing animals, such as the lemming, can live in the tundra all year round.

Until recently, human activity in the tundra was limited to the hunting, herding, and fishing economy of the indigenous people. This system was adapted to and had little impact on the natural ecosystem. Recent advances in technology have, however, opened the tundra to mineral exploitation and the establishment of military bases. The import of such energy-rich systems into a delicate natural environment can lead to dramatic alterations in the structure and processes of the natural environment (see Environmental Issue: Living on Permafrost—Do We Understand the Implications?, p. 316).

DESERT BIOMES

Deserts exist where available moisture is so low that all plants and animals must possess special drought-resistant adaptations in their bodies or their behavior. Deserts are mainly located around the tropics because of the high temperatures and calm descending air

found there. Deserts can also form in continental interiors, in any area of rain shadow, or along coasts that have an upwelling cold current offshore (see Chapter 6). The effects of low or irregular precipitation can, to some extent, be offset by groundwater storage or the concentration of runoff by local topography. Within the barren landscape, therefore, there may be small patches of complete and even lush vegetation. In general, however, and in comparison with all other land biomes, the desert supports few species and has a low energy-fixing ability.

All desert organisms have evolved mechanisms of coping with the dry conditions. Some succulent cacti take up water when it is available and store it for future use (Figure 12-26). Others find the very small amounts of water they need almost continuously through wet and dry seasons. A third group of plants complete their life cycle in the space of one wet season, or they simply lie dormant most of the year and grow only when it is damp. Studying these adaptations is fascinating in itself; it is also useful in selecting agricultural crops for arid and semiarid lands.

Figure 12-26 Succulent plants, such as these teddy bear cacti, survive dry periods by storing water in their tissues. They also have a tough outer skin that reduces transpiration.

The animal life of deserts is restricted by the low productivity of the plants to insects, reptiles, birds, and burrowing animals such as gerbils and gophers. Burrows as little as 50 cm (20 in) below the surface have a more or less constant temperature, escaping the daily highs and lows of the surface.

MOUNTAIN BIOMES

The biomes described so far are found in a narrow layer close to sea level. The atmospheric environment, however, changes much more rapidly with altitude than it does with latitude. **Mountain biomes** change dramatically with height. The drop in temperature that comes with a climb of 100 m (300 ft) is roughly equivalent to moving 160 km (100 mi) northward at sea level in North America. The changes are most marked in tropical areas, where a climber can go from tropical forest to an alpine environment by ascending 5000 m (15,000 ft). In the tropics, highland areas often provide a better climate for humans than the lowlands. The Andes of Peru and Colombia, for example, offer better living conditions and soils than the rainforests to the east or the coastal plain to the west.

Some of the vegetation belts found on mountains do not occur near sea level. Examples are the mountain forests of the western cordilleras of North America and, more spectacularly, the cloud forest that forms at high altitudes in tropical areas. Cloud forests form at the height where water vapor in rising air condenses into water droplets and cloud. The water creates a rich forest that contrasts with the dry scrub and grassland below.

OCEAN BIOMES

In the **ocean** the physical environment, the ocean water, is constantly moving and changing in composition and temperature. The continuous movement and mixing makes it difficult to define distinctive biomes within the oceans. Some patterns are, however, apparent. The temperature gradient from poles to equator also applies to the ocean. Superimposed on this gradient are patterns relating to shallow continental shelves and deep or open ocean, and the presence of warm or cold masses of water. Salinity variations are also important. A number of identifiable ecosystems can be defined in these intermingled and dynamic gradients (Figure 12-27).

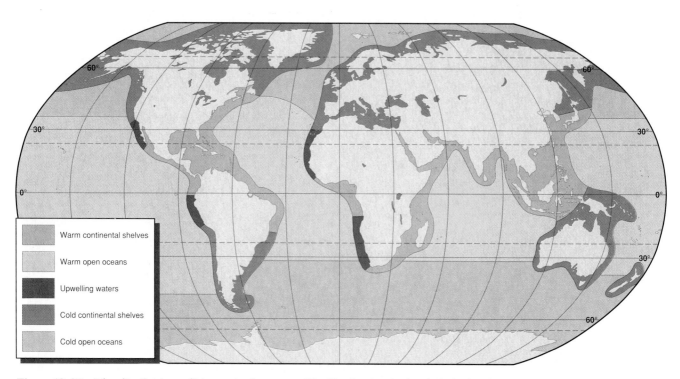

Figure 12-27   The distribution of biomes in the ocean. The distribution is closely linked to latitude, to the presence of deep and shallow water, and to upwelling currents.

SOILS AND HUMAN ACTIVITY

The species *Homo sapiens* is part of the biosphere's natural system, but our abilities to plan and organize and our self-awareness set us apart from other organisms. Many people—scientists, politicians, and ordinary citizens—are concerned that human impacts on the biosphere, already greater than our size or numbers would suggest, are reaching critical proportions. Many scientists now think that there are no completely natural ecosystems left in the world because human impacts on the atmosphere and ocean affect all places on the planet.

Human impact on the biosphere is most complete in cities, where natural ecosystem processes are confined to parks and gardens, and in intensive agricultural systems (Figure 12-28). The annual harvest of grain crops constantly restarts succession, so that most agricultural systems never develop further than their childhood. The homeostatic mechanisms of a complex food web and a store of nutrients are missing. Crop ecosystems are vulnerable to invasion from other species that compete with the crop plants for resources or feed on them. These species we call

(a)

Figure 12-28 Agricultural ecosystems. (a) The agricultural system shown here consists of one cereal crop. Its growth is subsidized by inputs of energy and nutrients, and further energy is used to harvest wheat growing in Berkshire, England, and transport it to market. (b) In less-developed countries, single crops are still grown but usually over smaller areas and with fewer subsidies of energy and artificial nutrients.

(b)

weeds or pests and try to eradicate with herbicides and pesticides. Other chemicals are added to the nutrient cycle in order to boost crop growth.

The apparently high productivity of agricultural systems (see Table 12-1) is heavily subsidized by energy and artificial nutrients. The energy subsidies include the use of fossil fuels to drive the tractors that prepare the seedbed and to pump water into irrigation canals. Energy is also used to make the fertilizers, pesticides, and herbicides that enter the biogeochemical cycles. Nitrogen fertilizers are made in factories by fixing nitrogen directly from the air; phosphorus may be mined to short-cut the natural process of weathering.

Human activities in agricultural systems have repercussions for other parts of the ecosystem. Excess fertilizers, particularly nitrates and phosphates, are washed out of the soil and may enter streams and lakes, causing rapid growth in the algae and plankton that live there. When the plankton die, the decomposers that feed on them use up the oxygen in the water, so other forms of life are suffocated. This group of processes is called *cultural eutrophication.*

Some pesticides, particularly DDT, contain organo-chlorine or chlorinated hydrocarbon compounds. The compounds become concentrated in the body tissue of successive organisms in the food chain. They are thought to be responsible for reducing the population of fish-eating birds such as the osprey and the bald eagle. The concentration of the chemicals in the bird's body is thought to interfere with the intake of calcium, which means the birds make fragile eggshells that are easily crushed by the incubating parent. DDT compounds are now widely banned but those released already are still present in the biosphere.

Homo sapiens is just one species in the biosphere dependent on Earth's resources for survival. The human abilities to plan and organize set us apart from other organisms. They should also tell us when human impact on an ecosystem has gone too far for the natural system to recover. At a local scale, however, the issues of cultural eutrophication and chemical poisoning described above were not widely foreseen. Humans are only just starting to recognize the enormity of our impact at a global scale through acid rain, destruction of the ozone layer, and the human input to climatic change.

The concept of conservation centers on sustaining resources so that an ecosystem or a species is not destroyed by its use. Some people feel that humans have a moral obligation to conserve other species. There are also more practical reasons. Each species is part of a food web and its extinction will alter the energy flows in an ecosystem from which humans might draw food. It is also part of the *gene pool,* the natural variation in genetic material from which humans draw medicines and domestic plants and animals. Humans should be aware that the biosphere will not look after itself and will not continue to meet our needs without careful husbandry. The biosphere is a living system of which we are only a part.

Humans have a greater impact on the biosphere than any other species. Humans affect natural ecosystems by disrupting flows of energy and nutrients. Deliberate impacts in one ecosystem may have unanticipated results in neighboring systems. The concept of conservation is based on the sustainable use of natural resources.

ENVIRONMENTAL ISSUE:

Deforestation—Does It Destroy the Environment?

Some issues involved in this debate are:

What is the economic value of forest ecosystems?

What is the ecological value of forest ecosystems?

What are the differing viewpoints of developed and developing countries?

Deforestation is one of the most long-standing human impacts on the biosphere. It started around the Mediterranean and in the Middle East in prehistoric times and is continuing today in the tropical rainforests. The original forests of the United States have been largely destroyed. Logging activities in the remaining forests of the Pacific Northwest are now partially controlled by legislation that protects forest bird species, but Congressional delegates are pressing for these restrictions to be lifted.

Forests are a major pool of carbon on the land, and through photosynthesis and respiration they help to regulate the concentration of carbon dioxide in the atmosphere. Trees intercept precipitation and return moisture to the air through transpiration and evaporation. Scientists suggest that up to 50 percent of the rainfall in the Amazon basin, Brazil, the world's largest rainforest, is recycled directly from evapotranspiration. Forests, especially tropical rainforests, are also diverse ecosystems that contain millions of plant and animal species.

Current concerns about deforestation center on the tropical rainforests. Such forests contain exotic and beautiful hardwood trees that can be cropped and sold for timber. They occupy large areas of land where water and sunlight, the two main ingredients for photosynthesis, are in abundant supply. The agricultural value of cleared land, however, is usually short-lived; the soil erodes rapidly once the forest is removed. The perceived value of the rainforests for logging and cropping means they are disappearing at an alarming rate. An area about five times that of Central Park (New York City) is cleared every hour.

There are three main agents of clearance in tropical forests: the commercial logger, the cattle rancher, and the small-scale farmer. The small-scale farmers, usually native or from neighboring countries, collectively account for more deforestation than the loggers and ranchers combined. Until the early 1960s, the small-scale farmer was typically a member of a native tribe who practiced rotating cultivation of small clearings in the forest. This had a minimal impact on the forest because it was practiced by a small number of people. From the 1960s onward, the indigenous peoples have been joined by people who migrated or have been moved to the marginal forestlands in the hope of a better life. The migrants move because they are poor, part of a rapidly growing population, and because they own no land. The newcomers practice a more settled form of agriculture, causing a widespread and permanent loss of forest cover. Deforestation by new settlers is concentrated in the frontier regions opened up by new roads, as shown in the satellite images. It is analogous to the widespread deforestation of Western Europe and the settlement of the American west.

In the developed world, conservationists have mounted a vigorous and emotional campaign to save the rainforests. Their arguments are countered by other groups of scientists, politicians, and industrialists. The opponents say, for example, that the tropical rainforests contain only a tiny fraction of the global carbon total. Thus, deforestation has only a minor effect on the global carbon balance and possible global warming. Disturbing the forest will certainly make many species extinct, but humans have used only a handful of the million or so species that we know about; we are unlikely to need many more. Deforestation gives opportunities to other species and for the evolution of new ones. The environmental concerns of well-nourished and well-housed populations in the developed world must be set against the grinding poverty that the exploitation of the rainforests is designed to offset.

Linkages

Forests are not only stores of genetic material, but also parts of ecosystems that trap energy from the sun, circulate water (**atmosphere-ocean system**), and convert soil minerals (**solid-earth environment**) into chemical foods. Human intervention in these ecosystem processes may cause disruption of the entire system.

Topics for Discussion

1. How important are forests in the hydrologic cycle and the global carbon balance?

2. How important is it to conserve species that we have not found yet and for which we have no use?

3. Can people in developed countries insist on conservation in the developing countries of the tropics?

4. Is forest conservation important enough to put people out of work in the logging industry?

(a)

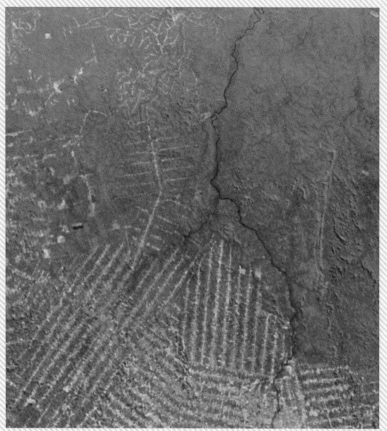

(b)

(a) Burning after forest clearance in the tropics.
(b) Clearing Amazon rainforest. The unbroken
forest is red; the clearings along roads are light
blue. This shows that forest clearance closely
follows the development of a road network.
(c) The extent of natural forest in North America
in 1620, when the first European settlers arrived,
and in 1990. About 5 percent of the original
forest remains.

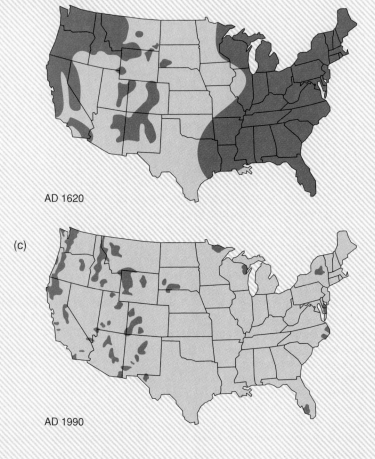

AD 1620

(c)

AD 1990

CHAPTER SUMMARY

1. Earth's plants and animals live in the biosphere, the thin layer of soil, water, and air at the planet's surface that can support life. The biosphere is formed by interactions among the ***atmosphere-ocean, solid-earth,*** and ***surface-relief environments.*** It is organized by the transfer of energy and chemical materials among these environments and living things. The ecosystem is an important framework for studying the biosphere. It can be applied at any spatial scale but may be difficult to delimit in practice.

2. Energy enters the ecosystem through the process of photosynthesis and is released by respiration. Photosynthesis creates energy-rich carbohydrates from water, carbon dioxide, and solar energy and produces oxygen (***atmosphere-ocean environment***). Photosynthesis is carried out by green plants (autotrophs) in water and on land. The autotrophs' energy is transferred to other organisms (heterotrophs) through the food web. The transfer of energy is not totally efficient.

3. Nutrients are chemical elements necessary for life. The supply of nutrients on Earth is finite and they are constantly recycled. Nutrients are exchanged among living things and the surrounding Earth environments. Most nutrients have one main reservoir, either the ***solid-earth*** (most micronutrients) or the ***atmosphere-ocean environment*** (most macronutrients).

4. Plants interact with each other and with their local environment. They grow in locations that provide an ideal or tolerable set of resources and in which they can compete successfully with other plants. Recurring ecological niches in the biosphere contain similar plant associations.

5. Over time, some plants modify their environment so that it is no longer possible for their offspring to live there. External conditions such as a landslide also create new environments for colonization. A new species takes over in the process of ecological succession. Succession stops when the environment is stable.

6. Ecological succession, evolution, and the changing positions of land and ocean cause changes in the distribution of species over time. Ecological succession leads to change in the ecosystem at a place as new species, better adapted to the environment, replace others. Evolution leads to new forms of life (***living-organism environment***). Movements of Earth's crust (***solid-earth environment***) distribute species around the world and, as in the case of the marsupials, may isolate them from competition with species that evolve later.

7. Biomes are regional ecosystems recognized and described by the growth-form characteristics of their dominant vegetation. Biomes are usually thought of as more-or-less uniform areas. Their environmental conditions, however, vary gradually over space and their boundaries are not easily recognizable on the ground.

8. Humans have a greater impact on the biosphere than any other species. Humans affect natural ecosystems by disrupting flows of energy and nutrients. Deliberate impacts in one ecosystem may have unanticipated results in neighboring systems. The concept of conservation is based on the sustainable use of natural resources.

KEY TERMS

biosphere   *354*

ecosystem   *354*

biome   *355*

photosynthesis   *355*

chlorophyll   *357*

autotroph   *357*

heterotroph   *357*

herbivore   *357*

carnivore   *357*

omnivore   *357*

food chain   *357*

decomposer   *358*

food web   *358*

net primary production   *359*

nutrient   *362*

biogeochemical cycle   *362*

ecological niche   *365*

ecological succession   *365*

homeostatic mechanism   *367*

evolution   *369*

tropical rainforest   *371*

tropical seasonal forest   *371*

midlatitude forest   *374*

CHAPTER REVIEW QUESTIONS

1. Draw a sketch showing how the major Earth environments interact to form the biosphere.
2. What are the similarities and differences between the flow of energy and the flow of nutrients in an ecosystem?
3. How do ecosystems change over time?
4. Describe how the process of evolution has affected the global distribution of living things.
5. Outline the different moisture and temperature requirements of the world vegetation regions.
6. What are the similarities and differences between deserts and the tundra?
7. Explain why the concept of biomes is not particularly helpful in a study of the oceans.
8. Explain the global distribution of tree growth.

SUGGESTED ACTIVITIES

1. Find out about the plants and animals living in a particular biome and construct a simple food web from your information.
2. Describe how ecological succession might affect ecosystems in your area.
3. Make a detailed study of a local ecosystem to show how different groups of organisms occupy different ecological niches.

CHAPTER
13
SOILS

THIS CHAPTER IS ABOUT:

- Soil development
- Soil components
- Soil characteristics
- Factors affecting soil formation
- Soil-forming environments
- Soils and human activity
- Environmental Issue: Desertification—Human or Climatic Disaster?
- Environmental Issue: Soil Erosion—Is It Inevitable?

Soil is a thin layer of weathered rock, dead plant and animal matter, water, and air, spread out across Earth's land surfaces. Soils differ from place to place because climate and the supply of raw materials varies around the world. Soil helps to dictate what plants grow where and the productivity of each country's agriculture.

Alfisols	Histosols	Spodosols
Andisols	Inceptisols	Ultisols
Aridisols	Mollisols	Vertisols
Entisols	Oxisols	Mountain areas

The interaction of the major Earth environments in the biosphere can be seen most clearly in the development of soil. **Soil** is a dynamic mixture of water, air, mineral, and organic materials that include decaying plant or animal tissue. It is spread out as a thin layer over Earth's surface. Soil provides physical support for green plants, the basis of the food web, and is the medium through which plants receive nutrients and water (Figure 13-1).

Physical geographers study soil because of its fundamental position in the ecosystem and its impact on human activities. Many scientific disciplines, such as chemistry, physics, and microbiology, contribute to an understanding of soils; and many other applied disciplines, such as agriculture, forestry, and engineering, make use of this knowledge. Within physical geography, biogeographers and geomorphologists study soils, emphasizing soil's importance in the biosphere and in the surface-relief environment.

SOIL DEVELOPMENT

Soils are the result of three interacting processes that bring different materials together. First, an exposed parent material is weathered to smaller fragments. Where the parent material is a rock surface that is weathered in place, the mass of smaller fragments is called *regolith*. Soils also develop from loosely deposited parent materials such as alluvium (from rivers), till (from ice), or loess (from wind action). Second, living and dead organic matter are added to the rock fragments. Third, both living and nonliving materials are moved, or **translocated,** in the soil, mainly by water. These three processes involve the interaction of all the major Earth environments. The details of these processes and their operation and interaction in the developed soil are discussed later in this chapter; this section outlines their interaction during initial soil formation.

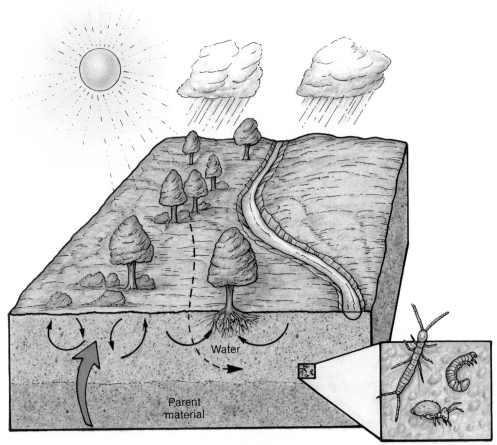

Figure 13-1   Soil in the ecosystem. Soils are a growth medium for green plants and are home to many decomposer organisms. Nutrients are weathered into the soil from the parent material, and water infiltrates after rain or snowmelt.

Not all weathered parent materials develop into soil. The transformation from a heap of inert mineral grains to an organized and living soil begins when living and dead organic materials are added (Figure 13-2). The initial input of organic matter is from lichens, mosses, or other simple organisms that can survive on the barren surface. These colonizing organisms modify their environment, making it a little more hospitable for other, more complex plants and animals. In particular they supply organic material, which is broken down and becomes part of the infant soil, providing energy for a decomposer food chain and nutrients that can then be recycled. Ecologic succession (see Chapter 12) ensures a gradual but continuous change of species over time and adds more organic material to the developing soil. Living soil organisms, particularly microorganisms and bacteria, break down the dead plant and animal tissue. They also help to move it physically around the soil. Water is the other main agent of translocation. Where precipitation is moderate to high, materials are washed from the upper part of the soil and deposited lower down; in arid regions materials may move upward as water is drawn to the surface by evaporation.

Continued weathering, addition of organic material, and translocation lead eventually to a developed soil. Such a soil is described by the characteristics of its **soil profile,** a vertical slice through the soil from the surface down to the parent material. The soil-forming processes, particularly translocation, tend to differentiate the soil profile into a series of horizontal layers, or **soil horizons** (Figure 13-3). Soil scientists dig profiles through the soil to examine the characteristics and spacing of soil horizons. Soil is not, however, made up of isolated two-dimensional slices at the Earth's surface. Soil is a three-dimensional body that varies continuously over the landscape. A **soil pedon** is a three-dimensional block of more or less uniform soil a few meters in size. Soil pedons are grouped together in *soil types* for the purpose of soil survey and mapping.

The soil horizons are given letters according to their composition and their position in the profile. Not all the horizons described in Figure 13-3 are present in every soil. The variability of soil profiles around the world is caused by variations in the soil-forming processes and the materials that they have to work on.

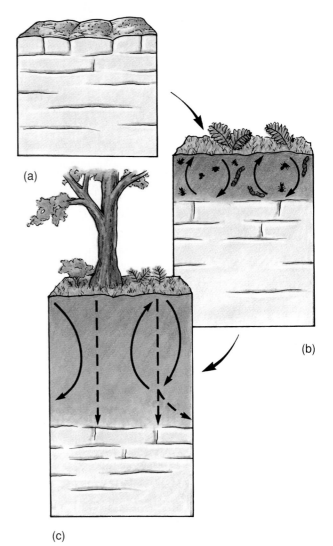

Figure 13-2 The stages of soil formation. (a) An exposed and weathered surface is colonized by simple organisms such as mosses and lichens. (b) More organic material is added to the soil through ecological succession, and weathering of the parent material continues. A decomposers food chain is started and the soil becomes deeper. (c) A developed soil, differentiated into soil horizons and in equilibrium with its surroundings.

SOIL COMPONENTS

Soil has four components—the mineral fraction derived from the parent material, organic material, air, and water.

MINERAL FRACTION

The mineral fraction is the remains of the weathered parent material, broken down to its individual grains. It is the relatively stable physical framework

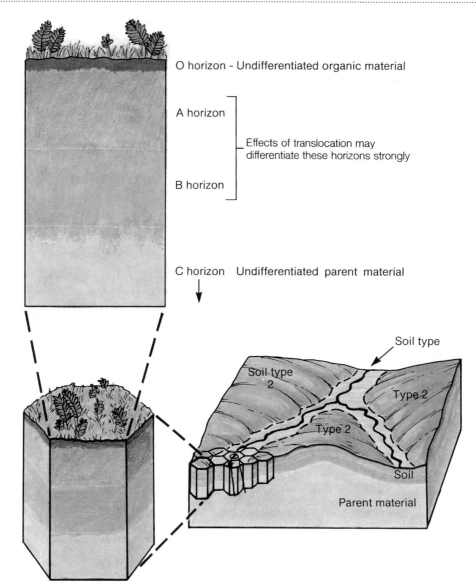

O horizon - Undifferentiated organic material

A horizon

Effects of translocation may
differentiate these horizons strongly

B horizon

C horizon   Undifferentiated parent material

Soil type

Soil type
2

Type 2

Type 2

Soil

Parent material

Figure 13-3   (a) A soil profile, differentiated into horizons by the translocation of
materials. (b) The soil pedon, a three-dimensional representation of the soil profile.
(c) Pedons fit together to form soil types and the soil landscape.

of the soil and in most cases is by far the largest
component. Its physical and chemical characteristics
are governed largely by those of the parent material.
The terms *sand, silt,* and *clay* are given to mineral
grains that fall into specific size ranges (Table 13-1).
Hard minerals within the parent material end up as
large durable sand grains; softer minerals weather to
smaller silt and clay fragments and to individual
chemical elements. The relative proportions of sand,
silt, and clay in the soil determine the important
property of soil textures discussed in the section on

TABLE 13-1 THE RELATIVE SIZES OF SAND, SILT, AND CLAY PARTICLES		
Particle	*Granules of Comparable Size*	*Size of Granules in MM*
Gravel	Hazelnuts	>2
Coarse sand	Builder's sand	0.2–2
Fine sand	Castor sugar	0.02–0.2
Silt	Confectioner's sugar	0.002–0.02
Clay	Putty	<0.002

soil characteristics. The very finest particles, **colloids,** are particularly important in keeping nutrients in the soil and making them available to plants.

ORGANIC MATTER

There are two sorts of organic matter in the soil—living and dead. The dead organic matter consists of the waste products excreted by organisms and the body tissues of dead plants and animals. This material is in various stages of decay as it is broken down by the decomposers, ready for the next stage of the biogeochemical cycle. The living organic matter comprises the millions of organisms that take part in the decomposition process and move materials around the soil. Organic matter in the soil usually represents about 2 to 6 percent by volume in the surface horizons but is much more important than this figure suggests.

The end product of the breakdown of dead organic material is **humus,** a structureless, dark brown or black material found close to the soil surface. It varies in type and distribution from one soil to another, depending on the sort of organic material and the conditions within the soil. The tiny organic particles bind to the small colloids of the mineral fraction to form the **clay-humus complex.** The clay-humus complex is important in controlling soil fertility, as is discussed in the section on soil chemistry.

SOIL WATER

Soil water, its dissolved elements, and suspended particles make up the **soil solution,** the liquid part of the soil. The soil solution, together with soil air, fills up the spaces or pores among the mineral and organic parts of the soil. It is the main agent of translocation, carrying dissolved chemical elements and small particles through the soil.

Water is held in the soil by the attraction of water molecules to each other and to soil particles. The same powers of cohesion hold a droplet of water together as it clings to the side of a glass. Soil water exists in one of three states in the soil. **Hygroscopic water** is held tightly, close to the surface of individual mineral grains (Figure 13-4). It is unavailable to plants because the attraction between the water and the mineral grain is greater than the pulling power of plants' roots. A soil in which the water is nearly all hygroscopic is said to be at *wilting point* because under this condition plants can obtain no moisture from the soil.

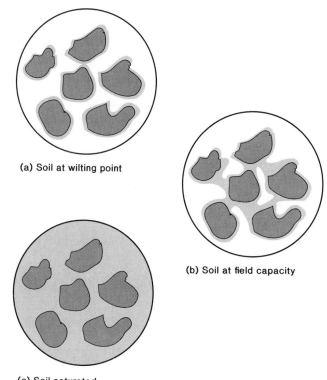

(a) Soil at wilting point

(b) Soil at field capacity

(c) Soil saturated

Figure 13-4 Water in the soil. (a) At wilting point the remaining water, mainly hygroscopic, is held tightly to the soil particles and cannot be drawn away by plant roots. (b) At field capacity capillary water fills the small pore spaces in the soil, and air fills the larger pore spaces. Capillary water moves slowly through the soil in response to pressure gradients between wet and dry areas. (c) In a saturated soil all pore spaces are filled with water, and air is excluded from the soil. A proportion of this water will normally drain from the soil as gravitational water, leaving the soil at field capacity.

Other water molecules can link up with the hygroscopic water attracted by the mineral grain. These molecules are held less tightly; they are called **capillary water,** they can move in any direction in the soil, and are available to plants. Capillary water remains in the soil because the combined attraction of the water molecules to each other and to the soil particles is greater than the force exerted by gravity. Capillary water moves from a wet to a dry area in the soil in response to variations in pressure similar to those found in the atmosphere. The most common movement is upward, toward roots and toward the soil surface where water is lost by evaporation and transpiration.

The third type of soil water is **gravitational water,** which is not held in the soil at all but moves freely under the influence of gravity. It is available to plants, but only temporarily. It is present immediately after rain but drains downward quickly. A soil with its maximum quota of capillary water but no excess gravitational water is said to be at **field capacity**—an ideal state for plant growth.

The amount of water held in each state of availability depends on the size of the soil particles and therefore the size of the pores between them. Water drains rapidly from the large pores in a coarse-grained, sandy soil, but it is held close to the soil particles in the smaller pores of a fine textured clay.

SOIL AIR

Soil air occupies the pore spaces not currently filled by water. A soil at field capacity contains a good balance of air and water. Air in the soil performs much the same function as it does in the free atmosphere. It brings oxygen to the soil organisms, including plant roots, and removes carbon dioxide. The composition of soil air differs from that of free atmospheric air because soil air is compartmentalized by intervening fragments of soil and water.

Soil forms as the result of three processes—weathering of a parent material, the addition of organic matter, and the movement or translocation of material in the soil profile. A soil has four components—the mineral fraction (***solid-earth environment***), organic matter (***living-organism environment***), water, and air (***atmosphere-ocean environment***).

SOIL CHARACTERISTICS

The nature of the four soil components and the processes that act on them produce a number of soil characteristics. The soil scientist uses these characteristics to find out how the soil is developing or to diagnose its ability to support plant growth. For example, a light-colored, acidic, sandy layer near the top of the profile, underlain by a rich brown clay horizon shows that material is being washed from the upper to the lower horizon. A soil with these characteristics is unlikely to support grain crops, but could be used for coniferous trees.

SOIL TEXTURE

Soil texture describes the relative proportions of sand, silt, and clay in the soil. Figure 13-5 shows the terms used internationally for soils with different textures. Soil texture affects the soil's ability to hold moisture and nutrients and make them available to plants. The presence of colloids, the smallest clay particles, is particularly important; so much so that the term clay is included in the name of any soil that contains more than about 25 percent clay. Soils with extreme textures, either very clayey or very sandy, are poor media for plant growth. The large grains and pore spaces of sandy soils mean that they can hold little capillary water or nutrients for plants. Clayey soils are compact and quite easily waterlogged. Much of the water is held tightly to the mineral grains and not available to plants. The ideal soil for plant growth is a loam, which is a mixture of sand, silt, and clay. Texture is a fundamental characteristic of the soil; it is not easily altered by management.

SOIL STRUCTURE

Soil structure describes how individual grains of soil are bound together into clumps. Soil aggregated in this way is more stable than a structureless mass of single grains and is less easily moved by wind or water. It is therefore a better medium for plant growth. The ideal structure for plant growth is small, crumblike aggregations; this gives plant roots ready access to both water and nutrients and allows plenty of air into the soil. The mechanisms that control soil structure are complex and incompletely understood. Unlike soil texture, though, soil structure *can* be influenced by human management.

The most important factor in promoting good structure is to maintain a reasonable level of organic matter in the soil, as humus helps to bind together the individual grains into aggregates. It is also helpful to avoid using heavy machinery, which creates large aggregates that block the movement of water, air, and plant roots.

SOIL CHEMISTRY

Soil chemistry describes the type and amount of chemical elements in the soil. Two related parts of soil chemistry are particularly important to soil's role in the ecosystem. These are soil fertility, which is the ability of the soil to supply nutrients to plants,

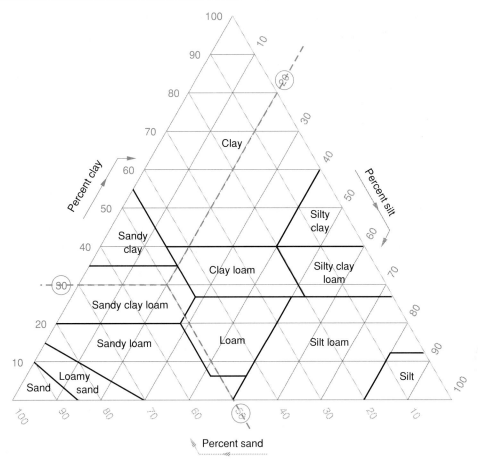

Figure 13–5   The percentage of sand, silt, and clay in the major soil texture classes. A soil with 20 percent silt, 30 percent clay, and 50 percent sand is a sandy clay loam, as highlighted on the diagram. In which texture class does a soil with 10 percent sand, 10 percent clay, and 80 percent silt fall?

and soil acidity, which affects the action and composition of soil microorganisms, the rate of breakdown of organic material, and the availability of nutrients in the soil.

Soil nutrients are held at three places in the soil. The first location is in the mineral fraction and unaltered organic material; these nutrients are not available to plants. The second site is the clay-humus co.nplex, a mass of small particles that is the main source of nutrients for plants. Finally, a smaller amount of available nutrients is found in the soil solution.

The tiny particles that make up the clay-humus complex carry electrical charges on their surfaces, which are mainly negative. Most nutrient ions released from weathering of parent materials are positively charged ions called *cations.* The cations are attracted to and held by the particles in the clay-humus complex. Negatively charged nutrient ions, called *anions,* such as sulfur and phosphorus, come from the breakdown of organic material. These are also held by the clay-humus complex at the smaller number of sites with positive surface charges.

Anions and cations are exchanged among the clay-humus complex, the surrounding soil water, and the plant roots. Some cations are more strongly attracted to the clay-humus complex than others. Hydrogen is held particularly strongly. Hydrogen ions from percolating rainwater, therefore, can dislodge other nutrients from the clay-humus complex. The dislodged nutrients dissolve in the soil solution and are usually carried farther down the soil profile. The clay-humus complex is, therefore, very important in maintaining soil fertility. Soils that contain colloids of clay and humus tend to have higher fertility than soils that contain only large grains. A soil's ability to supply nutrients is described by its **cation exchange capacity,** which measures the number of cations that can be held in a unit of soil.

Soil **acidity,** measured on the pH scale, reflects the concentration of dissolved hydrogen ions in the soil. Acidity is an important soil characteristic because it affects the chemical state and mobility of other nutrients and materials. Figure 13-6 shows how the pH readings of agricultural soils compare to those of common household products.

Acid soils occur where there is a poor supply of nutrient ions from the mineral fraction and organic material, or where rainfall is high and the nutrient ions are replaced by hydrogen. Acid deposition (see Chapter 4) from factories and power stations can also raise acidity. Under very acidic conditions, the action of the decomposer organisms slows down, and a deep and incompletely broken-down organic layer can accumulate. This deep organic soil is called peat.

Alkaline soils are typical of arid and semiarid regions. In these areas water is in short supply. Nutrients are not washed out of alkaline soils but can accumulate in concentrations that are dangerously high for plant growth.

SOIL COLOR

Color is one of the more obvious soil characteristics and is a good indicator of the processes operating in the profile. Most soils can be described as black, brown, orange, red, yellow, gray (including gray-blue

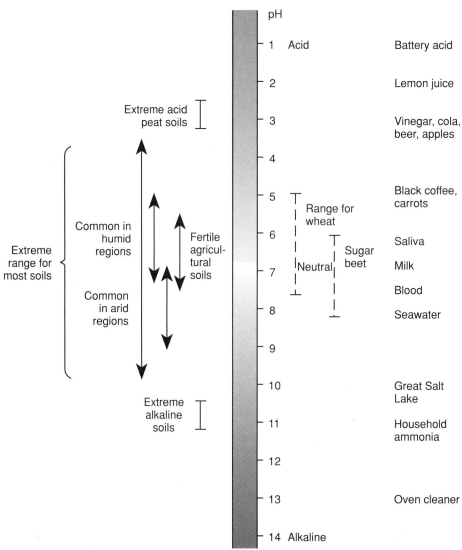

Figure 13-6 The acidity found in different types of soils and the tolerance range of some common crops. The pH values of some household goods are included for comparison.

or gray-green), and white. To make sure that everyone describes soil color in the same way, an international descriptive code has been set up based on a series of colored chips similar to those used to choose paint (Figure 13-7). Soil color is controlled by the distribution of organic material in the profile and by the state of any iron present. Soils with little organic matter are white or gray; those with much organic matter are brown or black. Where iron has been oxidized, the soils have a red tint. Where iron is not oxidized the dominant colors are gray, gray-blue, and gray-green; these colors are particularly common in waterlogged soils.

The soil-forming processes act on the soil components, creating different conditions of texture, structure, chemistry, and color. A biogeographer uses such conditions to understand the processes operating in the soil.

FACTORS AFFECTING SOIL FORMATION

The three processes of soil formation—weathering, the addition of organic matter, and translocation—vary in intensity from place to place in the biosphere and create different soil profiles. The soil-forming processes are controlled by the action and interaction of five factors—climate, parent material, topography, organisms, and time (Figure 13-8).

CLIMATE

Climate is the dominant factor in soil formation and affects all three soil-forming processes. First, it controls how fast parent material is weathered and how far the regolith is broken down to its constituent minerals. Second, climate is a major influence on the type and number of organisms in the ecosystem. This determines the type and amount of dead organic material added to the soil and how well it is broken

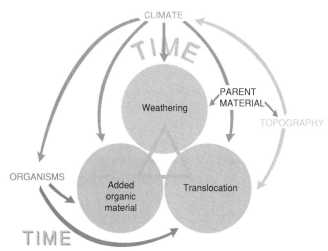

Figure 13-8 The effect of the soil-forming factors (climate, parent material, topography, organisms, and time) on the soil-forming processes (weathering, translocation, and the addition of organic matter). The soil-forming factors interact among themselves, as do the soil-forming processes. Virtually all elements of this diagram influence, or are influenced by, the others. Only the main linkages are shown here. Try filling in further linkages on an expanded version of this diagram.

(a)

(b)

Figure 13-7 (a) Munsell soil color chart. Each colored chip on the page has three index figures and a short written description that identify it uniquely. Separate pages take the scientist through shades of yellow, red, gray, and green. The chart is named for Albert Munsell, who developed the naming system in 1905 as a way of systematically describing the colors used for an atlas. (b) This handful of soil shows the considerable variations that are possible in the "brown" color of soil.

down. Third, climate determines the balance between precipitation and temperature, influencing the vertical translocation of water and soluble nutrients in the soil profile.

When rainfall is high and potential evapotranspiration (see Chapter 4) is low, rainwater moves down through the profile. Nutrient cations on the clay-humus complex are replaced by hydrogen ions and are washed out of the soil. This process is so common in soil development that it has been given the special name of **leaching.** Where the downward movement of water is very strong, the finest clay particles may also be washed down the profile. The physical transport of particles in suspension is termed **eluviation** (from the Latin *ex* or *e,* meaning out, and *lavere,* meaning to wash). **Illuviation** (from the Latin *il,* meaning in, and *lavere*) refers to the redeposition of materials elsewhere in the profile. In the humid mid-latitudes, leaching typically results in a nutrient-deficient or *eluviated* upper horizon and an enriched or *illuviated* lower horizon. The eluviated horizon is often called simply the E horizon.

Where rainfall is low and potential evapotranspiration is high, any water that filters into the soil will eventually be drawn upward to the surface or toward plant roots. Nutrients carried in the water are also drawn upward and deposited near the surface.

Climate controls virtually all soil characteristics and processes. The global distribution of soils, therefore, closely matches the distribution of major climatic environments (see Chapter 6) and world vegetation types (see Chapter 12). On a local scale, soils vary with the microclimates formed by altitude and aspect.

PARENT MATERIAL

The nature of the parent material affects the soil chemistry and the soil texture. Parent materials that contain a high proportion of nutrient cations, such as calcium, sodium, potassium, magnesium, and iron, maintain a reasonable soil fertility even if the nutrients are regularly washed out of the soil in rainwater. Under the same climatic conditions, parent materials without these cations, such as dune sand and sandstone, produce an acidic soil high in hydrogen ions but low in other nutrients. Sandstone is coarse grained and produces sandy soil with a low colloidal content that drains rapidly. Some igneous and metamorphic rocks weather to complex clay minerals that contribute to the clay-humus complex

and help to hold nutrient ions and water in the soil. In a given climate, parent materials that weather easily produce large amounts of regolith and eventually a deep soil. Resistant rocks produce shallow and often stony soils.

TOPOGRAPHY

The detail of the surface-relief environment at a specific site acts as a stage for the major soil-forming factors. The principal variable is slope, which affects the movement of both materials and water.

Soils that develop on steep slopes are subject to soil erosion through surface wash and creep. Water tends to follow the angle of slope and moves downhill on the surface or through the upper soil horizons. Vertical movement of water through the profile is reduced because of greater flow parallel to the slope. Soils at the foot of a slope collect both materials and water from above.

The waterlogged soils of wetlands allow water to accumulate on low-lying sites. They have low aeration, so they support specialized and sometimes unique biologic communities. Once drained, wetlands may produce fertile soils.

The details of surface relief at a site can lead to a considerable change in soil conditions over a short distance. This is well-illustrated by a soil *catena,* a sequence of soils that develops on a slope under uniform conditions of climate and parent material (Figure 13-9). The elevation and aspect of a site also affect soil processes by influencing the effects of climate.

SOIL ORGANISMS

Soil organisms are the living part of the soil ecosystem. Their presence and activity are controlled by their physical environment and by the flows of energy and nutrients they receive. Some soil organisms, particularly the bacteria and algae, live almost completely within the soil. They are part of the decomposers' food chain, gaining energy from the carbohydrate molecules in the dead plant and animal tissues that fall to the ground as waste material. Their action on the soil is mainly chemical, in that they break down the dead organic matter in place. Some of the larger soil fauna have a physical as well as a chemical effect on the soil. They dig and burrow and move both organic and inorganic material about. Earthworms, for example, literally eat their way through the soil. Up to 6 tons of soil per hectare (15 tons per acre) may pass through their

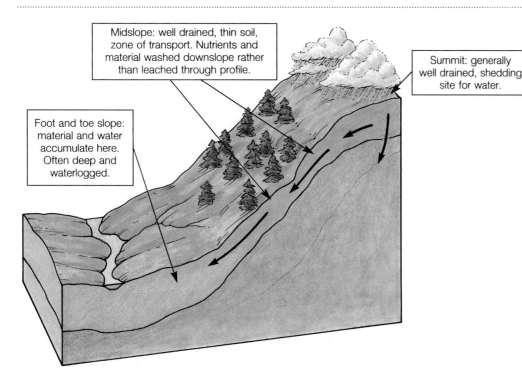

Midslope: well drained, thin soil, zone of transport. Nutrients and material washed downslope rather than leached through profile.

Summit: generally well drained, shedding site for water.

Foot and toe slope: material and water accumulate here. Often deep and waterlogged.

Figure 13-9 An example of a soil catena, in which soils at different positions on a gradient develop markedly different profiles. Differences in soil development due to topography are important controls on the local-scale distribution of vegetation in an ecosystem.

bodies in a year. The higher plants growing at the soil surface also influence the soil by adding organic matter and penetrating the soil with their roots. Grasses probe the soil with thousands of small roots; trees dig deeper with fewer, larger, roots.

TIME

The soil-forming processes of weathering, addition of organic material, and translocation operate over time. Beginning with unaltered parent material, a soil typically develops through the continued action of the soil-forming processes over hundreds and thousands of years, to a mature profile (see Figure 13-2). If undisturbed, the developed soil exists in a dynamic equilibrium with its environment. It will change only if the soil-forming processes alter substantially in response to change in the controlling factors described here.

Over time, the influence of each of the factors listed above changes. Initially the interaction between climate and parent material—between the atmosphere-ocean environment and the solid-earth environment—is most important. Through the process of weathering, this interaction controls the nature of the regolith and how quickly it accumulates. Soon the rock surfaces are colonized by living things and the living-organism environment starts to contribute, providing the decomposers and the beginnings of a nutrient cycle.

The surface-relief environment influences proceedings throughout. It provides a location for the regolith to accumulate or be deposited and for soil to form. Its main influence, however, is in controlling the movement or translocation of materials in water through the soil. As the soil develops, weathering becomes less important because the parent material is protected beneath the soil. The main processes are then the addition, incorporation, and cycling of organic material and the translocation of materials up and down the profile. Climate, soil organisms, and topography therefore have a greater day-to-day influence on changes in the developed soil than the nature of the parent material, although the parent material continues to influence soil development because it determines the texture and the availability of nutrient cations.

SOIL-FORMING ENVIRONMENTS

Figure 13-10 shows how the subjects examined in this chapter fit together and how this study of soils will be completed. The three soil-forming processes are controlled by five factors, one each from the atmosphere-ocean, solid-earth, surface-relief, and living-organism environments, all acting over time, the fifth factor. Together, the soil-forming processes and the factors that influence them determine the basic nature of the soil. The initial result of their interaction is a soil made up of four components—mineral

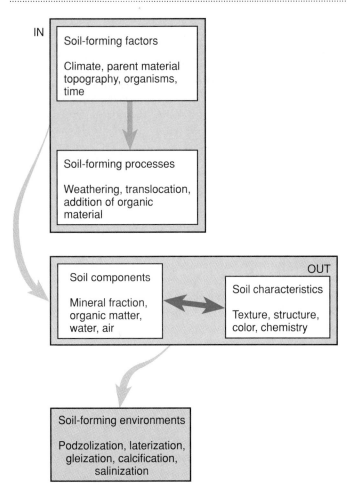

IN

Soil-forming factors

Climate, parent material
topography, organisms,
time

Soil-forming processes

Weathering, translocation,
addition of organic
material

Soil components

Mineral fraction,
organic matter,
water, air

OUT

Soil characteristics

Texture, structure,
color, chemistry

Soil-forming environments

Podzolization, laterization,
gleization, calcification,
salinization

Figure 13-10   The linkages among the topics studied in this
chapter. The soil-forming processes and soil-forming factors
work on the soil components, producing a range of
conditions recognizable through the soil characteristics. The
different types of conditions can be summarized by a limited
number of soil-forming environments.

fraction, organic matter, water, and air. In a well-developed soil, the soil-forming processes have worked
on these components to produce soil characteristics—
texture, structure, color, and chemistry. The components and characteristics of soil develop together and
are interdependent.

There is clearly a tremendous potential for variety
in soil characteristics around the world. In practice,
though, the soil-forming factors and processes fall naturally into a limited set of soil-forming environments.
There are five major soil-forming environments; podzolization, laterization, gleization, calcification, and
salinization, each of which is described below. Soil scientists and physical geographers can examine the characteristics of a soil in the field, and through their

knowledge of soil-forming processes, they can work
backward through the sequence shown in Figure 13-
10 to understand its origin.

PODZOLIZATION

Podzolized soils are acidic and strongly leached.
Organic material, nutrients, and fine particles of clay
are washed from the upper horizons of the soil and
deposited lower down. A podzolized soil typically has
a grayish-white E horizon and a darker, brownish B
horizon that may have a touch of red or yellow (Figure 13-11). Podzolized soils are typical of the northern boreal forest biome. The needle-leaf trees and
hardy shrubs that grow here produce acidic organic
material which is low in nutrients. Few decomposers
can live in the acidic conditions and severe cold, and
the organic material is broken down very slowly.
Podzols also form under less extreme climatic conditions on parent materials such as sandstone that are
nutrient deficient and free draining.

The Pilgrims came across podzolized soils at their
early settlements in Plymouth County, Massachusetts.
Their attempts at plowing mixed the organic surface
horizon with the leached horizon beneath, creating a
poor soil. Many of these farmers abandoned their
land for the promise of a new life in the west.

LATERIZATION

Laterization is the dominant soil-forming environment in the humid tropics and subtropics. The hot,
wet climate is ideal for rapid chemical weathering and
leaching. The most striking features of a lateritic soil
are its depth, its lack of horizons, and its deep red or
yellow color (Figure 13-12). Heavy rainfall encourages strong leaching, and virtually all the nutrient ions
are washed out of the upper horizons. The nutrients
may be washed out of the soil completely or deposited as a hard layer lower down the profile.

There is virtually no organic matter in lateritic
soils despite the fact that they frequently support
lush vegetation. Compared to weathering and
translocation, the influence of organic matter on soil
formation is minimal. The soil organisms work
rapidly in the hot wet climate, and the nutrients
they release from the litter are almost immediately
taken up by plants or leached out of the soil. There
is little humus in the soil.

The lateritic soil described here is typical of laterization in the humid tropics. Laterites also occur in a
more moderate form in the subtropics. In the United
States, the red soils of the old cotton belt—Georgia,

Figure 13-11   A profile from a podzolized soil in France. Note the bleached E horizon close to the soil surface, and the richly colored illuviated B horizon.

Figure 13-12   A profile through a laterized soil in Kenya. The soil shows the typical bright red color and lacks horizons.

North and South Carolina, and Alabama—are the result of weak laterization. Podzolization and laterization are both dominated by the vigorous downward translocation of materials through the soil profile. In the case of laterization, this is accompanied by rapid breakdown and recycling of organic material, plus rapid and complete weathering.

GLEIZATION

Gleization occurs in soils that are waterlogged. Saturated profiles are most common in areas of moderate to high rainfall, but gleization is not confined to this type of climate. It develops in soils that have a high clay content, are at collecting sites in the landscape, or develop where the subsurface drainage is impeded for any reason. Gleization is common in podzolized soils where leaching has created an impermeable layer of clay or iron in the B horizon. Waterlogged soils are not an attractive environment for most soil organisms because the soils contain little oxygen. Organic material, therefore, tends to accumulate at the surface. A gleyed soil is characteristically blue-gray in color because iron in the soil is not oxidized. Iron may be oxidized along root channels and other entry points for air and the soil may be mottled with red or orange (Figure 13-13).

Figure 13-13   A gleyed soil from Brazil, showing the typical gray color.

CALCIFICATION

In contrast to the soil-forming environments discussed so far, **calcification** takes place in arid and semiarid regions, where available rainfall better matches potential evapotranspiration. There is still, however, some downward movement of water, particularly after rainfall or snowmelt. This gentle leaching displaces calcium carbonate from the upper horizons and redeposits it at depth as a *calcic* horizon. The other nutrient ions remain in the upper horizon. The calcic horizon is recognizable in the profile by its light color. The degree of differentiation between the A horizon and the calcic horizon and their depths depend on the strength of leaching.

Calcification is typical of the soils found under midlatitude grasslands. Compared to podzolization and laterization, it is a moderate process and creates fertile soils. Most of the nutrients are retained in the upper horizons, the climatic regimes are favorable to soil organisms, and organic material is readily incorporated into the profile. Figure 13-14 shows an example of calcified soil and the ecosystem that it supports.

SALINIZATION

The final major soil-forming environment is **salinization,** which, as the name suggests, occurs when salts, such as compounds of sodium, calcium, and magnesium, accumulate in the profile. Salinization is dominated by *upward* translocation of materials in water. Salinization is typical of arid regions where evaporation is high; it is particularly common in low-lying or flat areas where subsurface drainage is poor. In this environment, moisture evaporates from the surface and the upper horizons creating a moisture gradient between the surface and the lower horizons. Capillary water is drawn up from supplies lower in the profile by capillary action in much the same way as kerosene moves up the wick of a lamp. This water carries dissolved salts upward, and when it evaporates, it leaves salts concentrated near or at the surface (Figure 13-15).

Salty soils in their natural state are virtually useless for agriculture. The soil chemistry can, however, be improved by irrigation and drainage, which flush the salts out of the soil. Ironically, irrigation without adequate drainage only worsens the problem and is a

(a)

(b)

Figure 13-14 (a) A calcified prairie soil from Texas. The soil has a loamy texture, and the deep brown color shows its high organic content. This example has formed over calcareous deposits. (b) The typical grassland community supported by calcified soils east of El Paso, Texas.

Figure 13-15 Salt-affected soil from Botswana. The upper horizons and soil surface are dominated by salt.

major cause of salinization in arid areas. The added water has three effects. First, the water itself contains traces of salts and these are added to the soil. Second, irrigation ensures a continuous supply of water for evaporation, so that salts continue to move up the profile and accumulate near the surface. Third, irrigation may raise the water table to the extent that the plant roots are in a saline solution. Under these conditions normal biological processes are reversed. Instead of drawing water from the soil, plant roots may lose water to the soil solution, causing the dessication and death of the crop.

Each of the soil-forming environments summarizes a particular range and combination of conditions and processes in the soil. They represent the five extreme soil types that make up Earth's surface. In practice, soils vary gradually and continuously within and among the soil-forming environments, creating a detailed mosaic of soil types at Earth's surface, each with slightly different characteristics. Given the importance of soils in the ecosystem and their impact on human activity, it is important that physical geographers make some sort of order out of this variation and communicate information about the nature and distribution of soils in the landscape. The geographer does this through soil classification. The Reference Section (p. 413) outlines the soil classification used in the United States.

The factors controlling soil formation are climate (**atmosphere-ocean environment**), the nature of the parent material (**solid-earth environment**), the effects of **surface relief**, the action of microorganisms (**living-organism environment**), and the time these factors are in operation. The soil-forming factors combine to create five soil-forming environments—podzolization, laterization, gleization, calcification, and salinization.

SOILS AND HUMAN ACTIVITY

Soils are a fundamental part of the biosphere. They help to dictate what plants will grow where and in what quantity, thus deciding what type of ecosystem will develop at a place. Through their impact on the ecosystem, soils also decide the size and distribution of the energy supply available to humans in food crops. Efficient use and conservation of soils has become a key issue in physical geography as the world's population grows and demands for food increase.

Only those soils formed by calcification or weak podzolization can support extractive agriculture in which little or no organic material returns to the soil. Even these fertile soils need occasional periods of rest and recuperation. Fertile soils cover only 11 percent of Earth's ice-free land surface, and their distribution is a major constraint to the distribution of human populations. Optimistic agriculturalists predict that up to 24 percent of Earth's ice-free surface could be cultivated, but this would mean bringing more marginal land into production (see Environmental Issue: Desertification—Human or Climatic Disaster? p. 408).

The most widespread impact of cultivation on soil is soil erosion. Many people associate soil erosion with agriculture in less-developed countries or with historical events such as the American Dust Bowl of the 1930s. It is seen as a problem that can now be mitigated by modern technology in developed countries. This is not the case; soil erosion occurs in all continents. In the developed world, modern farming practices tend to exacerbate rather than solve the problem. Soil is renewed naturally at a rate of about one ton per hectare per year. Erosion rates are currently estimated at between 18 and 100 times this rate. In the United States, the average annual soil loss is 18 tons per hectare. In states with intensive agriculture, such as Iowa and Missouri, the rate is over 35 tons per hectare.

The Dust Bowl on the Great Plains of the United States in the 1930s is a prime example of soil erosion on a massive scale and provides a useful lesson to contemporary soil scientists. It was caused by three factors. First was the agricultural economy of the time. During the First World War (1914–1918), international wheat prices soared, which provided an incentive to extend the acreage of wheat on the Great Plains. The extension was achieved by investment in increased mechanization, particularly the use of tractors and combine harvesters. In time this created a vicious cycle in which continued high wheat production was needed to pay off the loans for the machinery, especially when wheat prices fell after the war. Second, as a result of relentless cropping to maintain profits, the soil was deeply plowed and lost its organic matter. The result was a loose topsoil susceptible to erosion. The third factor was the climate of the area; the Great Plains are naturally dry, with strong winds and a tendency for severe drought. A group of drought years between 1926 and 1934 led to massive wind erosion of the topsoil.

Since this period, a policy of soil conservation has been introduced to the United States. Soil erosion can be physically slowed or halted by minimally disturbing the soil surface in plowing, by using crops that have a high vegetation cover, by covering the soil surface with mulch, and by plowing along the contours or using terraces on steep slopes (Figure 13-16). The real problem, however, lies in persuading the global agricultural community that this long-term investment in soil fertility is preferable to higher profits or a better standard of living in the short term (see Environmental Issue: Soil erosion—Is It Inevitable? p. 410).

In summary, healthy soil is vital to the functioning of the ecosystem and to the capture of energy for human beings. The limited distribution of fertile soils at Earth's surface restricts the distribution and activity of humans. Given the current and future demands for food production, it would seem appropriate to conserve

(a)

(b)

(c)

(d)

Figure 13-16   Methods of controlling soil erosion. (a) Beans planted in wheat straw in Kansas. (b) Strip cropping, in which different crops are planted in narrow parallel belts. (c) Building terraces to prevent erosion on slopes. (d) Plowing along the contours instead of up and down the slopes reduces water erosion.

and nurture this limited resource. The trend, however, seems to be in the opposite direction, with 6 million hectares (roughly the size of West Virginia) going out of production in the world each year because of soil erosion, salinization, and other forms of soil degradation. Cropping becomes uneconomical on a further 20 million hectares each year for the same reason.

Soils are an important resource for humans. Soil quality partly determines the type and amount of plants that can be grown. Soil erosion is a common problem on cultivated land. Overproduction can lead to desertification in dry areas.

ENVIRONMENTAL ISSUE:

Desertification—Human or Climatic Disaster?

Some issues involved in this controversy are:

What are the causes of desertification? How does desertification affect human activities?

Desertification is the degradation of land to the extent that it is unable to support vegetation and becomes desertlike. Estimates of the world area affected by desertification range from 20 to 32 million sq km (7.7 to 12.4 million sq mi) affecting both the developed and developing world.

The human cost of desertification can be high. Harrowing media reports in the early 1970s and again in the 1980s showed millions of people at risk from starvation in the Sahel region of Africa, the area just south of the Sahara desert. The famine was the result of desertification combined with poverty, misguided politics, and poor agricultural economics. In the developed world, the cost of desertification is less tragic, but it affects the ability to produce food. It decreases crop yields, and desertified land must eventually be taken out of production. Salt contamination of soils in the southwestern United States and the drastic wind erosion of the Dust Bowl are both forms of desertification.

The causes of desertification are complex and still not fully understood. The main discussion centers around whether desertification is caused by climatic change or by human mismanagement. The favored theory is that desertification is a long-term process caused by mismanagement of land but exacerbated by drought conditions. In the absence of drought, poor management may lead to desertification anyway. In the absence of poor management, the land could survive short-term or infrequent drought.

The two main characteristics of desertification are degradation of vegetation and degradation of soil. Degradation of vegetation occurs in two ways. First, the overall biomass may be reduced by cutting of firewood, overbrowsing of shrubs, or overgrazing of grasslands, for example. Second, the natural vegetation can be changed to a less productive form. For example, overgrazing may remove some palatable species of grass and allow grasses and shrubs that are more tolerant of grazing to take over. This reduces the grazing quality of the land, putting increased grazing pressure on other areas.

Degradation of vegetation occurs in developing countries for a number of reasons that are rooted in the struggle for development or simply survival. First, the population of these countries is growing rapidly and an increase in agricultural production is needed to feed the people. This brings marginal lands, such as dry areas, into cultivation or increases the number of animals kept there. Second, there has been a gradual change from nomadic grazing to more settled and permanent herding, often based around new wells and boreholes. This concentrates grazing in a smaller area. Third, to increase production of agricultural goods for export, productive land is overcultivated and marginal lands are used for cash crops instead of the subsistence agriculture to which they are suited.

The degradation of vegetation leads eventually to the degradation of soil. Typically the soil is exposed for longer periods and is not bound together by vegetation. It is therefore vulnerable to erosion by wind and by water. Degradation of soils and vegetation are chronic, long-term effects of the human mismanagement of resources. Drought simply tips the balance in favor of disaster.

Linkages

Desertification links the **atmosphere-ocean environment** processes of climate change with soil-forming processes and vegetation cover in the **living-organism environment.** Human intervention alters the delicate natural balance.

Topics for Discussion

1. How can the human causes of desertification be stopped?
2. Can currently degraded lands be salvaged and brought back into production?
3. How can areas at risk from desertification be recognized?

(a) Overgrazing by livestock in arid lands such as this area in Kuwait may cause desertification.
(b) Different levels of grazing either side of this fence line in Harding County, New Mexico have had dramatically different effects on the vegetation.

ENVIRONMENTAL ISSUE:

Soil Erosion—Is It Inevitable?

Some of the issues in this debate include:
   What causes soil erosion?
   How far can soil management reduce soil erosion?

Soil erosion is a natural process in the surface-relief environment. Cultivation hastens soil erosion by loosening the topsoil and exposing it directly to wind and water. The finest and the lightest mineral particles are removed, as is much of the organic matter. This loss reduces the nutrient content and the nutrient and water-holding capacity of the soil. The soil particles are eventually deposited elsewhere, which can lead to silting in irrigation and water conservation works.

Soil erosion in the American Dust Bowl affected all segments of society in the 1930s. Farmers lost their crops and their land, city-dwellers dealt with choking dust-storms, and the nation as a whole lost food resources, all in a period of considerable economic depression. Not surprisingly, there was widespread support throughout society for measures to reduce soil erosion. Farmers, aware of the risks to future productivity, took the advice of soil erosion control agencies along with the government subsidies that allowed them to maintain their incomes while changing farming practices. With no repeat of the drought years, the worst effects of soil erosion on the Great Plains were brought under control quite rapidly, and soil erosion dropped down the political agenda.

From the 1970s onwards, however, soil scientists made a concerted effort to change the existing policy and legislation on soil erosion. They demonstrated that much public money was wasted in short-lived conservation measures in which a farmer could enroll land in a conservation scheme for a short time (one or two years) and then return immediately to the original, eroding practice without penalty. Further, much of the most erodible land was also highly productive and was never enrolled in these schemes.

To the individual farmer, soil conservation was simply a waste of money. The major off-site costs of soil erosion, such as silting and contamination of water supplies, were not borne by the agricultural community. In contrast, the cost of soil conservation measures was the landowner's responsibility.

Mounting criticism of soil conservation policy led to new legislation in 1985 in the form of the Conservation Title of the Food Security Act of 1985. The Sodbuster provisions in this legislation allowed highly erodible land to be brought into production only if a conservation scheme was implemented. Failure to implement such a scheme barred the farmer from various government benefits. This and similar schemes were not as effective as hoped by conservationists because not all farmers receive government funds. The Conservation Reserve Provisions pay farmers to retire land from production. This scheme had limited success to start with because the rents were too low to attract the most productive land. More productive and erodible land has been enrolled now that the rents are higher, but the scheme is an expensive use of public money.

The Conservation Compliance Provisions are potentially the strongest pieces of legislation in the Conservation Title. These provisions demand that all farmers on erodible land should develop a conservation plan, to have been implemented by 1 January 1995. The initial criteria of acceptable rates for residual erosion were, however, *increased* by staff of the Soil Conservation Service in 1990. If implemented, these new rates would allow some farmers to comply with the regulations while still supporting a soil loss of 10 tons per hectare per year. The legality of the revised rates has been challenged in court.

One major lesson from this legislation is that even the soil conservation agencies are unwilling to "get tough" and force the agricultural community, and its consumers, to bear the real costs of soil erosion. In many countries, agricultural communities have a certain degree of exemption from government constraint, and this situation is economically and politically difficult to reverse. It is further complicated by conflicting government priorities, for example, the desire to increase productivity and to decrease soil erosion, all at minimum cost.

Linkages

Soil erosion links the **atmosphere-ocean environment** processes with the **surface-relief environment** and the soil-forming processes in the **living-organism environment.** It is a natural process exacerbated by human activities.

Topics for Discussion

1. Who should pay for soil conservation measures?
2. Who should pay for the off-site damage caused by soil erosion?
3. How compatible are the desires to improve productivity and to control soil erosion?

Erosion from a corn field.

(a)

Dust storm approaching a small western Kansas town.

(b)

Soil accumulation on a deserted homestead in Dallas County, Texas in 1937. The accumulations are approximately 5 feet high due to soil collecting in weeds in the fence rows.

(c)

CHAPTER SUMMARY

1. Soil forms as the result of three processes—weathering of a parent material, the addition of organic matter, and the movement or translocation of material in the soil profile. Soil has four components—the mineral fraction (*solid-earth environment*), organic matter (*living-organism environment*), water and air, (*atmosphere-ocean environment*).

2. The soil-forming processes act on the soil contents, creating different conditions of texture, structure, chemistry, and color. A biogeographer uses such conditions to understand the processes operating in the soil.

3. The factors controlling soil formation are climate (*atmosphere-ocean environment*), the nature of the parent material (*solid-earth environment*), the effects of *surface relief*, the action of microorganisms (*living-organism environment*), and the time these factors are in operation. The soil-forming factors combine to create five soil-forming environments—podzolization, laterization, gleization, calcification, and salinization.

4. Soils are an important resource for humans. Soil quality partly determines the type and amount of plants that can be grown. Soil erosion is a common problem on cultivated land. Overproduction can lead to desertification in dry areas.

KEY TERMS

CHAPTER REVIEW QUESTIONS

1. Draw an annotated diagram to show how Earth environments interact to form soil.

2. Outline the stages in the development of a soil profile.

3. Explain the difference between soil structure and soil texture and describe the importance of each as a soil characteristic.

4. Using the photographs of soil profiles in this chapter, show how useful soil color is in showing what processes are going on in the profile.

5. What are the similarities and differences between the soil-forming environments of podzolization and laterization?

6. Outline the relative importance of each of the soil-forming processes and factors (climate, parent material, etc.) in each soil-forming environment. Create a table showing your results.

7. Describe some of the impacts of human activity on soils.

SUGGESTED ACTIVITIES

1. Discuss what physical geography gains from and contributes to the study of soils.

2. Examine soil profiles exposed in road cuts in your area. Describe their visible characteristics, such as color, number, position, and clarity of horizons. Determine which soil-forming environment is dominant in the profile.

3. Use a soil map and the Reference Section on soil classification to learn about the nature and distribution of soils in your area. What relationships can you find among soil, topography, and land use?

REFERENCE SECTION

GLOSSARY

The terms included in the glossary are those listed as Key Terms at the end of Chapters 1–13. The chapter in which they are first defined is indicated in parentheses.

A

ablation. The process of ice thinning when melting and evaporation exceed accumulation. (10)

absorption. The retention of radiated energy, such as insolation, instead of its scattering or transmitting. The process of absorption transforms the radiated energy to another form, often heat energy. (3)

abyssal plains. Wide, relatively flat areas on ocean floors found at depths between 3500 and 4000 meters that are covered by layers of fine sediment. (7)

acidity. The number of hydrogen ions present in a solution; defined by a pH value of less than 7. (13)

adiabatic expansion. The cooling of a rising parcel of air without loss of heat to its surroundings. As the air rises, it expands in air of lower density and the heat energy it contains is spread through its greater volume, lowering the temperature. (4)

adiabatic lapse rate. The rate at which temperature falls in a rising body of air. (4)

aeolian processes. Wind-based processes. (10)

air mass. A large body of air in contact with the ground and having relatively uniform internal conditions of temperature and humidity. (5)

albedo. The proportion of insolation that is reflected; it varies according to the reflecting surface. (3)

alluvial fan. A fan- or cone-shaped stream deposit. (9)

alpine glacier. Ice masses that are confined by mountain relief, forming valley and cirque glaciers. (10)

altitude. Height above sea level. (3)

anticyclone. A weather system in which there is high atmospheric pressure at the center. The winds diverge in clockwise patterns in the Northern Hemisphere (counterclockwise in the Southern Hemisphere). (5)

aquifer. A permeable rock layer that has sufficient porosity to hold water and allow it to be withdrawn for use. (9)

argon (Ar). One of the three major atmospheric gases; the least chemically active. (2)

aridity. Dry conditions at a place with low precipitation or with an evaporation rate that exceeds the precipitation. (4)

aspect. The direction in which a slope faces. (6)

atmosphere. A thin envelope of gases, dust, and water droplets surrounding Earth, formed largely of nitrogen, oxygen, and argon, and divided into layers by temperature. (2)

atmosphere-ocean environment. The major Earth environment that includes the gaseous atmosphere and ocean waters and the interactions between them. (1)

atmospheric pressure. The force exerted by the weight of the atmosphere. (2)

atom. The smallest particle of an element that can exist alone or combine with atoms of other elements. (2)

autotroph. A plant that is able to manufacture carbohydrates in the process of photosynthesis. (12)

B

barrier beach. An elongated beach form lying parallel to the coastline. (11)

basal sliding. The process of glacier movement in which meltwater lubricates movement of the glacier base. (10)

baseflow. The groundwater contribution to streamflow. (9)

base level. The lowest level of continental erosion, usually sea level, but may be inland lakes. (8)

beach. Depositional landforms in the coastal zone. (11)

bed load. The coarsest fraction of a stream's load, usually moves in contact with the stream bed. (9)

biogeochemical cycle. A pathway along which nutrients move between the living and nonliving parts of the ecosystem. (12)

biologic weathering. Process of rock disintegration that involves the action of living organisms. (8)

biome. A regional ecosystem in which the climatic conditions, resultant soil development, and dominant growth-form of vegetation are more or less uniform. (12)

biosphere. The living-organism environment, created by the interacting margins of the atmosphere-ocean, solid-earth, and surface-relief environments. (12)

boreal forest. Coniferous forest in cold continental interiors. (12)

boundary current. A surface ocean current that flows south-to-north on the western sides of oceans and north-to-south on the eastern sides. (3)

braided stream. A stream channel in which the waterflow divides around channel bars. (9)

C

calcification. A soil-forming process in which gentle downward percolation of rainwater washes calcium carbonate into the lower horizons but leaves other nutrients near the surface. (13)

calorie. One calorie is the amount of heat needed to raise the temperature of 1 gram of water 1° C at sea-level pressure. (3)

capacity. *See* stream capacity.

capillary water. Molecules of water that are loosely held in the soil against the force of gravity. Capillary water travels through the soil in response to pressure gradients and is the main source of water for plants. (13)

carbonation. The process of chemical weathering in which rainwater (a weak carbonic acid) reacts with calcium carbonate (limestone). (8)

carbon dioxide. A minor atmospheric gas capable of absorbing long-wave radiation; important in photosynthesis. (2)

carnivore. An organism, usually an animal, that eats only meat. (12)

cation exchange capacity. A measure of the number of nutrient cations that can be absorbed by a unit of soil, commonly used as an indicator of soil fertility. (13)

cavern. A cave (underground cavity) system in limestone rock. (9)

chelation. A process of chemical weathering that is dependent on organic matter and acids. (8)

chemical weathering. The process of rock disintegration that involves chemical reactions. (8)

chlorophyll. The green pigment in plant leaves that receives sunlight in the first step of photosynthesis. (12)

cirque. An amphitheater-shaped rock basin eroded by a glacier, usually formed just below high-mountain peaks, which often contains a lake or the glacier itself. (10)

cirriform cloud. High-level cloud composed of ice particles. (4)

clay-humus complex. A mixture of tiny particles of clay and humus in the soil. The particles have surface electrical charges that attract and hold nutrient ions. (13)

cliff. A steep rock face, common in coastal zones where wave attack is strong and the rocks are cohesive. (11)

climate. The long-term condition of the atmosphere as measured by average conditions and the range of weather experienced. (2)

climate change. A change in climatic conditions at a location that involves a shift of the average conditions of temperature, rainfall, or other climatic factors. (6)

climatic environment. The surrounding conditions that produce the measured characteristics of a region's climate, including heat and water balances, surface type, and weather systems. (6)

climograph. A diagram that includes graphs and that illustrates the general climatic conditions of a place. (6)

cloud. Condensed masses of water droplets, ice particles, or both, in the atmosphere above the ground. (4)

cloud cluster. A group of cumulonimbus clouds that forms near the equator. Cloud clusters are responsible for a high proportion of tropical rainfall. (5)

coastal zone. The zone from just below low tide to just above high tide where marine processes have an important influence on landforms. (11)

cold front. A front at the leading edge of an advancing cold air mass. (5)

cold ice sheet, glacier. An ice sheet or glacier where the ice remains at very low temperatures, tens of degrees below freezing. (10)

colloid. The smallest fragment of mineral material in soil. Colloids form a major part of the clay-humus complex. (13)

competence. *See* stream competence.

compound. A substance formed by the bonding of different elements, for example, water (H_2O). (2)

condensation. The process by which water vapor changes to liquid water. (4)

conduction. The transfer of heat energy from a hot to a cool substance by molecular contact and without movement within or between the substances. It is most effective in solids. (3)

continent. A large area of land, standing above sea level by virtue of the presence of relatively light sial rocks. (7)

continental rise. An area on an ocean basin's margin with a surface slope up toward the continent. It is a wedge-shaped deposit composed of debris washed from the continent that thins on the ocean side. (7)

continental shelf. The top surface of a wedge of sediment formed on a continental margin, usually covered by less than 130 meters of water. (7)

convection. The transfer of heat energy within a substance by circulatory movement resulting from differences in temperature (and density). It is most effective in liquids and gases. (3)

convergent plate margin. A plate margin where plates are colliding and plate material is destroyed. (7)

converging winds. Winds that blow toward a common center, causing air to pile up in the middle and rise. (3)

coral reef. A rock structure composed of limestone that is formed just below sea level by colonies of coral animals. (11)

core. Innermost layer of Earth's rocks. (7)

Coriolis effect. The deflection of winds, to the right in the Northern Hemisphere and to the left in the Southern Hemisphere, resulting from Earth's rotation. (3)

corrasion. The mechanical erosion of rock as other rock fragments move across it in running water, flowing ice, wind, or seawater, abrading the rock. (9)

corrosion. Denudation of rock by chemical processes active in a moving agent such as running water. (9)

counter-radiation. Long-wave radiation from the troposphere gases and clouds back to the Earth's surface. (3)

crevasse. Crack in surface ice that may be 50 meters deep. (10)

crust. The outer layer of the solid-earth environment, composed of sial (continental crust) and sima (denser ocean-floor crust). (7)

cumuliform clouds. Separate, heaplike clouds formed by the convection of bubbles of air. (4)

cyclone. A weather system in which there is low atmospheric pressure at the center. The winds converge in counterclockwise patterns in the Northern Hemisphere and clockwise patterns in the Southern Hemisphere. (5)

D

decomposers.   Organisms that gain their energy by digesting dead plant and animal tissues in the detrital food chain. Most decomposers are small, or microscopic, organisms living in the soil. (12)

deflation.   Wind action that removes unconsolidated clay, silt, or sand from a land surface. (10)

delta.   An alluvial deposit at a river mouth, where fluvial and marine processes interact. (11)

density.   The concentration of matter; mass per unit volume (for example, kg/m^3). (2)

deposition.   The dropping of particles transported by running water, flowing ice, wind, or the sea into low-energy environments. (8)

desert biome.   An ecosystem that develops in areas where potential evapotranspiration far exceeds input from precipitation. (12)

desert pavement.   An area of bare stones forming a crust in an arid environment. (10)

dew.   Condensation of water vapor in contact with the ground when the vapor is cooled below saturation point. (4)

dew point.   The temperature at which condensation begins in a cooling parcel of air; the maximum temperature at which a relative humidity of 100 percent is reached. (4)

discharge.   The flow of water in a stream, measured in volume per unit of time. (9)

dissolved load.   The portion of a stream's load carried in solution. (9)

divergent plate margin.   A plate margin where two plates move apart and new plate material forms in the resulting gap. (7)

diverging winds.   Winds that blow outward, or away from, a common central area. (3)

doline.   A surface depression on limestone rock in a karst landscape. (9)

downburst.   A strong downdraft of air in a thunderstorm. (5)

drainage divide.   The perimeter of a watershed or drainage basin. (9)

drumlin.   A streamlined, elongated hill formed of glacial till. (10)

dune.   A deposit of windblown sand forming a mound or ridge. (10)

E

earthquake.   A shock, or series of shocks, generated by sudden movement in Earth's crust or upper mantle. (7)

easterly wave.   A tropical weather system on the equatorial margins of subtropical anticyclones that is characterized by a wavelike feature in the isobars. (5)

ecologic niche.   The set of environmental conditions in which an organism can survive and compete successfully with other organisms. (12)

ecologic succession.   The growth and development of an ecosystem over time. (12)

ecosystem.   An organized system made up of plants, animals, and the nonliving environment to which they are linked by flows of energy and materials. (12)

element.   A chemical substance that has a distinct formula. (2)

El Niño current.   A warm ocean current that flows south off Colombia and Ecuador between Christmas and Easter, and that occasionally extends farther south for longer periods (the enhanced El Niño current). (6)

eluviation.   The physical removal of particles in suspension from one part of the soil to another, usually from upper to lower horizons. (13)

emergent coastal landforms.   A coastline that shows evidence of the sea level falling relative, or the land rising relative, to sea level. (11)

environment.   The sum total of conditions in which an organism lives. (1)

environmental lapse rate.   The rate at which temperature decreases with increasing height in the troposphere. (3)

epeirogenesis.   Movements of Earth's crust that produce broad upward or downward warping of the surface. (7)

equable climate.   A climatic environment in which there is little difference between summer and winter temperatures. (6)

equator.   The line of latitude (0°) that lies equidistant from the poles and is at right angles to Earth's axis. (1)

equinox.   The days (approximately March 21 and September 21) when the sun is overhead at the equator and all parts of the world have 12 hours of daylight and 12 hours of night. (3)

erosion.   The wearing away of land surface and removal of rock debris by agents such as running water, flowing ice, wind, or the sea. (8)

esker.   A narrow winding ridge of sand and gravel formed in a stream flowing beneath ice. (10)

estuary.   A river mouth that widens toward the sea and is influenced by tidal changes. (11)

evaporation.   The process by which a liquid, such as water, is changed to a gas, such as water vapor. (4)

evapotranspiration.   The total transfer of liquid water to water vapor at Earth's surface: evaporation plus transpiration. (4)

evolution.   The gradual development of new species. (12)

exfoliation.   The process by which layers of rock break into sheets when exposed at the surface. (8)

F

fall.   A type of rapid, downward mass movement in which blocks of rock move freely without requiring lubrication. (8)

fault-block mountain.   A mountainous area where the rocks have been uplifted between faulted boundaries. (7)

faults.   A form of rock deformation in which the layers are broken and displaced. (7)

field capacity.   The condition at which a soil contains the maximum amount of capillary water but no excess gravitational water. (13)

fjord.   A narrow inlet of the sea in a mountain region formed by the drowning of a glaciated valley. (11)

floodplain.   The floor of a river valley, adjacent to the channel, that is covered by floodwater. (9)

flow.   A type of mass movement in which water or ice lubricates the regolith being moved. Varieties include earthflows, mudflows, and avalanches. (8)

fog.   Condensation in the air in contact with the ground so that visibility is reduced to under one kilometer (0.62 mi). (4)

folds.   A form of rock deformation in which the rock layers are contorted, but not broken. (7)

food chain.   A linked sequence of plants and animals in the ecosystem through which energy is transferred. (12)

food web.   A number of interlinked food chains. (12)

front.   A narrow zone in the atmosphere between two contrasting air masses. (5)

frost.   Condensation on ground that is cooled below freezing point. (4)

frost wedging.   A form of physical weathering in which the alternate freezing and thawing of water in a rock exerts pressure and splits the rock apart. (8)

G

general circulation model.   A mathematical model based on a two-dimensional grid of surface climate observations, sometimes including vertical layers as well. Such models demand very powerful computers and are used to predict future climate. (6)

Geographic Information System (GIS).   A computer database used to collect, store, retrieve, analyze, and display spatial data. (1)

geologic (rock) cycle.   The circulation of mineral matter in Earth's crust and upper mantle that produces different rock types. (7)

geostrophic wind.   A wind that flows parallel to the isobars as a result of a balance between the pressure gradient force and the effect of Earth's rotation. (3)

gleization.   A soil-forming environment characteristic of waterlogged and poorly aerated soils. Little oxygen is available and this restricts the presence and activity of soil organisms. (13)

global warming.   A trend toward rising temperatures in the atmosphere-ocean environment. (6)

gradation.   The sum total effect of processes that act to wear down land areas. (8)

gravitational water.   Soil water that is not held in the soil but moves freely under the influence of gravity. (13)

greenhouse effect.   The relationship between Earth's atmosphere and insolation in which the atmosphere is transparent to incoming short-wave radiation, but traps outgoing long-wave heat radiation. (3)

ground ice.   Ice that occurs in soil or rock, either filling pore spaces or forming larger wedges or masses. (10)

groundwater.   Water that accumulates in aquifer rocks below the water table. It is the main source of stream baseflow. (9)

gully.   A small valley excavated by flowing water, usually on unvegetated land, in which water flows occasionally. (9)

gyre.   A circular flow of ocean surface water, combining several currents. (3)

H

Hadley cell.   The vertical convectional cell in the tropical atmosphere, incorporating the surface trade winds, ascending air above the inter-tropical convergence zone, northerly flow aloft, and descending air in the subtropical high-pressure zones. Named after G. Hadley (1735). (3)

hail.   The form of atmospheric precipitation that is composed of lumps of ice, which may have an internal structure of concentric layers. (4)

hammada.   A flat, bare-rock surface in a desert that may be partly covered by a lag deposit. (10)

heat.   Heat is the total molecular movement in a substance: the greater the molecular movement, the greater the heat energy and the higher the temperature. (3)

heat balance.   The balance between insolation and terrestrial radiation in the atmosphere. (3)

heave.   A form of mass movement in which wetting and drying, or freezing and thawing, of the ground surface causes it to rise and fall. It causes patterned ground on flat surfaces and creep on slopes. (8)

herbivore.   An animal that eats only plants. (12)

heterotroph.   Describes an organism that cannot photosynthesize and must rely on autotrophs and other heterotrophs for its energy supply. (12)

homeostatic mechanisms.   The mechanisms by which an ecosystem maintains a stable equilibrium with its environment by adjusting to any disruption in flows of energy or materials. (12)

humidity.   The amount of water vapor present in the atmosphere. (4)

humus.   A structureless brown or black material that is the final product of the breakdown of organic material in the soil. The tiny organic particles join with clay colloids to form the clay-humus complex. (13)

hurricane.   A tropical cyclone that occurs in the northern Atlantic Ocean. (5)

hydrologic cycle.   The circulation of water from ocean to atmosphere, from atmosphere to the land surface, and across the land surface back to the ocean. (4)

hydrolysis.   A form of chemical weathering in which water combines with rock minerals to form clay minerals and sand particles. (8)

hygroscopic water.   Water held tightly to the surface of individual mineral grains in the soil. It is unavailable to plants because the attraction between the water and the mineral grain is greater than the "sucking power" of a plant's roots. (13)

I

ice abrasion.   The erosion of rock by ice, in which rock fragments contained in the ice scrape the rocks beneath and along the sides of the glacier or ice sheet. (10)

ice age.   A time when ice sheets formed over the polar areas and spread out to cover high-latitude regions. (6)

ice cap.   A small ice sheet. (10)

ice plucking.   The erosion of rock by ice, in which the basal ice freezes around blocks in well-jointed rock and pulls them free. (10)

ice sheet.   A large continuous layer of ice that may cover all or most of a continent. (10)

ice wedge.   Ice that fills cracks in frozen soil in permafrost conditions. (10)

igneous rocks.   Rocks that are formed by the solidification of molten rock after it has migrated from the point of melting to the point of cooling. (7)

illuviation.   The redeposition of materials in a soil following their transport in water. (13)

infiltration.   The process by which water enters the soil. (9)

insolation.   The flow of radiant energy from the sun, mainly in shorter wavelengths. (3)

instability in the atmosphere.   The state of a rising body of air when it remains warmer than the surrounding air and continues to rise. (4)

interception.   The process by which precipitation is prevented from direct impact on the ground by its vegetation. (9)

internal deformation.   The process of crystal deformation under pressure that causes ice masses to move. (10)

Inter-Tropical Convergence Zone.   The line, close to the Equator, where northeast and southeast trade winds converge. (3)

ions.   Atoms or molecules that have a positive or negative charge; they occur in solutions and in the outer atmosphere. (2)

island arc.   An arc-shaped line of volcanic islands formed on one side of an ocean trench where two oceanic plates collide. (7)

isobar.   A line drawn on a map joining places of the same atmospheric pressure. (3)

isoline.   A line used in cartography that joins points of equal value. (1)

isostasy.   Changes in the height of Earth's continental surface as a result of adding, or removing, loads such as sediment or ice. (7)

isotherm.   A line drawn on a map joining places of the same atmospheric temperature. Isotherms are usually surface lines and may be corrected to a sea-level value. (3)

J

jetstream.   Strong wind near the top of the troposphere. (3)

K

kame.   A landform resulting from deposition along the sides or margins of a glacier or ice sheet by glacial meltwater. (10)

karst landscape.   A landscape on limestone rock that is characterized by surface depressions and underground caverns. (9)

knickpoint.   A change of slope on a stream course, in which a gentler upper gradient gives way to a steeper lower one. (9)

Köppen classification of climate.   An empiric climate classification based on the average monthly and annual temperature and moisture characteristics that delineate natural vegetation regions. (6)

L

landform.   The shape of Earth's continental surface identified as specific features such as mountains, volcanoes, river valleys, coastal cliffs, etc. (7)

lapse rate.   The rate of temperature change with height—a vertical temperature gradient. (3)

latent heat.   The amount of heat absorbed or released when a body changes its state or phase without any change of temperature in the body. In water, heat is absorbed as latent heat during the melting of ice and evaporation of liquid water; it is released during condensation and freezing. (4)

laterization.   A soil-forming environment in which strong leaching and aggressive weathering create a deep, red soil that is acidic and nutrient-deficient. (13)

latitude.   The angular position of a place north or south of the equator. Used with longitude to give a precise location to all places on Earth's surface. (1)

lava flow.   A layer of molten igneous rock moving across Earth's surface. (7)

leaching.   The transport of materials in solution through the soil profile, typically a downward washing of nutrients in rainwater. (13)

lightning.   A visible flash that results from electrification within a thunderstorm. (5)

living-organism environment.   The major Earth environment that is composed of plants and animals. It is closely linked with, and dependent on, the other major environments. (1)

load.   *See* stream load.

loess.   A deposit of silt and clay that has been carried by wind. (10)

longitude.   The angular position of a place east or west of the prime meridian, an arbitrary reference line running from North Pole to South Pole. Longitude is used with latitude to give a precise location to all places on Earth's surface. (1)

longshore drift.   The movement of sand and pebbles along the shore when waves approach at an angle to the shore. (11)

M

magma.   A mobile mixture of liquid and gaseous minerals that cools to form an igneous rock. (7)

mangrove.   Plant that is able to send roots down through shallow water and have a growth of leaves above the water. (11)

mantle.   The layer of the solid-earth environment between the crust and core. It is mostly solid, with the weak asthenosphere near its upper boundary. (7)

mass movement.   The movement of rock and unconsolidated regolith downhill, largely by the influence of gravity. (8)

meandering channel.   A winding stream channel in which the outer sides of bends erode, causing steep banks, while deposition occurs on the inside banks. (9)

Mediterranean forest and scrub.   A dry and scrublike vegetation that develops where winters are mild and humid and summers are hot and dry. This vegetation type is called chaparral in the southwestern United States. (12)

meridian.   Imaginary reference line around the globe from the North to the South Pole that joins places of the same longitude. (1)

metamorphic rock.   Rock that is formed by the alteration of minerals under extreme heat or pressure, but without complete melting or migration. (7)

midlatitude anticyclone.   A high-pressure weather system that occurs in middle latitudes in association with cyclones or as a larger "blocking" form. (5)

midlatitude cyclone.   A low-pressure weather system that occurs in middle latitudes. Most have fronts as the result of converging contrasting air masses, but some do not. (5)

midlatitude forest.   A forest type dominated by deciduous trees. Midlatitude forest is the natural vegetation of most of western Europe and parts of the eastern United States. It has mostly been cleared for agriculture. (12)

midlatitude grassland.   Grassland vegetation found in the drier parts of the midlatitudes; the natural vegetation of the American prairies and the Eurasian steppes. Most of this biome has been replaced by arable agriculture. (12)

mineral.   Naturally occurring combinations of chemical elements bonded together in orderly crystalline structures. (7)

molecule.   A combination of atoms held together by electric bonds. (2)

moraine.   A landform produced by the deposition of glacial till. Examples include terminal moraines and ground moraines. (10)

mountain biomes.   The ecosystems associated with high mountains. (12)

N

natural environment.   The set of physical and biologic conditions that surround human beings at Earth's surface. (1)

net primary production.   The amount of energy the autotrophs fix into an ecosystem over and above their own needs. (12)

nitrogen (N).   The most common gas in Earth's atmosphere. (2)

nutrients.   The chemical elements that are needed for life. (12)

O

occluded front.   A front in a midlatitude cyclone where the cold front has caught up with the warm front and lifted the entire warm sector off the ground. It forms at the end of the life cycle of a midlatitude cyclone. (5)

ocean.   A major water storage area on Earth's surface, fills an ocean basin. (2)

ocean basin.   A structural depression in Earth's crust that is filled by water and floored by mafic igneous rock. (7)

ocean biomes.   Living environments in the oceans. (12)

ocean ridges.   Raised areas along divergent plate margins where ocean floor is created. (7)

ocean trench.   An elongated deep in the abyssal plain formed along a convergent plate margin; the deepest part of an ocean basin. (7)

omnivore.   An organism that eats both plants and animals. (12)

orogenesis.   The process of mountain formation. (7)

outwash plain.   A depositional landform created by the meltwater deposition of sand and gravel in front of an ice sheet or ice cap. (10)

overland flow.   The surface movement of water derived from precipitation, usually in a shallow sheet. (9)

oxidation.   The process of chemical weathering in which oxygen dissolved in water reacts with iron and other rock minerals to form oxides. (8)

oxygen (O_2).   The second gas of the atmosphere; an important product of photosynthesis. (2)

ozone (O_3).   A minor atmospheric gas, present near the surface, but concentrated in the upper stratosphere, where it absorbs ultraviolet rays. (2)

P

parallels of latitude.   Imaginary reference lines that form concentric circles around the globe and join places of the same latitude. (1)

patterned ground.   Polygonal patterns of stones on the surface in permafrost areas, produced by frost heaving. (10)

percolation.   The process by which water moves downward through soil and into rock via pores and joints. (9)

periglacial environment.   The area that lies in front of an ice sheet in polar latitudes, where the climate and landform processes are controlled by very low temperatures. (10)

permafrost.   Permanently frozen ground, occurs in very cold parts of the world. (9)

permeability.   The rate at which liquids or gases will pass through a substance; often used in relation to water passing through soil or rock. (9)

photosynthesis.   The manufacture of carbohydrates from water and carbon dioxide in the presence of sunlight. (12)

physical geography.   The study of the distribution of weather, climate, landforms, plants, and animals, together with an understanding of the factors causing such distributions. (1)

physical weathering.   Processes that cause rocks to disintegrate by changes of temperature and other forms of mechanical stress. (8)

pingo.   A small, circular hill underlain by an ice mass. (10)

plate.   A large, virtually rigid section of lithosphere. (7)

plate tectonics.   The idea that subsurface convection currents cause plates to move and thereby change the positions and relief of the continents and ocean basins. (7)

podzolization.   A soil-forming environment characterized by acid conditions and excessive downward movement of water, causing a leached and nutrient-deficient upper horizon and an illuviated lower horizon. (13)

polar easterlies.   Winds that blow from the east polewards of 60° North and South. (3)

pollution.   Materials that are added to the ocean or atmosphere by human activities and that degrade the quality of the medium. (2)

porosity.   The proportion of void space in a rock or soil. The voids may be filled with air or water. (8)

pressure gradient.   The difference in pressure per kilometer between high and low centers of atmospheric pressure. (3)

processes.   The forces at work to accomplish change in Earth environments. (1)

R

radiation.   The transfer of energy by means of electromagnetic waves. (3)

rain.   The form of atmospheric precipitation that is composed of large drops of liquid water. (4)

reduction.   The process of chemical weathering in which oxygen is removed from minerals, usually in conditions where groundwater contains little oxygen. (8)

reflection.   Radiation that bounces off a substance without changing its own wavelength or affecting the substance. (3)

regolith.   A layer of loose rock fragments, sand, and clay particles formed as the result of rock weathering or deposition following mass movement, streamflow, ice movement, or wind action. (8)

relief.   The physical form of Earth's surface. It is made up of a range of landforms of varying size, height, and shape. (7)

remote sensing.   The use of aerial photographs or satellite images to gather information about Earth's environments. (1)

river.   A stream of water flowing in a channel from high to low land. (9)

rock.   A compact, consolidated mass of mineral matter; the minerals may be interlocked following crystallization or rock fragments cemented together. (7)

rock cycle.   *See* geologic cycle.

Rossby wave.   The path of the polar-front jetstream forms these wave patterns as it circles the globe in midlatitudes. There are generally four to six Rossby waves around the globe. (3)

S

Sahel.   The zone along the southern edge of the Sahara that has a semiarid climate and is subject to prolonged periods of drought. (6)

salinity.   The proportion of dissolved salts in water, measured as parts per thousand. (2)

salinization.   A soil-forming environment in which water is drawn to the soil surface, depositing salts in the upper horizon and at the surface. (13)

saltation.   The movement of sand when the wind causes particles to bounce across the surface. (10)

salt marsh.   A vegetation community that develops on mudflats in intertidal conditions; the plants tolerate saline and fresh water at high tide and exposure to air at low tide. (11)

saturated air.   Air that contains the maximum water vapor for a specific temperature and pressure. Any further addition will be balanced by condensation. (4)

savanna.   Grassland vegetation found in the tropics. (12)

scale.   The mathematical relationship between a distance on the ground and its representation on a map. (1)

scattering.   The dispersal of radiation in all directions after hitting a substance; does not change the wavelength of radiation or affect the substance. (3)

sea ice.   The formation of ice by the freezing of the sea surface. (10)

sedimentary rock.   Rock that is formed from products of the breakup of other rocks, from organic debris, or from minerals precipitated out of solution; hardened sediment. (7)

sensible heat.   Heat energy in the atmosphere that changes atmospheric temperature when it is transferred by convection or conduction. (4)

shield.   A section of continental crust mainly composed of metamorphic rocks older than 570 million years. (7)

shore platform.   A coastal landform that is either horizontal or sloping gently seaward and that is formed by wave action, weathering, or deposition of calcium carbonate. (11)

slide.   A form of mass movement that involves movement of a rock section along a slide plane. (8)

slope.   An inclined surface. Landform slopes are individual units having an area, slope angle, and orientation. (8)

smog.   Fog that is caused or enhanced by human pollutants. Originally *"smoke-fog,"* it is now used for reductions in visibility caused by vehicle exhausts and other sources. (2)

snow.   The form of atmospheric precipitation that is composed of ice crystals. (4)

snow line.   The altitude of the lower limit of permanent snow cover. (6)

soil.   The dynamic layer of water, air, mineral, and organic materials at Earth's surface in which most green plants grow. (13)

soil horizon.   A horizontal layer in the soil differentiated by its contents, characteristics, and processes from other layers. (13)

soil pedon.   A conceptual three-dimensional unit of more-or-less uniform soil used as a mapping unit. (13)

soil profile.   A vertical slice through a soil from the surface down to the parent material. (13)

soil solution.   The liquid part of the soil, consisting of soil water, dissolved elements, and suspended particles. (13)

soil structure.   A description of how individual grains of mineral material are bound together into clumps or peds. (13)

soil texture.   The relative proportions of sand, silt, and clay in a soil. (13)

solid-earth environment.   The major Earth environment that is composed of rocks, including the core, mantle, and crust, and in which movements are caused by internal sources of energy. (7)

solstice.   The days in the year when the sun is overhead at the Tropic of Cancer (approximately June 21) and Tropic of Capricorn (approximately December 21). (3)

solution.   The process of chemical weathering in which matter is changed from a solid or gaseous state into a liquid one by combining with a solvent such as water. (8)

specific heat.   The amount of heat required to change the temperature of one gram of a substance by one degree Celsius. (3)

spring.   A point at which groundwater flows out on the surface. (9)

stability in the atmosphere.   The state of a rising body of air when it becomes cooler than the surrounding air and falls back toward the ground. (4)

storm wave.   A high-energy sea wave that is generated in the midlatitude storm belts. (11)

stratiform cloud.   Layer clouds forming a structureless gray sheet that covers the sky. (4)

stream capacity.   The total load that a stream can carry at a particular velocity. (9)

stream channel.   The landform in which a stream flows, defined by its bed and banks. (9)

stream competence.   The largest particle that a stream can move at a particular velocity. (9)

stream load.   The material that a stream carries, including dissolved, suspended, and bedload fractions. (9)

structure.   An organized group of connected items. (1)

submerged coastal landforms.   A coastline that results from the rise of sea level or the sinking of land; for example, valley mouths that are drowned. (11)

subsidence.   The mass movement process in which the land surface is lowered by removal of underlying rock or soil. (8)

subtropical high-pressure system.   Semipermanent areas of high pressure centered over the Atlantic, Pacific, and Indian oceans between 20° and 30° North and South. (3)

surface creep.   The movement of sand by the wind in a rolling and sliding mechanism caused by wind drag and particle collisions. (10)

surface-relief environment.   The major Earth environment in which atmospheric processes interact with the rocks at the surface of the solid-earth environment. (1)

suspended load.   The materials that a stream carries in suspension. (9)

swell wave.   A wave in the sea characterized by regular occurrence and a long wavelength in relation to its height. (11)

T

temperature.   The measure of heat contained in a substance. (3)

temperature gradient.   The rate of temperature change with distance or height (*see* lapse rate). (3)

temperature inversion.   A lapse rate that increases (instead of decreases) with increasing altitude for a section of the troposphere. (3)

terrestrial radiation.   Long-wave radiation from Earth's surface (heat). (3)

thermocline.   The zone in the oceans separating surface warm water and deep cold water. (2)

thunder.   The sound generated by the discharge of electricity when lightning occurs in a thunderstorm. (5)

thunderstorm.   Single or multiple cumulonimbus clouds that produce heavy rain, thunder, lightning, and strong downdrafts of air. (5)

tidal current.   A flow of water in response to tidal changes, either in-and-out of a river mouth or around an island. (11)

tidal range.   The difference in height between high-tide and low-tide levels (11)

tide.   The rise and fall of the ocean water level resulting from the gravitational attraction of the sun and moon on Earth. (11)

till.   Sediment deposited by glaciers and ice sheets. (10)

tornado.   A small, violent rotating storm that forms in extremely active thunderstorms and is characterized by a funnel-shaped vortex descending from the cloud base to the ground. (5)

trade winds.   Surface winds in the tropics, blowing from the northeast in the Northern Hemisphere and the southeast in the Southern Hemisphere. (3)

translocation.   The movement of materials in the soil, primarily by water, but also through the action of soil organisms. (13)

transpiration.   The process by which plants pass water vapor into the atmosphere. (4)

transport.   The movement of rock fragments, particles, and dissolved matter by mass movements, rivers, glaciers, the wind, and the sea. (8)

tropical cyclone.   A major cyclonic weather system in the tropics, known locally as a hurricane or typhoon. (5)

tropical rainforest.   Forest that develops in the constantly hot and wet climate around the equator, characterized by evergreen trees, a high primary production, and high species diversity. (12)

tropical seasonal forest.   Mixed deciduous and evergreen forest that marks the transition from tropical rainforest to the drier conditions of tropical grassland. (12)

troposphere.   The lowest layer of the atmosphere that is in contact with the ground and the layer in which weather-producing movements take place. (2)

tundra.   A biome found in the cold conditions near the poles and at high altitudes, with a typical vegetation of mosses, lichens, and some hardy herbs, shrubs and stunted trees. (12)

typhoon.   A tropical cyclone that occurs in the western Pacific Ocean. (5)

U

U-shaped valley.   A deep valley that has been eroded by a valley glacier. (10)

V

volcano.   A landform produced by the eruption of lava or ash through a vent. (7)

W

warm front.   A front at the leading edge of an advancing warm air mass. (5)

warm glacier.   A glacier in which the ice is close to melting point nearly all year. (10)

water budget.   Comparison of the inputs (precipitation) and outputs (evaporation, runoff) of water at a location to provide an indication of its water availability. (4)

watershed.   The area drained by a river and its tributaries; the catchment or drainage basin. (9)

water table.   The upper level of groundwater saturation in an aquifer. (9)

water vapor.   The gaseous form of water that forms a variable gas in the atmosphere and absorbs long-wave radiation. (2)

wave.   A disturbance of a body or medium that moves from point to point. In the sea, the water surface is deformed by an alternate rising and falling motion generated by wind action on the surface. (11)

wave-induced current.   A flow of water toward the coast that results from sea waves entering shallow water. (11)

weather.   The short-term (few hours or days) changes in atmospheric conditions, such as temperature, rain, cloudiness, and wind. (2)

weather forecast.   An attempt to predict weather at a place for the next few hours or days. (5)

weathering.   The breakdown and decomposition of solid rock and rock fragments in response to atmospheric processes. (8)

weather system.   Distinct and repeated patterns of weather, such as a cyclone. (5)

westerlies.   Winds that blow from the west. (3)

whaleback forms.   Elongated, smoothed domes and basins formed by glacial abrasion. (10)

wind.   A horizontal movement of air relative to Earth's surface. (3)

wind abrasion.   The process of erosion in which windborne sand particles wear away rock. (10)

Y

young folded mountain.   A mountain system composed of relatively recent (up to 200 million years old) rocks that have been subject to compression and folding at a convergent plate margin. (7)

MAP PROJECTIONS

A map should show the spatial relationships of features at Earth's surface as accurately as possible in terms of distance, area, and direction. These requirements can be met when the cartographer is dealing with a relatively small area, such as a county or a small state in the United States. Over this area, Earth's curvature is minimal and the surfaces can be mapped onto a piece of flat paper with little distortion. However, at a global scale, the problems are acute; the skin of a sphere will simply not lie flat without being distorted in some way. No flat map can portray shape, distance, area, and direction over the spherical globe accurately. The different portraits of Earth shown in the diagrams below are different map projections, each of which attempts to minimize overall distortion or to maximize the accuracy of one measure of space. There are no right or wrong map projections, because they all contain distortion of one sort or another. In virtually all projections, linear scale is particularly not constant across the map, meaning that measurement of long distances will be inaccurate.

The simplest way to imagine constructing a map projection is to think of a transparent globe with a

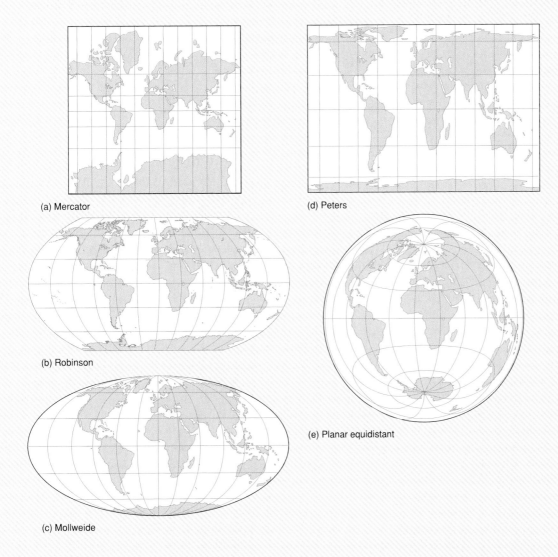

(a) Mercator

(b) Robinson

(c) Mollweide

(d) Peters

(e) Planar equidistant

(a–e) Map projections in common use.

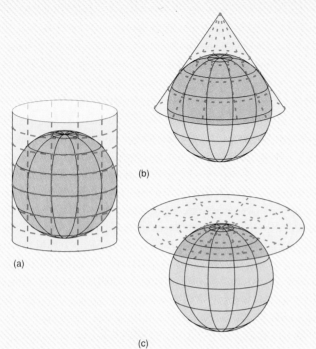

(a)

(b)

(c)

Methods of constructing map projections: (a) cylindrical, (b) conical, and (c) planar. The cylindrical and conical projections are unwrapped from the globe to give a flat map.

lightbulb inside it and a sheet of paper touching the globe in one of the three ways shown above. The outline of the continents and the lines of latitude and longitude are silhouetted on the paper to form the map. In fact, most projections are not constructed as simply as this. They are calculated mathematically and may be a compromise among a number of methods and designed to preserve the best features of each. There are three main desirable properties in a map and different projections attempt to preserve each one.

Equal-area projections, as the name suggests, maintain the areas of the continents in their correct proportions across the globe. They are used to depict distributions such as population, zones of climate, soils,

or vegetation. The Robinson projection used for world maps in this book is an equal-area projection. The scale distortion in *conformal maps* is equal in the two main directions from the projection's origin but increases away from the origin. This means that conformal projections maintain the correct shape over small areas, but the outline of continents and oceans mapped over a larger area is distorted. *Azimuthal maps* are constructed around a point or *focus.* Lines of constant bearing or compass direction radiating from the focus are straight lines on an azimuthal map. Distortion of shape and area is symmetrical around the central point and increases away from it. The azimuthal projection is used most often in geography to portray the polar regions, which are distorted in projections designed for lower latitudes. Azimuthal maps can also be equal-area or conformal.

It is mathematically impossible to combine the properties of equal area and reasonably correct shape on one map. The commonly used Mercator projection is a conformal map and demonstrates graphically how poor this projection is for showing geographical distributions. The Mercator projection was developed in 1569 by a Flemish cartographer, Gerardus Mercator, to help explorers and navigators. It has the important property, for navigation, that a straight line drawn anywhere on the map, in any direction, gives the true compass bearing between these two points. However, the size of land masses in the mid- and high latitudes (toward the poles) is grossly distorted in this projection. Alaska, for example, appears to be the same size as Brazil, although Brazil is actually five times as large. Greenland appears similar in size to the whole of South America. This distortion is partly because the meridians, which actually come together toward the poles, are shown with uniform spacing throughout the Mercator map. Despite its unsuitability for most geographical purposes, the Mercator is still very widely used.

MAP SYMBOLS

The following symbols (p. 428) are used on topographic maps produced by the United States Geological Survey.

TOPOGRAPHIC MAP SYMBOLS

BOUNDARIES

National

State or territorial

County or equivalent

Civil township or equivalent

Incorporated-city or equivalent

Park, reservation, or monument

Small park

LAND SURVEY SYSTEMS

U.S. Public Land Survey System:

Township or range line

    Location doubtful

Section line

    Location doubtful

Found section corner; found closing corner

Witness corner; meander corner

Other land surveys:

    Township or range line

    Section line

Land grant or mining claim; monument

Fence line

ROADS AND RELATED FEATURES

Primary highway

Secondary highway

Light duty road

Unimproved road

Trail

Dual highway

Dual highway with median strip

Road under construction

Underpass; overpass

Bridge

Drawbridge

Tunnel

BUILDINGS AND RELATED FEATURES

Dwelling or place of employment: small; large

School; church

Barn, warehouse, etc.: small; large

House omission tint

Racetrack

Airport

Landing strip

Well (other than water); windmill

Water tank: small; large

Other tank: small; large

Covered reservoir

Gaging station

Landmark object

Campground; picnic area

Cemetery: small; large

RAILROADS AND RELATED FEATURES

Standard gauge single track; station

Standard gauge multiple track

Abandoned

Under construction

Narrow gauge single track

Narrow gauge multiple track

Railroad in street

Juxtaposition

Roundhouse and turntable

TRANSMISSION LINES AND PIPELINES

Power transmission line: pole; tower

Telephone or telegraph line

Aboveground oil or gas pipeline

Underground oil or gas pipeline

CONTOURS

Topographic:

    Intermediate

    Index

    Supplementary

    Depression

    Cut; fill

Bathymetric:

    Intermediate

    Index

    Primary

    Index Primary

    Supplementary

MINES AND CAVES

Quarry or open pit mine

Gravel, sand, clay, or borrow pit

Mine tunnel or cave entrance

Prospect; mine shaft

Mine dump

Tailings

SURFACE FEATURES

Levee

Sand or mud area, dunes, or shifting sand

Intricate surface area

Gravel beach or glacial moraine

Tailings pond

VEGETATION

Woods

Scrub

Orchard

Vineyard

Mangrove

COASTAL FEATURES

Foreshore flat

Rock or coral reef

Rock bare or awash

Group of rocks bare or awash

Exposed wreck

Depth curve; sounding

Breakwater, pier, jetty, or wharf

Seawall

BATHYMETRIC FEATURES

Area exposed at mean low tide; sounding datum

Channel

Offshore oil or gas: well; platform

Sunken rock

RIVERS, LAKES, AND CANALS

Intermittent stream

Intermittent river

Disappearing stream

Perennial stream

Perennial river

Small falls; small rapids

Large falls; large rapids

Masonry dam

Dam with lock

Dam carrying road

Intermittent lake or pond

Dry lake

Narrow wash

Wide wash

Canal, flume, or aqueduct with lock

Elevated aqueduct, flume, or conduit

Aqueduct tunnel

Water well; spring or seep

GLACIERS AND PERMANENT SNOWFIELDS

Contours and limits

Form lines

SUBMERGED AREAS AND BOGS

Marsh or swamp

Submerged marsh or swamp

Wooded marsh or swamp

Submerged wooded marsh or swamp

Rice field

Land subject to inundation

SOIL CLASSIFICATION

Soil classification in the United States follows the *United States Comprehensive Soil Classification System,* also called Soil Taxonomy, developed by the Soil Survey staff of the United States Department of Agriculture since the 1950s.

Soil Taxonomy divides the soils of the world into eleven *soil orders,* primarily on the basis of closely defined diagnostic horizons. Each horizon is recognized by its position in the profile and by characteristics such as thickness, color, chemistry (particularly the type and amount of nutrients held in the soil), texture (particularly the amount of illuviated clay), and structure. The maps show the distribution of the soil orders across the world and in the United States. The soil orders are further divided into 47 suborders,

then into about 230 great groups, 1200 subgroups, 6000 families, and 13,000 series. Each series can be further subdivided into a few minor variants or phases. Soil maps used at a local scale show the distribution of soil series and phases; the descriptions at this level have stayed more or less the same through different classifications.

The nomenclature of Soil Taxonomy is an important and occasionally bizarre characteristic of the system. Each syllable of the soil's name, at any level, provides information about its characteristics. As far as possible the root of each syllable is drawn from Latin or Greek so that it will be recognized in many languages. The table shows how this system works for the first division into soil orders. Each soil order

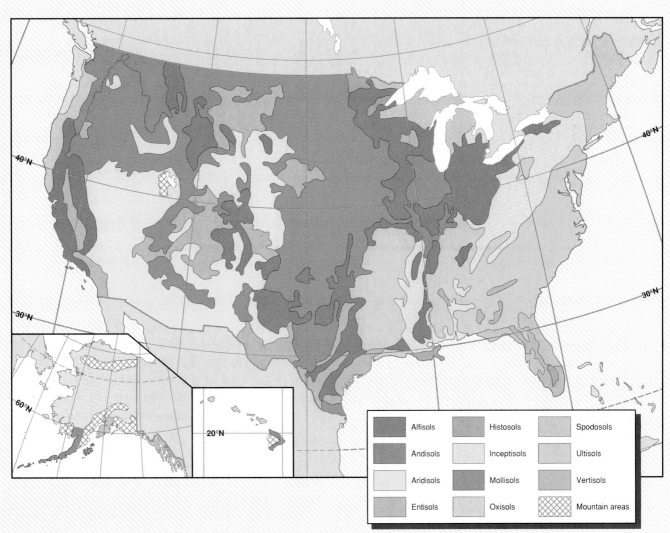

Alfisols **Histosols** **Spodosols**
Andisols **Inceptisols** **Ultisols**
Aridisols **Mollisols** **Vertisols**
Entisols **Oxisols** **Mountain areas**

The distribution of Soil Taxonomy's 11 soil orders in the United States.

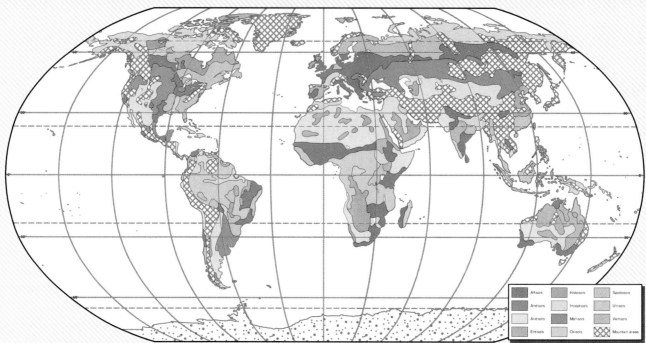

The global distribution of Soil Taxonomy's 11 soil orders.

TABLE	Derivation of the Names of the Soil Orders in Soil Taxonomy					
ORDER	**SYLLABLE**	**DERIVATION AND MEANING**		**ORDER**	**SYLLABLE**	**DERIVATION AND MEANING**
Entisol	Ent	(No derivation); recent soil		Spodosol	Od	*Spodos* (Greek) wood ash; ashy (podzol) soil
Andisol	And	*Ando* (Japanese) volcanic		Aridisol	Id	*Aridus* (Latin) dry; dry soil
Inceptisol	Ept	*Inceptum* (Latin) beginning; young soil		Mollisol	Oll	*Mollis* (Latin) soft; soft soil
Vertisol	Vert	*Verto* (Latin) to turn; inverted soil		Alfisol	Alf	(No derivation); aluminum (Al) and iron (F) soil
Histosol	Hist	*Histos* (Greek) tissue; organic or bog soils		Ultisol	Ult	*Ultimus* (Latin) last; ultimate (of leaching)
				Oxisol	Ox	*Oxide* (French) oxide; oxide soils

contains one descriptive syllable a linking 'i' or 'o,' and the syllable 'sol' from the Latin *solum,* meaning "soil." Names for the lower levels of the hierarchy are constructed by adding more descriptive syllables at the beginning of the name, with additional descriptive words at the family level. The series is given a local geographical name.

Entisols

Entisols show little or no evidence of pedogenic processes. They are found on steeply eroding slopes, floodplains that are regularly swamped with new material, or in any place where the normal soil-forming processes have not had a chance to act. This includes newly exposed surfaces where it is simply a matter of time before a soil develops and soils on old but resistant parent materials such as sand dunes, where soil formation is slow or nearly nonexistent.

Andisols

The soil order Andisols was introduced to Soil Taxonomy in 1990 to describe soils that develop on parent materials of volcanic origin, including ash, tephra, and solid rocks such as basalt. These soils were previously described as Entisols. They are found on or near active volcanoes and their global distribution therefore mirrors the distribution of volcanic activity. The main distinguishing feature is the nature of the mineral fraction. They also tend to have high organic content and a high cation-exchange capacity. Over time, Andisols may develop to Ultisols or Oxisols.

Inceptisols

Inceptisols are young soils that have reached a more advanced state of development than Entisols. They can be considered an early stage in the development of a "mature" soil. Although they have some degree of horizonation, weathering is incomplete and they still contain primary minerals released from the parent material but not yet broken down. Inceptisols are as varied an order as the Entisols and they have a wide geographic distribution.

Vertisols

Vertisols have a high clay content; this means that they contract and crack open when dry. Material from the surface falls into the cracks and is eventually wetted by rain. The soil is dampened in the cracks and therefore expands from the bottom upward. This creates pressures and tensions that result in the upward movement of the soil and the gradual inversion of the profile. For this reason, Vertisols are also known as "self-swallowing" soils. The pressures of expansion and contraction tend to produce a typical microrelief, a few centimeters high and a meter or so wide, of alternating depressions and ridges. In common with the Entisols and Inceptisols, the Vertisols have a wide geographical distribution that reaches all climatic environments.

Histosols

The Histosols are the organic or peat soils, composed mainly of organic materials in various stages of decay. They form where organic matter accumulates more quickly than it can be broken down. This is typical in waterlogged conditions, which promote acidity and inhibit activity in the detrital food chain. Histosols will develop at any site, under any climatic regime, where these conditions are met, although they are clearly most common in areas of moderate to high precipitation. The most typical Histosols are the blanket bogs and peats of northwestern Europe. In North America, the largest areas of Histosols are in Alaska and Minnesota.

Spodosols

Spodosols form most readily on freely draining parent materials under conditions of high rainfall and heavy leaching. In general, they are the product of podzolization and were called podzols under earlier classification systems. A diagnostic feature is the presence of a B horizon in which organic material, aluminum, and (usually) iron have been redeposited from the E horizon. Spodosols can develop in any humid climate, but are most often found under coniferous forest vegetation.

Aridisols

Aridisols are the shallow soils of arid areas. The most important factor affecting their formation is the lack of moisture. The natural vegetation is typically sparse and shrubby so there is little organic input to the soil and a high risk of wind erosion. Although the low precipitation precludes much downward translocation, in some profiles the high pH encourages the solution and movement of silica, which may be redeposited at deeper levels. In other profiles, a calcic horizon may develop as water moves up through the profile from a parent material rich in calcium.

Mollisols

The Mollisols are the soils of the world's grassy plains, prairies, and steppes and are the most extensive soil order in the United States. They are productive soils, presently used to grow grain and raise livestock. The dominant soil-forming environment is calcification. The diagnostic characteristic is the presence of a thick dark brown or black A horizon rich in organic material. The organic content decreases with depth.

Alfisols

The final three soil orders, Alfisols, Ultisols, and Oxisols, all develop under damp to wet conditions. They are differentiated primarily by the temperature regime under which they form. Alfisols are usually found in the coolest conditions. They are an intermediary stage in soil genesis between the Mollisols and the Spodosols. In the United States, they typically have, or have had, a vegetation of deciduous forest. Because they combine high nutrient content with moderate temperatures and moisture regimes, these soils are fertile and have been used extensively for agriculture.

Ultisols

Ultisols are found in the warmer temperatures of the mid- to low latitudes. They are strongly weathered and leached and are therefore nutrient-poor and acidic. A subsurface horizon with high amounts of clay particles is usually present. Ultisols are an intermediary step from Alfisols to Oxisols and are often found with one or another of those soil orders in the landscape. The difference between Alfisols and Ultisols is in their degree of leaching and their nutrient content; their profiles are otherwise similar. Without management, Ultisols are poor agricultural soils, mainly because the release of nutrients by weathering is more or less equal to their removal by leaching, and the nutrients in the organic material are rapidly recycled.

Oxisols

The Oxisols are the result of intense laterization. They are the extremely weathered and leached soils of the tropics and do not occur in the continental United States. Oxisols are the end process of a pedogenic regime starting at Entisols and moving through Inceptisols, Alfisols, and Ultisols. In general, they have very little horizonation.

CREDITS

ILLUSTRATIONS

DIPHRENT STROKES: 1-2, 1-12A, 3-20, 3-22, 3-30, 6-8, 6-10, 6-25, 7-15, 7-36, 10-6B, 11-1B, 11-17B, 13-5.

LINE ART

1-7 & 1-8: From Fred M. Shelley and Audrey E. Clarke, *Human and Cultural Geography.* Copyright © Wm. C. Brown Communications, Inc., Dubuque, Iowa. All Rights Reserved. Reprinted by permission.

6-5C: Redrawn from T. J. Chandler, *The Climate of London,* © 1965 Hutchinson. Used by permission of the author; **6-30A-D:** Modified from *Earthquest,* Vol. 5, No. 1, with the following sources: A and B, J. T. Houghton, et al., *Climate Change: The IPCC Assessment,* Cambridge University Press, Cambridge, 1990; C, N. J. Shackleton and N. Opdyke, *Quarternary Research 3,* 39–55, 1973; D, T. J. Crowley, in *Journal of Climate 3,* 1282–1292, 1990; **6-34:** C. Lorius, et al., "Antarctic Ice Core: Carbon Dioxide and Change Over the Last Climatic Cycle," *EOS,* Vol. 68, pp. 681–684, 1988, copyright by the American Geophysical Union.

7-5: B. Isaks, et al., "Seismicity and the New Global Tectonics," *Journal of Geophysical Research* 73:5855–5899, 1968, copyright by the American Geophysical Union; **7-29:** Redrawn by permission from M. J. Selby, *Earth's Changing Surface.* Copyright 1985 Oxford University Press, Oxford, England; **7-40:** Adapted with permission from Michael Summerfield, *Global Geomorphology,* © 1991 Longman Group Limited.

8-7: Redrawn with permission from N. M. Strakhov, *Principles of Lithogenesis,* Oliver & Boyd, © 1967, Longman Education; **8-8:** Partly based on M. A. Carson and M. J. Kirkby, *Hillslope Form and Process,* Cambridge University Press, Cambridge, England, 1972; **8-10A,B & 8-12A-C:** From D. J. Varnes, *Slope Movement Types and Processes,* TRB Special Report 176, Transportation Research Board, National Research Council, Washington, DC, 1978. Used by permission of the National Academy of Sciences; **8-13:** Reprinted from "Recent Development of Mountain Slopes in Scandinavia" by A. Rapp from *Geografiska Annaler,* 1960, 42:71–200, by permission of Scandinavian University Press; **8-19:** Redrawn with permission from R. U. Cook and J. C. Doornkamp, *Geomorphology in Environmental Management.* Copyright 1990 Oxford University Press, Oxford, England; **8-21:** From A. J. Parsons, *Hillslope Form,* copyright 1988 Routledge, London. Used by permission.

9-19: Modified from W. B. Bull, "The Alluvial Fan Environment," *Progress in Physical Geography,* 1:222–270, © 1977 Edward Arnold. Used by permission of Edward Arnold (Publishers) Limited; **9-32:** Reprinted from "A Cycle of Sedimentation and Erosion in Urban River Channels" by

M. G. Wolman from *Geografiska Annaler,* 1967, 49A:385–395, by permission of Scandinavian University Press.

10-11: Redrawn with permission from M. J. Selby, *Earth's Changing Surface.* Copyright 1985 Oxford University Press, Oxford, England; **10-29:** Redrawn with permission from J. T. Hack, "Dunes of the Navajo Country," *Geographical Review* 31:240–263, copyright 1941 The American Geographical Society.

11-6: Redrawn with permission from J. L. Davies, *Geographical Variation in Coastal Development,* copyright 1980, Oliver & Boyd, United Kingdom; **11-8:** Modified from J. S. Pethick, *Introduction to Coastal Geomorphology,* © 1984 Edward Arnold. Used by permission of Edward Arnold (Publishers) Limited; **11-12:** After R. J. Small, *The Study of Landforms,* Cambridge University Press, Cambridge, 1970; **11-15:** Redrawn from Stephen P. Leatherman, *Barrier Island Handbook,* University of Maryland, 1988. Used by permission of the author; **11-16:** Redrawn from Stephen P. Leatherman, *Barrier Island Handbook,* University of Maryland, 1988. Used by permission of the author; **11-27:** Adapted from Michael Summerfield, *Global Geomorphology,* © 1991 Longman Group UK.

PHOTOS

Table of Contents: 1: Natural Environment Research Council, United Kingdom; **2:** Michael Bradshaw; **3:** Michael Bradshaw; **4:** © James H. Karales/Peter Arnold, Inc.; **5:** NASA; **6:** United States Department of Defense; **7:** Michael Bradshaw; **8:** Phototone, Pebbles 2. 3. 46; **9:** Michael Bradshaw; **10:** © John Carnemolla/APL/Westlight; **11:** © John Shaw/Tom Stack & Associates; **12:** National Zoo; **13:** Hari Eswaran, USDA

Chapter 1: Opener (starfish): © Jeffrey L. Rotman/Peter Arnold, Inc.; **Opener (lava):** © Penisten/APL/Westlight; **Opener (stones):** © Matt Meadows/Peter Arnold, Inc.; **Opener (web):** © Carl R. Sams II/Peter Arnold, Inc.; **1-1A:** © Gregory Ochocki/Photo Researchers, Inc.; **1-1B:** © Anna E. Zuckerman/Tom Stack & Associates; **1-1C:** © Fred Maroon/Photo Researchers, Inc.; **1-1D:** © Bruce Berg/Visuals Unlimited; **1-1E:** © Norman Benton/Peter Arnold, Inc.; **1-3A:** © J. Lotter/Tom Stack & Associates; **1-3B:** © IFA/Peter Arnold, Inc.; **1-3C:** © Thomas Kitchin/Tom Stack & Associates; **1-3D:** © Milton Rand/Tom Stack & Associates; **1-4:** Ruth Weaver; **1-5,1-6:** Courtesy of A. Smith, University of Plymouth; **1-11 A,B:** NASA; **1-12B:** NASA, Michael Helfert; **1-13, 1-14 B-D, 1-15B:** Natural Environment Research Council, United Kingdom; **1-16 A,B:** Courtesy of Mark Abbott, Oregon State University

Chapter 2: Opener (background): © Jeffrey L. Rotman/Peter Arnold, Inc.; **Opener (inset):** © Scott Blackman/Tom Stack & Associates; **2-1:** NASA; **2-5:** © Greg Vaughn/Tom Stack & Associates; **2-8:** © Carl Purcell/Photo Researchers, Inc.; **p. 36 (top):** © Patrick Watson; **p. 36 (bottom):** NASA

Chapter 3: Opener (background): Phototheque MonTresor; **Opener (inset):** © Kevin Schafer/Peter Arnold, Inc.; **3-6:** USDA; **3-7:** © Jim Zuckerman/Westlight; **3-11:** NASA;

3-19: NASA, Gene Feldman; **3-25, 3-34:** Michael Bradshaw; **p. 71:** © Patrick Downs/Los Angeles Times Photo

Chapter 4: Opener (background): © James H. Karales/Peter Arnold, Inc.; **Opener (inset):** © Craig Aurness/Westlight; **4-4:** © Mack Henley/Visuals Unlimited; **4-9A:** © William James Warren/Westlight; **4-9B:** © Kevin Schafer/Peter Arnold, Inc.; **4-9C:** © Kent Wood/ Peter Arnold, Inc.; **4-9D:** © John Gerlach/Visuals Unlimited; **4-11:** NASA, Michael Helfert; **4-13:** Michael Bradshaw; **4-15:** NASA, Bill Rossow; **p. 93 (both):** Tom Carroll, NOAA

Chapter 5: Opener (background): © H. David Seawell/Westlight; **Opener (inset):** © Charles Campbell/Westlight; **5-7:** © Howard B. Bluestein/Photo Researchers, Inc.; **5-8:** © Keith Kent/Peter Arnold, Inc.; **5-9:** © Science VU/Visuals Unlimited; **5-13:** NOAA; **5-15:** NASA; **5-17:** NASA, Michael Helfert; **5-19:** National Oceanic and Atmospheric Administration, National Weather Service; **p. 119 (left):** S. J. Williams, U.S. Geological Survey; **p. 119 (right):** © Steve Starr/Saba; **p. 120 (top):** NOAA; **p. 120 (bottom):** © F. Rossotto/Tom Stack & Associates

Chapter 6: Opener (background): © Ron Watts/Westlight; **Opener (inset):** © Landform Slides; **6-2:** United States Department of Defense; **6-14:** Paul Bradshaw; **6-17:** USDA; **6-35:** Michael Bradshaw; **p. 165:** © Spencer Swanger/Tom Stack & Associates

Chapter 7: Opener (background): © Penisten/APL/Westlight; **Opener (inset):** © Bill Everitt/Tom Stack & Associates; **7-1:** Ruth Weaver; **7-10:** © Greg Vaughn/Tom Stack & Associates; **7-14 A,B:** U.S. Geological Survey; **7-16:** J. P. Lenney; **7-21, 7-22A:** Michael Bradshaw; **7-23 A,B:** © Landform Slides; **p. 190 (inset):** © Mark C. Burnett/Photo Researchers, Inc.; **p. 120 (background):** U.S. Geological Survey; **7-26 A,B:** U.S. Geological Survey; **7-28B:** © Kevin Schafer/Tom Stack & Associates; **p. 213 (top right):** © Larry Davis/Los Angeles Times Photo; **p. 213 (top left):** © Jim Mendenhall/Los Angeles Times Photo; **p. 213 (bottom):** © Lacy Atkins/Los Angeles Times Photo; **7-30:** © J. P. Lenney; **7-32:** World Ocean Floor authors Bruce C. Heezen and Marie Tharp, Date (1977) © by Marie Tharp 1977. Reproduced by permission of Marie Tharp.; **7-42:** NASA

Chapter 8: Opener (background): © Craig Aurness/Westlight; **Opener (inset):** © F. Gohier/Photo Researchers, Inc.; **8-4:** © Landform Slides; **8-5:** USDA; **8-11:** © Landform Slides; **8-16:** © M. Timothy O'Keefe/Tom Stack & Associates; **8-22 A-D:** Michael Bradshaw; **p. 245:** U.S. Geological Survey

Chapter 9: Opener (background): © Matt Meadows/Peter Arnold, Inc.; **Opener (inset):** USDA; **9-6:** Science VU/Visuals Unlimited; **9-11:** © Landform Slides; **9-12A:** © Tim Hauf/Visuals Unlimited; **9-12B:** © Dick Poe/Visuals Unlimited; **9-13, 9-15, 9-16A:** USDA **9-16B:** Michael Bradshaw; **9-17:** © Neil Rabinowitz; **9-20:** © Peter Kresan; **9-26 A,B:** Landform Slides; **p. 275 (inset):** © M. Timothy O'Keefe/Tom Stack & Associates; **p. 275 (background):** U.S. Geological Survey; **p. 282 (both):** Courtesy of EOSAT, Lanham, Maryland

Chapter 10: Opener (background): © John Carnemolla/APL/Westlight; **Opener (inset):** NASA, Michael Helfert; **10-2:** U.S. Geological Survey; **10-6A:** NASA, Michael Helfert; **p. 293 (inset):** © Verna R. Johnston/Photo Researchers, Inc.; **p. 293 (background):** U.S. Geological Survey; **10-7:** © Landform Slides; **10-8A:** © Kim Heacox/Peter Arnold, Inc.; **10-8B:** © Landform Slides; **10-12:** © J. P. Lenney; **10-15:** © Landform Slides; **10-16:** Michael Bradshaw; **10-18:** © Landform Slides; **10-19A:** © Fred Bruemmer/Peter Arnold, Inc.; **10-19B:** © Steve McCutcheon/Visuals Unlimited; **10-21:** USDA; **10-26, 10-28:** © Landform Slides; **10-33:** NASA; **p. 317 (both):** © Steve McCutcheon/Visuals Unlimited

Chapter 11: Opener (background): © John Shaw/Tom Stack & Associates; **Opener (inset):** NASA, Michael Helfert; **11-1A:** Michael Bradshaw; **11-4:** © Scott Blackman/Tom Stack & Associates; **11-9 A,B:** © Ted Clutter/Photo Researchers, Inc.; **11-10:** USDA; **11-11:** © Peter Gilday; **11-13:** Courtesy of Joann Mossa, University of Florida; **p. 336 (both), p. 337 (inset):** Jeffrey M. Greene; **p. 337 (background):** U.S. Geological Survey; **11-17A:** NASA, Michael Helfert; **11-19A:** © TSA/Tom Stack & Associates; **11-19 B-D, 11-23:** NASA; **11-25:** © Thomas Kitchin/Tom Stack & Associates; **11-28B:** © Alex MacLean/Landslides; **p. 349 (both):** NASA

Chapter 12: Opener (background): © Carl R. Sams II/Peter Arnold, Inc.; **Opener (inset):** © Alex S. MacLean/Peter Arnold, Inc.; **12-2A:** © Tom & Pat Leeson/Photo Researchers, Inc.; **12-2B:** Courtesy of Dr. Charles Hogue, curator of Entomology, Los Angeles County Museum of Natural History; **12-2C:** © Chip & Jill Isenhart/Tom Stack & Associates; **12-2D:** © Richard Thom/Visuals Unlimited; **12-7:** National Zoo; **12-8:** C. J. Tucker/NASA; **12-11B:** © E. R. Degginger/Animals Animals/Earth Scenes; **12-12B:** © John D. Cunningham/Visuals Unlimited; **12-14B:** © Robert W. Hernandez/Photo Researchers, Inc.; **12-17B:** © Kevin Schafer/Tom Stack & Associates; **12-18A:** © Jany Sauvanet/Photo Researchers, Inc.; **12-18B:** © Michael Dick/Animals Animals; **12-18C:** © Zig Leszczynski/Animals Animals; **12-19A:** © Rafael Macia/Photo Researchers, Inc.; **12-19B:** © Kevin Schafer/Peter Arnold, Inc.; **12-20:** © Tom McHugh/Photo Researchers, Inc.; **12-21A:** © Jim Zuckerman/Westlight; **12-21B:** © James L. Shaffer; **12-22A:** © Glenn Oliver/Visuals Unlimited; **12-23A:** © Tim Davis/Photo Researchers, Inc.; **12-23B:** © John Cancalosi/Peter Arnold, Inc.; **12-23C:** © Joe McDonald/Visuals Unlimited; **12-24A:** © Kevin Magee/Tom Stack & Associates; **12-24B:** © Alex S. Maclean/Peter Arnold, Inc.; **12-25,12-26:** © Stephen J. Krasemann/Photo Researchers, Inc.; **12-28A:** © Holt Studios Int'l/Photo Researchers, Inc.; **12-28B:** © Neil Rabinowitz; **p. 385:** © Randall Hyman/Stock Boston; **p. 386:** NASA

Chapter 13: Opener (background): © Tony Stone Images; **Opener (inset):** USDA; **13-7 A,B:** USDA; **13-11, 13-12, 13-13:** Hari Eswaran, USDA; **13-14A:** USDA; **13-14B:** © Jack Ryan/Photo Researchers, Inc.; **13-15:** Hari Eswaran, USDA; **13-16 A,B:** USDA; **13-16C:** © Tim Hauf/Visuals Unlimited; **13-16D:** © David M. Dennis/Tom Stack & Associates; **p. 409 (top):** © Rudolf Arndt/Visuals Unlimited; **p. 409 (bottom):** © Larry Brock/Tom Stack & Associates; **p. 411 (top):** Science VU/Visuals Unlimited; **p. 411 (middle):** USDA; **p. 411 (bottom):** USDA

INDEX